DATE DUE

AUG 26 2004	
GAYLORD	PRINTED IN U.S.A.

Pilots, Man Your Planes!

THE HISTORY OF

NAVAL AVIATION

WILBUR H. MORRISON

Hellgate Press

Central Point, Oregon

PILOTS, MAN YOUR PLANES!
© 1999 Wilbur H. Morrison
Published by Hellgate Press

HELLGATE PRESS
a division of PSI Research
P.O. Box 3727
Central Point, Oregon 97502-0032

(541) 479-9464
(541) 476-1479 fax
info@psi-research.com e-mail

Editorial and Interior design: Editype, Jackson, MI
Cover design: Steven Burns

Library of Congress Cataloging-in-Publication Data
Morrison, Wilbur H., 1915–
   Pilots, man your planes! : the history of naval aviation / Wilbur
H. Morrison.
      p.   cm.
   Includes bibliographical references and index.
   ISBN: 1-55571-466-8 (cloth)
   1. United States.   Navy– –Aviation– –History.   I. Title.
VG93.M664   1999
359.9'4'0973– –dc21   99–10941
                                                                      CIP

Printed and bound in the United States of America
First Edition 10 9 8 7 6 5 4 3 2 1
Printed on recycled paper when available.

Dedicated to the memory of
Admiral John H. Towers

Acknowledgments

I have been writing about naval aviation since 1954, when I joined the public relations staff of Douglas Aircraft's El Segundo Division in California. My military background was with the Air Force in World War II as a Twentieth Air Force bombardier-navigator. Therefore, my exposure to naval aviation was somewhat of a shock—I knew almost nothing about it. But in my ten years with the Douglas Aircraft Company I learned to admire the men and women who—along with my own Air Force pioneers—created military airpower. Their story has never been adequately told in a single volume of history. I tried to tell it earlier with *Wings Over the Seven Seas, U.S. Naval Aviation's Fight for Survival,* which I began to write in 1964, and finished ten years later with its publication. In all frankness, I failed to do justice to the broad scope of naval aviation through the years. At the time, my expertise as a historian was too limited. I am proud of my 1983 book, *Above and Beyond: 1941–1945,* about carrier aviation in World War II, but it covers only a part of naval aviation's history.

Now, after years of exhaustively exploring that history, using Douglas's voluminous files to start with (now largely destroyed following the company's merger with McDonnell) and access to Lockheed California Company's files—where I worked as military publicity manager for a number of years—I feel confident that I have covered naval aviation's history from its beginnings in 1910 through the Persian Gulf War in 1991 in a professional manner.

I am deeply indebted to both Douglas and Lockheed for exposing me to thousands of Naval Aviation's personal stories, which otherwise would have been lost, and grateful that I had the foresight to add them to my personal files while I was employed by these companies. The best of them are included in *Pilots, Man Your Planes!*

I would like to extend my special thanks to the late General Curtis E. LeMay and the late Admiral James S. Russell, and Admiral Ulysses S.G. Sharp (Retired). Retired Rear Admiral James D. Ramage provided extensive recollections about World War II and the Vietnam War.

Others who have given me exceptional assistance include: Rear Admiral Philip Anselmo, Vice Admiral Lyle G. Bien, Colonel Steve Furimsky (Marine Corps, Retired), Colonel (Retired) James Bush, Captain R.C. Knott, Captain Sandy Russell, Captain Robert Arnold (Retired), Captain Jack Tanner, the late Edward H. Heinemann, Patrick J. Carney and Raul R. Fernandez Jr., director and former assistant director, respectively, of the library at Camp Pendleton, Stewart Slack, and Seth Lewis, who made final changes in Stewart Slack's drawings.

Wilbur H. Morrison
Fallbrook, California

Foreword

This is Wilbur Morrison's third book on naval aviation, and it covers the entire history in depth, with much of the subject previously treated superficially by other writers.

Like his two previous books, *Wings Over the Seven Seas: U.S. Naval Aviation's Fight for Survival,* in 1974, and *Above and Beyond: 1941–1945,* in 1983, it does not rely on archival material. It is basically about the men and women who developed naval aviation through the years and fought for it in peace and war.

Through extensive interviews, Morrison tells how well—or badly—naval aviators performed in all wars in this century. Most importantly, for future generations, he evaluates the lessons learned—often the hard way—with suggestions on how these mistakes can be avoided in the future. Morrison's interviews with naval aviators high and low form the backbone of the text in a way that no archival research could ever provide. This is their story—and one that can be read with pride by all Americans.

James D. Ramage
Rear Admiral, USN (Retired)

Introduction

Despite the growing importance of aviation technology in the military services, men and women of courage and vision still determine an organization's success or failure. Their achievements have been particularly noteworthy in the United States Navy. Naval aviators since 1910 have always had to fight for their survival in peace and war despite government apathy, interservice rivalry, and the lack of vision of many officers in their own service. Fortunately, a few prophetic leaders emerged in each generation to provide the necessary direction so that naval aviation could expand despite its adversarial environment.

The acceptance of aviation by high officials of the United States Navy at first was frustrated by surface admirals who refused to concede that the battleship had outlived its usefulness. In the mid-1990s, the United States Navy is dominated by submariners, in almost all positions of authority, who continue to oppose attack carriers as the Navy's first line of battle.

For those unfamiliar with Navy terminology, the following glossary is included. When "V" is used, it means a Navy's heavier-than-air unit or aircraft. When "M" is added, it stands for Marines. The letter "F" stands for fighter, "B" for bomber, "TB" for torpedo bomber, "SB" for scout bomber, and "Y" and "X" for early stages of an experimental aircraft. Letters assigned to airplanes identify the plane's manufacturer. "D" was for Douglas, "B" for Boeing, and "M" for Martin. Secretary of Defense McNamara ordered such identification dropped in the 1960s.

This is the story of naval aviation and of the men and women who fought for it through its obscure beginnings in 1910 through the end of the Persian Gulf War in 1991, and Desert Fox in late 1998. Such a story could not be found in historical archives. It had to come from the people who created naval airpower. It has been my privilege to know many of them. These people were particularly valuable because of their vivid recollections of the early pioneers whom I had never met, although I was born in 1915 and most naval aviators belong to my generation. Many of their achievements are little known because so few books have been devoted exclusively to naval aviation's history. Hopefully, *"Pilots, Man Your Planes!"* will help to fill some of that void.

Contents

Part I

Men of Vision

1

Naval Aviation's Fight for Recognition

Although flying machines had an irresistible allure for some people in the United States prior to the twentieth century, it was nothing compared to the enthusiasm expressed by Europeans for each new development in aviation.

Military leaders in Europe early recognized airpower's potential in warfare. When the gas balloon was first flown in France during 1782, and the first human ascension was made the following year in a captive balloon by Jean Pilâtre de Rozier and his passenger the Marquis d'Arlendes, the French army was quick to realize the usefulness of airpower. Balloons were used first during the Napoleonic Wars in 1794 for observation of enemy lines.

In contrast, American military officials failed to appreciate the impact of aerial machines on war operations. Except for a few men of vision, this was particularly true of the United States Navy. For years its primary offensive weapon had been the battleship because it could stand off thousands of yards and "slug it out" with long-range projectiles.

In many nations of Europe in the late nineteenth century, and in the United States, bombing experiments were conducted from captive and free balloons. Such uses gave rise to concern after they demonstrated during the American Civil War what could happen if bombs were dropped upon innocent men, women, and children instead of being used just for observation of enemy lines.

Although bombing from balloons was in its infancy, it was evident that bombs eventually could be transported indiscriminately over battlefields, towns, and cities. This was a sobering thought as early as 1890, because it seemed only a matter of time before balloons were used in a manner that was too horrible to contemplate.

A possible further extension of their use to destroy great cities caused an understandable fear. This was expressed by the Hague Peace Conference in 1891. In a broad statement of policy, the conference declared that the discharge of aerial projectiles in war would be outlawed.

Even greater concern was aroused when the first successful metal dirigible was designed in 1897 by David Schwarz, a Hungarian. It took off from Berlin's Tempelhof Field powered by a 16-horsepower Daimler engine. It flew for several miles before it crashed due to a fuel leak.

Assistant Secretary of the Navy Theodore Roosevelt—who later coined the slogan "Speak softly and carry a big stick"—correctly sensed the potential of airplanes in naval warfare as early as 1898. His inquisitive mind was intrigued by the successful flight on May 6, 1896, of a small, steam-powered model airplane, which weighed only 30 pounds and had a 13-foot wingspread. War with Spain was imminent, and Roosevelt wrote a memorandum to the Secretary of the Navy describing the successful flights of the model airplane designed by Samuel P. Langley, Secretary of the Smithsonian Institution.

Roosevelt was determined to protect the United States from all forms of aggression. He envisioned some kind of a bombardment airplane that would give potent emphasis to his "big stick" philosophy. He proposed that a study be made of aeronautics and the airplane.

"I have assigned Commander C.H. Davis, Director of the Naval Observatory in Washington, D.C., to head a special board to undertake a study," Roosevelt wrote.

After extensive analysis and investigation, the board reported that airplanes were useful as a means of scouting and reconnaissance. It also said they might serve as a fast means of communications between isolated bases. It concluded significantly that airplanes could be used as engines of offense for dropping high explosives into enemy camps and fortifications.

The potential for mass destruction of cities became more apparent on July 20, 1900, when Germany's Count Ferdinand von Zeppelin flew the first of his many airships with their wire-braced, aluminum hulls covered with cotton cloth and sixteen gas cells of hydrogen. Von Zeppelin had made his first balloon ascent during the American Civil War.

Although Orville and Wilbur Wright achieved a historic advance in aviation on December 17, 1903 by flying the world's first powered, heavier-than-air flying machine at Kitty Hawk, Americans took little notice. The event wasn't even covered by the press until later. Even then it was greeted largely with nonchalance. The Wrights made four flights that day, the longest almost a minute in duration, covering 850 feet. They had worked for seven years testing scale models in a wind tunnel. Their successful powered flights used a lightweight 26-horsepower engine that gave more power at less weight than any other engine at the time. Their first airplane was built at a cost of only $1,000.

Five years later the Wright brothers were still improving their aircraft designs. During a demonstration of their airplanes in Europe, Wilbur flew 62 miles at an average altitude of 361 feet.

After years of testing new models, Orville was seriously injured in September 1908, when the first United States Army air trials were held at Fort Myer, Virginia. His passenger, Lieutenant Thomas Selfridge, was killed, becoming the first American casualty of powered flight.

By July 1909, Orville was fully recovered from his injuries, and he insisted upon making the twenty practice flights needed to get the Wright Flyer qualified for acceptance by the United States Army.

A 17-year-old teenager, who would later carve a special niche in the aeronautical world, was present on July 30 when Orville made his final tests. Donald W. Douglas was fascinated as the Wright flying machine circled the field at speeds averaging 42 miles an hour. His eyes were bright with excitement as he admired the graceful, curving surfaces of the Flyer's wings as they reflected the sun from their white surfaces. Witnessing the miracle of powered flight for the first time, with the staccato rattling of the Flyer's engine turning the large propellers at such speeds they were just a blur, Douglas was almost overwhelmed with emotion.

During the final test flight, Orville Wright and Lieutenant Benjamin Foulois took off from Fort Myer for Alexandria. After completing their round trip of 10 miles at an average speed of 42 miles an hour, Douglas and other spectators roared their approval.

It was a great day for aviation. After the Wright biplane satisfactorily completed all contractual terms, it was purchased for $30,000 and became the world's first successful military aircraft.

Douglas often chuckled later about one Army requirement that specified the Wright Flyer must have the capability of being transported in a mule-drawn wagon.

A short time later, Douglas's father took him to a flying exhibition at Morris Park Racetrack alongside the Harlem River. There he saw Glenn Curtiss make a demonstration flight by flying a mile in one of his pusher-type machines, which was purchased by the New York Aeronautical Society.

The air age surged forward on July 25, 1909, after Frenchman Louis Blériot flew across the English Channel in his monoplane powered by a three-cylinder engine driving a two-bladed propeller. He became the first human being to travel across the Channel other than by sea, and the trip took only thirty-seven minutes.

In the United States, the Wright brothers kept improving their airplanes. Orville had already logged hundreds of flights, and two days after Blériot's flight he set a new flight duration record of one hour, one minute and forty seconds.

Glenn Curtiss captured the International Cup of Aviation on August 28, 1909, along with a $5,000 prize, by outflying the previous winner, Louis Blériot, who emerged in second place. Curtiss flew 12.42 miles in 15 minutes, 50.6 seconds. Count de Lambert, another participant, described Curtiss's flight in these words, "The day on which man in his primitive form crawled out of the water and found he could move and live on land was no more an epoch than this."

Now that the heavier-than-air flying machine was a reality, the whole concept of naval warfare at sea was altered, although battleship "diehards" refused for years to concede the superiority of the airplane.

Imaginative men in the Navy posed the question, Why not deliver exploding projectiles by airplanes instead of lobbing them at targets from deck and turret guns? They argued further that this method vastly increased the range of attack, added a potent element of surprise, and afforded less danger to one's own forces. In retrospect it seems inconceivable that these men should have been so castigated for their vision.

Lieutenant George C. Sweet was the first naval officer to ride in an airplane. In 1909, he told his superiors, "Aviation should be developed quickly in the Navy because the governments of France and Germany have given substantial encouragement to their aeronautical inventors. Our Navy should not fall behind in aviation activity," he said, "but should strive to lead other navies in its development and fully exploit the airplane."

Commander F.L. Chapin, Naval Attaché in Paris, was authorized to attend an air meet at Rheims, France. He reported later, "Impressive advances have been made in Europe in the science of aviation. I recommend that battleships be equipped with airplanes launched by catapult while ships are underway."

He discussed the possibility of using airplanes for bombing attacks against enemy ships, particularly at night when the ships could not see to fire back at low-flying aircraft. However, his words were largely ignored in Washington.

If the Navy lacked vision, a civilian aviator by the name of Glenn H. Curtiss made up for some of it. In 1910, during a speech in New York City, he said, "Battles of the future will be fought in the air. The airplane armed with guns and bombs will decide the destiny of nations." It would be 10 years before another champion of airpower—Brigadier General William "Billy" Mitchell of the United States Army—would say the same thing.

Curtiss was a man of action. A newspaper needled him into conducting a series of aerial bombing tests to prove his advanced theories. "I accept the challenge of the *New York World,*" he said. "I'll personally conduct a series of aerial bombings on Lake Keuka."

During July 1910, Curtiss used flags to form the outline of warships in the shallow water at the lake near Hammondsport, New York. Then he repeatedly used lead pipes to simulate bombs as he dropped them within the flagged target areas.

A brilliant engineer, Captain Washington I. Chambers, had been the assistant to the Secretary of the Navy's Aide for Materiel in 1909. When the Navy Department decided to follow air developments more closely the following year, he was the logical man for the job. He was intimate with most of the air-minded civilians and officers of that day.

Thrill seekers crowded airports throughout the United States as bombing experiments multiplied at air meets. Underpowered airplanes were difficult enough to fly without the problem of releasing bombs. Carrying only a pilot and a few gallons of fuel, the planes were barely able to stagger off the ground. The pilot was forced to keep both hands and feet against the controls while busily leaning left or right against special shoulder yokes to actuate the ailerons for turning. Safety belts had not even been thought of, so the risks were serious enough without complicating the pilot's job by having him release bombs. Airmen realized early that some sort of mechanical device was needed for releasing bombs from airplanes.

If a bomb could be lifted into the air, how should it be dropped safely? Obviously it was dangerous to carry it in the pilot's lap, or in a sack tied to the flimsy airframe. An accidental explosion was always a dreaded possibility during a rough cow pasture takeoff.

Young daredevils traveled from city to city performing aerial acrobatics in their fragile machines. Many died in spectacular crashes as crowds willingly paid high prices to see these wonders of the air age.

Most professional men in uniform ridiculed the whole idea. "Aerial bombardment will never qualify as a major threat in modern warfare whether on land or sea," they said. Little did they realize that these experiments, crude as they were by later standards, foreshadowed changes in warfare that would revolutionize military thinking for decades to come.

Although aerial bombardment had been outlawed by an international agreement according to terms of the Second International Peace Conference at the Hague in 1907, the United States was the only first-class power to sign the declaration.

The Italian government ignored the agreement when its fliers dropped explosives on Arab troops at Benghazi and Derna in 1911. The cans of nitroglycerine hurled at defenseless troops did little damage because of their limited power, but the action was a forerunner of worse to come.

The United States Secretary of the Navy, George L. Meyer, formed the Aeronautical Reserve, a new organization of private citizens for advancing aeronautical science as a means of supplementing the national defense. He announced on September 26, 1910 that Captain Washington I. Chambers, assistant to the Aide for Materiel, had been designated as the officer to whom all correspondence on aviation should be referred. This was the first time a provision was specifically made for aviation in the United States Navy.

A few days later the General Board, of which Admiral George Dewey was president, recommended more positive action. On October 1, he wrote the Navy Secretary that in recognition of "the great advances which have been made in the science of aviation and the advantages which may accrue from its use in this class of vessel, the problem of providing space for airplanes or dirigibles be considered in all new designs for scouting vessels."

The chief of the Bureau of Steam Engineering, Captain H.I. Cone, in a letter to the secretary reminded him that "the rapid improvement in the design and manipulation of airplanes and the important role they would probably play made it necessary to requisition an airplane for the USS *Chester* and the services of an instructor to teach one or more officers to fly the machine.

Secretary Meyer approved the recommendation and directed that whoever was appointed keep Captain Chambers fully advised of the results of all experiments.

The appointees, Naval Constructor William McEntee and Lieutenant N.H. Wright, joined Chambers on October 22 as official Navy observers to the international Aviation Tournament at Belmont Park, New York.

As a result of their observations, the chief of the Bureau of Construction and Repair suggested to Secretary Meyer that steps be taken to obtain one or more airplanes to develop their use for naval purposes. He recommended that in the absence of specific funds for their purchase, specifications for the battleship *Texas* be modified to require the contractors to supply one or more aircraft as a part of their obligation.

A civilian pilot, Eugene Ely, made the first takeoff from a ship on November 14, 1910 when his 50-horsepower Curtiss plane used a wooden platform built on the bow of the USS *Birmingham*. The ship was at anchor in Hampton Roads and Ely landed safely on Willoughby Spit.

Two weeks later Glenn Curtiss followed up this achievement by writing to the Secretary of the Navy that he would train one naval officer without charge as one means of assisting "in developing the adaptability of the aeroplane to military purposes."

Secretary Meyer approved the recommendation and Lieutenant T.G. Ellyson was ordered to report on December 23 to the Glenn Curtiss Aviation Camp at North Island, San Diego, California. After undergoing flight training, Ellyson became the Navy's first officer to become an aviator.

Eugene Ely followed up his feat of taking off from a Navy ship in November by making a landing on the USS *Pennsylvania* at 11:01 A.M. on January 18, 1911. He used a specially built platform on the armored cruiser, which was anchored in San Francisco Bay. He took off at 11:58 and returned to a nearby airfield. His flights from Navy warships demonstrated the feasibility of using airplanes on ships, and Secretary Meyer followed these experiments with great interest.

With Lieutenant Ellyson assisting in another test to adapt airplanes to the needs of the Navy, the first successful hydroplane flight was made by Curtiss at North Island, San Diego. Curtiss used his standard biplane with a single main float in place of the tandem triple floats used in earlier tests. During two flights he proved conclusively that the single float was superior to the earlier triple floats. It became the standard for Navy hydroplanes up to World War I.

The first funds for naval aviation were appropriated on March 4, 1911, and assigned to the Bureau of Navigation. Twenty-five thousand dollars was specifically allocated for "experimental work in the development of aviation for naval purposes."

The new naval Flying Corps was placed in charge of Captain Chambers, who reported for duty with the General Board. When space was not available for his offices they were moved to the Bureau of Navigation.

Eugene Ely prepares to land on the cruiser *Pennsylvania* in
San Francisco harbor. (Courtesy of U.S. Navy.)

Meanwhile, the Wright Company made a formal offer on March 9 to train a pilot for the Navy if it purchased an airplane for $5,000. This condition was later rescinded, and Lieutenant John Rodgers was ordered to Dayton, Ohio, where he became the Navy's second flying officer.

Chambers now used part of the congressional appropriation to order two Curtiss biplanes on May 8. One, the Triad, was equipped with pontoons for operations at sea. It had a metal-tipped propeller that allowed it to reach a speed of 45 mph. A two-seater, its side-by-side arrangement provided controls for both a pilot and a passenger. It was designated the A-1 and became the Navy's first airplane.

Lieutenant John H. Towers reported for duty June 27, after training at the Curtiss School in Hammondsport, New York. He became the Navy's third officer qualified to fly.

Towers demonstrated the Navy's new A-1 on Lake Keuka at Hammondsport on July 1 during a five-minute flight, achieving an altitude of 25 feet. Lieutenant Ellyson flew as a passenger with Curtiss on a later flight that evening, and then made two flights by himself. Two days later Ellyson completed the Navy's first night flight and landed without the aid of lights.

Chambers was now ordered to temporary duty at the Naval Academy to establish an aviation experimental station. Greenbury Point was selected and the following September it became the Navy's first naval air station.

Captain W.I. Chambers, left, and Lieutenant Ellyson in the A-1.
(Courtesy of U.S. Navy.)

Glenn Curtiss demonstrated the amphibious characteristics of the Navy's A-1 on July 10 by taking off from land in the Triad, lifting its wheels while in the air, and making a landing in the water on its hull.

This was also the year that Lieutenant Riley E. Scott, U.S. Coast Artillery, invented a bombsight that incorporated the first bomb rack. It weighed sixty-four pounds, and the sight had a small telescope. After many flights against simulated targets, the system proved amazingly accurate, although low altitudes and low airspeeds made its task easier.

These achievements were acclaimed as outstanding by the Army Ordnance Department, and further development was authorized. Meanwhile, Captain Chambers kept a watchful eye on these experiments.

Captain Bradley A. Fiske, now in charge of the War Plans Division of the General Board, went on record in 1911 as favoring the airplane as the best weapon for the defense of the Philippines. He recommended that four air stations be set up on Luzon, each with at least 100 planes, to be used to attack troops of an invading army. Admiral George Dewey agreed with this brilliant plan, but somehow it was put aside without serious further consideration. A more favorable reaction might have changed the course of history in later years.

A shipboard launching device was demonstrated by Ellyson at Hammondsport on September 7, when he made a successful takeoff from an inclined wire rigged from the

Rear Admiral Bradley A. Fiske in 1912.
(Courtesy of U.S. Navy.)

beach down to the water. In describing the test, he said, "The engine was started and run at full speed and then I gave the signal to release the machine.... I held the machine on the wire as long as possible as I wanted to be sure that I had enough headway to rise and not run the risk of the machine partly rising and then falling.... Everything happened so quickly and went off so smoothly that I hardly knew what happened except that I did have to use the ailerons and the machine was sensitive to their actions."

Ellyson wrote the Department of the Navy a week later that fliers needed special clothing, which, he made clear, he hoped the Navy would purchase. He described the attire as a light helmet, with detachable goggles, or a visor, with a covering for the ears but with holes so the engine could be heard. He cited the need for a leather coat lined with fur or wool, leather trousers, high rubber galoshes, and gauntlets. He further specified that some type of life preserver was needed.

Captain Chambers wrote Glenn Curtiss on October 17 about the need for improved power plants. He discussed heavy oil (diesel) engines, and turbine engines similar in principle to those that, thirty years later, made jet propulsion engines practicable. "In my opinion," he said, "this turbine is the surest step of all, and the aeroplane manufacturer who gets in with it first is going to do wonders."

Experiments with airborne wireless transmission were conducted at Annapolis by Ensign C.H. Maddox and Lieutenant Towers in the A-1. This system, using a trailing wire antenna reeled out after takeoff, failed to transmit properly. The wire was too weak and no definite results were obtained.

Meanwhile, the Bureau of Ordnance continued its experiments to develop a shipboard launching device. Captain Chambers urged that a catapult for launching airplanes be designed similar to the one that launched torpedoes.

Congress became concerned early in 1912 about the lack of American airpower, so $65,000 was authorized for Naval Aviation—an insignificant and tragic amount compared with the $16 million allotted to aviation among the European powers. Unfortunately the Navy Department refused to even spend this amount.

An early but limited interest in the helicopter was shown on March 11 when the Secretary of the Navy authorized the expenditure of not more than $50,000 to develop models of a helicopter design by Commodore F.E. Nelson of the USS *West Virginia*.

At the Michelin competitions in Paris during August of 1912 $50,000 was posted for distribution among winners of various events. Army Lieutenant Scott turned in an impressive performance with his precision bomb rack, dropping three bombs at a time on a rectangular target. He scored eight hits with 15 bombs released from an average altitude of 2,264 feet, winning a $5,000 prize.

The Navy's General Board showed interest in all of these new tactics as they recommended the Naval Air Service to join the fleet at Guantanamo as quickly as possible.

Captain Chambers arranged for two of his top aviators, Lieutenant Towers and Lieutenant B.L. Smith of the Marine Corps, to establish a camp there during January 1913. Smith was the Marine Corps' second aviator. (Lieutenant Alfred A. Cunningham preceded him as Naval Aviator number five.)

Their Curtiss and Wright floatplanes were hardly suited for attacks at sea, but Towers made the Navy's first attempt at bombing stationary targets on February 5. An enlisted technician at the aviation camp rigged a simple bombing tube with crosshairs but without special lenses for these flights.

A few days later, while flying scouting, observation, and gunfire-spotting missions, the two aviators caught up with Atlantic Fleet warships and splattered flour bombs on their spotlessly clean decks.

The results of these experiments proved that aerial bombardment at that period was limited and ineffective because of a complete lack of high-explosive bombs, bomb racks, and bombsights. The airplanes also were too underpowered and unreliable.

Captain Chambers's fertile imagination had been at work for years to develop a seaplane catapult. At the Washington Navy Yard on November 12, 1912, Lieutenant Ellyson in an A-3 was successfully catapulted, and four years later Chambers's invention became standard equipment on capital ships of the Atlantic Fleet. The *North Carolina* was the first battleship to be equipped with his compressed-air system. The inventor was awarded a medal for his invention by the Aeronautical Society, and for his "unusual achievements in being the first to demonstrate the usefulness of the aeroplane in navies, in developing a practical catapult for the launching of aeroplanes from ships, in assisting in the practical solution of the hydro-aeroplane by the production in association with others of the flying boat...through his perseverance and able efforts in advancing the progress of aeronautics in many other channels...."

Among Navy men whose imagination was captured by possible strategic uses of aircraft was Captain Bradley Fiske, a dynamic and forceful speaker on the subject of naval airpower. He saw the airplane as a means of extending the range and accuracy of his favorite weapon, the torpedo. In 1912, he took out a patent covering the weapon and the technique of dropping it.

After the Navy authorized the conversion of the old battleship USS *Texas* as a target ship for bombing tests, Captain Chambers weighed the proposal cautiously, then declined the invitation. "Naval aircraft now available cannot carry large enough bombs to inflict serious damage to a capital ship," he said.

In view of the facts, it was a wise decision. Such tests were delayed for eight years until heavy bombers became available after World War I.

The British conceived the idea of flying in formation in 1912. They suggested military airplanes could be maneuvered in flight similar to cavalry. Lieutenant Hugh Dowding of the Royal Air Force subsequently led a flight of twelve airplanes past a reviewing stand in an aerial display of basic cavalry formations. Thus, formation flying was conceived as a means of mutual protection and to expedite in-flight movement of a group of aircraft under leadership of a squadron commander. Bombing planes were the first to use such tactics during World War I.

Despite the brilliant achievements of the Wright brothers and Glenn Curtiss, the dedicated efforts of Captain Fiske and Captain Chambers, and the enthusiastic men of the small Naval Air Service, the United States continued to fall behind in aeronautics. This was due to lack of wide public support, continual patent litigation among pioneer aircraft inventors and manufacturing companies, and a lack of facilities for aerodynamic experimentation. In truth, the skill of professional aircraft technicians was superior to the abilities of most airplane designers and engineers of the period.

The small Naval Air Service, composed of only eight planes, included three Curtiss hydro-aeroplanes, two Wright hydro-aeroplanes, a Burgess, and two Curtiss flying boats. Despite the size of the organization, experiments continued and Lieutenant Towers set a new world record for endurance on October 6, 1912—six hours, ten minutes and thirty-five seconds. Also, serious tests to develop tactics against submarines were conducted.

When the Massachusetts Institute of Technology established a much-needed course in aerodynamics in May 1913, the Navy was asked to designate a qualified officer to teach the course. A young naval constructor, Jerome C. Hunsaker Jr., was given the assignment. He soon became an outstanding figure in international aviation, and aided research in the United States through his trips abroad. It was through his efforts that a vital wind tunnel was erected at Cambridge.

The Navy's General Board unfortunately was only an advisory agency. It frequently came up with challenging new ideas that were not in line with the thoughts of the other admirals. For example, in December 1912, they endorsed, in vain, a new offensive weapon to be known as the "aeroplane destroyer," with a speed of eighty miles per hour and equipped with machine guns and small bombs.

Chambers, meanwhile, requested that an office of aeronautics be created. His request was not considered seriously, although he was placed in charge of a special board that recommended fifty planes for the fleet, with as many spares, along with six additional planes for an advanced base ashore. He also recommended three dirigibles, to give the Navy a truly mobile air service.

The Navy Appropriations Act for fiscal 1914 recognized that flying personnel should be compensated for their added risks, and an increase of fifty percent in pay and allowances was approved for officers on duty as fliers of heavier-than-air craft. A limit of thirty officers was established, but no naval officers above the rank of lieutenant commander—or major in the Marine Corps—could be a pilot.

Pilot safety improved during 1914, as new aircraft were equipped with a compass, altimeter, inclinometer, and an airspeed indicator. A radio and a generator were recommended as necessary accessories, but did not become standard parts of such aircraft for several years.

On October 7, the Secretary of the Navy appointed a board of officers—with Captain Chambers as senior member—to draw up a "comprehensive plan for the organization of a Naval Aeronautic Service." Its report—submitted after twelve days of deliberation—emphasized the need for expansion and for integration of aviation with the fleet. This was the first comprehensive program for an orderly development of Naval Aviation. Its recommendations included the establishment of an Aeronautic Center at Pensacola for flight and ground training, and for the study of advanced aeronautical engineering; the establishment of a central aviation office to coordinate the aeronautical work of the various Navy bureaus; and the assignment of a ship to train aviators in operations at sea and to make practical tests of the equipment necessary for such operations. The plan called for the assignment of one aircraft to each major combatant ship, and the expenditure of $1,297,700 to implement the program.

Chambers, despite his untiring efforts on behalf of Naval Aviation, was still not appreciated by the Department of the Navy. In late 1913, top officials decided to rid themselves of this outspoken individual, using the excuse that he did not have sufficient sea service as a captain for promotion to admiral. Chambers was forced to accept involuntary retirement. Although he was removed from his post, a change in orders permitted him to remain on duty for another six years, but not in a position of responsibility.

Mark Bristol, like Chambers, was not a pilot, but he was an equally strong-minded individual. After he replaced Chambers on December 17, 1913, he became an early proponent of the aircraft carrier—particularly after the British converted the old cruiser *Hermes* into a carrier for ten planes.

The Office of Aeronautics, under Captain Bristol as its director, was transferred January 7, 1914, from the Bureau of Navigation to the Division of Operations in the office of the Navy Secretary. Three days later, Navy Secretary Josephus Daniels announced that "the science of aerial navigation has reached that point where aircraft must form a large part of our naval forces for offensive and defensive operations."

In less than twenty-four hours after receiving orders on April 20, an aviation detachment of three pilots, twelve enlisted men, and three aircraft under Lieuten-

Early Pensacola officers. Taken in March 1914, this photo shows the men who developed Naval Aviation. From left: Lieutenant V.D. Herbster, Lieutenant M.D. Mellvain, Lieutenant P.L. Bellinger, Lieutenant R.C. Saufley, Lieutenant John H. Towers, Lieutenant Commander H.C. Mustin, Lieutenant B.L. Smith, Ensign de Chevalier, and Ensign Stolz. (Courtesy of U.S. Navy.)

ant Towers sailed from Pensacola on board the USS *Birmingham* to join Atlantic Fleet forces operating off Tampico. They were ordered to assist the United States Marines who had occupied Veracruz, Mexico, to protect American interests in Mexico's civil war.

A second aviation detachment from Pensacola, with one pilot and three student pilots and two aircraft, was dispatched the following day. Lieutenant P.N.L. Bellinger was in charge on board the USS *Mississippi* as it headed for Veracruz.

Once the need for scouting services diminished there, the aviation detachment resumed routine flight operations.

Aviation was formally recognized July 1, 1914, with the establishment of an Office of Naval Aeronautics in the Division of Operations under the Secretary of the Navy.

Meanwhile, in Europe, the stage was set for a fierce world conflict. Aerial bombardment would not be a decisive factor, although the knowledge acquired in those years—and the devotion of a small coterie of dedicated individuals—would do much to save the free world twenty-five years later.

Austrian Archduke Francis Ferdinand and his wife were assassinated in Sarajevo on June 28 by Serbian nationalist Gavrilo Princip. Austria, a key member of the Central Powers—which included Austria-Hungary, Germany, Bulgaria, and Turkey—declared war on Serbia a month later. Germany formally declared war on Russia on August 1, on France two days later, and invaded Belgium August 4. Great Britain declared war on Germany on the date of the Belgium invasion. With Great

Britain's allies France, Russia, Belgium, Serbia, Romania, Montenegro, Portugal, Italy, and Japan joining the fray, the scene was set for a great war. Its underlying causes were imperial, territorial, and economic rivalries that had defied diplomatic settlement.

The Germans moved against Russia in August, defeating her armies on the eastern front at the Battle of Tannenberg. The French resisted attempts by the German armies to sweep through France after Belgium was quickly overrun. At the first battle of the Marne in September, French troops stopped the Germans only 25 miles from Paris. Trench warfare bogged down the German armies for the rest of 1914.

During the early months of the conflict, airmen flew back and forth across the battle lines making notes of enemy troop movements and taking photographs. For a time there was a gentleman's agreement that neither side would fire upon the other's airplanes.

Such a situation could not last. The inevitable happened when a pilot took a gun along and shot at an enemy plane.

No one knows who dropped the first bomb, but there are many accounts of French reconnaissance pilots strafing soldiers in the trenches or marching along roads.

It was not until October 1914 that the first planned bombing raid was flown, when England's Royal Naval Air Service sent Lieutenant C.H. Collett to bomb Germany's airship sheds at Düsseldorf. A month later, three British Avro-504 biplanes bombed the Zeppelin airship works at Friedrichshafen. The Germans retaliated, their aviators hurling small bombs upon the cities of Dover and London.

These bombings, though limited in scope, heralded the start of the air war, despite agreement by most major European powers to refrain from aerial bombardment of any kind.

Although hand grenades were dangerous weapons to take on a flight, they were light and easy to hurl over the side of the cockpit. Ground troops, understandably annoyed at this new tactic, retaliated by firing at every airplane that came within range regardless of its nationality. This forced aviators to increase their altitudes, and some even resorted to sitting on cast-iron stove lids for protection.

After several disastrous accidents with hand grenades, pilots began looking for a bomb that could be accurately dropped from higher altitudes and explode on contact.

The first answer to the problem was finned dart aerial bombs. They were small and ineffective because of their limited yield and fragmentation capability. They were carried in a wicker basket on the cockpit floor between the airmen's legs. Observer-bombardiers often dumped them over the side by the basketful.

Some aircrews stowed their darts in the loops of the rounded turtledeck aft of the rear cockpit. This gave the rear-seat man the chance to fire his free-swinging machine gun and drop darts on troops in the trenches. Some enterprising crews even tried to destroy a gas-filled barrage balloon by running their wheels across the top, then blowing it up with a handful of darts. It was not a practice for the fainthearted!

In the United States, naval aviators took the war in Europe seriously. In 1914, Lieutenant V.D. Herbster, and Lieutenant B.L. Smith of the Marine Corps, conducted experiments at Indian Head Proving Ground in Maryland. Using a new technique, with bombs armed by a windwheel mechanism, they flew into the wind at 1,000 feet while the bombardier waited until the aircraft was just short of the target.

Herbster and Smith both agreed, with a nervous chuckle, that the drops were successful if the bombardier managed to disengage his fingers from the windwheel!

Although the bombs tumbled end over end at first, they later straightened out and landed at a good angle. Hits were scored on small targets, but it was decided to improve the trajectory by adding stabilizing fins and contact fuses.

Shortly thereafter Captain Bristol established a bombing course for aviators at Pensacola patterned after these experiments. Specifications were drawn up for the manufacture of larger bombs with delayed-action fuses for use against submarines and ships.

After the start of the war in Europe on July 28, 1914, and despite recommendations by the Chambers board, the Navy had only twelve airplanes for training. This prompted Bristol to plead for two planes for each of the sixteen battleships in the fleet, a request receiving Fiske's strong backing. On November 23, Bristol was finally named Director of Naval Aeronautics. With this new authority, he appealed to the General Board for assistance in building up his air service.

After thorough study, the board agreed that the situation was deplorable, claiming aircraft were the eyes both of armies and navies, and finding there was no limit to their offensive capabilities. While the board urged the Department of the Navy to make $5 million available to aviation, they pointed out that only a third of previous allocations had been spent.

No action was immediately forthcoming, but Lieutenant Towers was ordered to London, Lieutenant Smith to Paris, and Lieutenant Herbster to Berlin to study European military aviation.

In Europe, following the British and French bombing raids, the German Navy decided to use the huge zeppelin airships in organized attacks on Allied cities. Ten bombs were dropped on Antwerp in Belgium on September 28, 1914, but the first large-scale bombing effort was not flown until Christmas Eve, when bombs plunged out of the sky on the city of Dover, England.

Germany had thirty airships available. Some of them had been constructed originally for the commercial transport lines of Delag, the German civilian carrier, prior to the war. The five-ton load-carrying capability of each zeppelin gave the Germans a decided advantage, and they moved swiftly to capitalize on it. They routinely bombed British cities against no opposition because the British had no airplane capable of reaching their altitudes to attack them. During the summer of 1915, Germany increased its airship raids, hoping to bring England quickly to its knees by destroying morale at home. Although raids were made regularly on several cities, including London, the raids were more of a nuisance than anything else.

British scientists finally developed more powerful engines, and fighters were built that could fly above the vulnerable zeppelins. Commander A.W. Bigsworth of the Royal Flying Corps led an attack on the LZ-9 near Dunkerque, and much to the zeppelin commander's shock, flew above it and dropped four bombs. The bombs caused the huge bulk to collapse, although it was able to limp back to its base at Ghent.

At a debriefing, Bigsworth told his superior, "Now that we can reach the airships, we need incendiary bullets and electrically-fired explosive rockets."

Flight Lieutenant J.K. Warneford destroyed the first German airship near Ghent on June 6, 1915. His tiny Morane monoplane soared majestically above the zeppelin, and when he was in proper position he dove down. The bag seemed to fill the sky as he dropped his bomb. He pulled back hurriedly on the stick as the Morane protested vigorously at such rough treatment. Then an awesome explosion rocked his airplane, throwing it into a violent tailspin.

Warneford was frantic during recovery, and for several seconds he thought the end had come. Finally he managed to level the wing, casting a quick glance at the zeppelin falling behind him with huge flames licking the sky as he headed for home base.

Concentrated bombing attacks by the Royal Naval Air Service, the Flying Corps, and French aviators against airship bases in Belgium forced their abandonment.

Night raids by the "zeps" were no longer effective. British defensive tactics became so successful that they were shooting down zeppelins regularly.

Most American naval aviators disliked dirigibles, but Captain Bristol avidly promoted them; it was one of the few times when this capable officer showed poor judgment.

German submarines began to blockade Great Britain starting in February 1915. Among the worst casualties of the U-boat war was the sinking of the *Lusitania* on May 7, resulting in the loss of 1,198 men, women, and children, including 128 Americans.

On the Western Front, during the battle of Ypres during April and May, the Germans used poison gas with deadly effect in their attempt to overcome the stout resistance of Allied troops.

On the Eastern Front, German and Austrian armies launched a "great offensive" to conquer all of Poland and Lithuania. By September 6 the Russians had lost a million men, and the Central Powers emerged victorious.

After the Allied "Great Fall Offensive" on the Western Front failed, the front lines remained the same as they had been at the start of the war.

In the United States the Naval Appropriations Act of 1916 was approved on March 3, 1915. At last it recognized the dire need for expansion of the Naval Air Service. A million dollars was approved, but equally important, the act authorized the formation of a committee, later known as the National Advisory Committee for Aeronautics. Two of its twelve members were Navy men and it soon began its historic contributions to aeronautical research.

Early advocates of aviation like Bristol, Chambers, and others had long fought for adequate compensation for men who risked their lives flying. The new act recognized these risks and increased the amount previously authorized to fifty percent of base pay, along with survivor's benefits for qualified aviators, amounting to a year's pay for next of kin. It was a welcome reward for the Navy's forty-eight eligible officers and its ninety-six enlisted men, plus the twelve officers and twenty-four enlisted men of the Marine Corps.

The Navy signed its first contract for lighter-than-air craft on June 1. The Connecticut Aircraft Company in New Haven was ordered to build one nonrigid airship. It was later designated the DB-1.

A week later Josephus Daniels, Secretary of the Navy, wrote a letter to Thomas A. Edison to mobilize science in the interests of naval aviation. "One of the imperative needs of the Navy, in my judgment," he said, "is machinery and facilities for utilizing the natural inventive genius of Americans to meet the new conditions of warfare...." Edison agreed and the Naval Consulting Board was established with a group of civilian advisers.

While the war bogged down in the trenches on the continent of Europe, the dire need for improvements in air weapons was apparent on both sides. Ordnance engineers went to work designing better bombs and releasing and sighting equipment.

The need for an aerial bomb rack that would operate efficiently was particularly urgent. An isolated instance in the early days pointed to the critical need for any workable device. During a flight by two German aviators over Allied lines, the observer-bombardier pulled the safety pin on a hand grenade. He moved quickly to hurl it over the side. To his horror, the grenade slipped from his grasp and rolled out of sight under the floorboards. Panic seized him. He knew the grenade was set to explode within seconds from the time it left his hand.

The pilot, sensing something was wrong, but not knowing just what, gaped in astonishment as his bombardier stood up in the cockpit and saluted an attacking Allied aircraft just before the plane disintegrated.

Bradley A. Fiske, finally a rear admiral, paid the price for outspoken criticism of his superiors by being replaced in 1915 as aide for operations. On July 1 he was appointed President of the Naval War College at Newport, where he could no longer promote naval aviation except through letters.

Bristol, meanwhile, recommended 120 planes and two dirigibles for the fleet and twenty-eight planes for naval defense districts. He said his request was based on the assumption that twenty-four fleet divisions of four planes each and three reserve divisions were a minimum requirement. He also asked for promotion slots for his airmen, who had suffered by merely being in the fledgling air service. Both Bristol and his able assistant Lieutenant Commander H.C. Mustin also pleaded in vain for construction of carriers, pointing out that the British Royal Navy had five nearly completed.

Pusher-type planes now were giving way to tractor designs, although many pushers continued to be built. Except for bombers, virtually all planes were small, and had relatively low-powered engines. A big order for a firm was six or eight airplanes a year, and they were priced between $10,000 and $15,000 each.

While Italian and German aviators forged ahead, Allied fliers often had to delay operations until shipments of bombs could be delivered.

Impatient ordnance men and pilots conceived a substitute for explosive bombs and made nonexplosive darts for dropping over the front lines. Shaped like a stonemason's star drill, and equipped with small guide vanes, bombardiers tossed these ugly weapons over the side in bundles. They were deadly during an attack on a marching column of troops, or a platoon at rest, because they inflicted horrible wounds. They often killed infantrymen because the eight- to twelve-inch missiles easily pierced their steel helmets.

Russian-built Sikorsky four-engine bombers carried their first loads over German lines during February 1915. Seventy-three of them attempted to fly from Russia for delivery to Allied airfields, but only three completed the trip. Engine troubles plagued the Illya Mourometz bombers, and crash landings were common in hayfields and cow pastures.

The Sikorsky bombers that did get into combat were never shot down. Their five machine guns—capable of firing at every conceivable angle—made them impregnable to German pursuit planes. The planes even had a tail gunner. Such a position was a vast improvement on the typical rear-seat gunner, whose range of trajectory was limited by an aircraft's tail.

The Sikorsky bombers would have posed an extreme threat to Germany if production had not been halted because of the Russian revolution in 1917.

Italian Caproni bombers maintained a production lead until 1917, when German Gothas and British Handley-Pages went into service. In the first two years of the war, French Caudron bombers also proved useful, along with the smaller Farmans and the Breguet Michelins.

In early 1916 Captain Bristol realized he could no longer fight the Navy bureaucrats in Washington. They were obviously out to get him removed from office. He beat them to it by requesting a transfer to command the *North Carolina*. He was removed as director of Naval Aeronautics on March 4, 1916, and both the title and the office ceased to exist. After his request to transfer to the *North Carolina* was approved, he was given the new title of Commander of the Air Service. Theoretically Bristol retained operational supervision over all aircraft, air stations, and the further development of aviation in the Navy. But this was window dressing by bureaucrats anxious to mollify their critics, who were charging them with failure to develop aviation.

After Bristol's reassignment in March, he wrote a parting report in which he sternly reminded the Navy's top brass that there were only twenty-four officers at Pensacola, and only ten qualified pilots.

At first Bristol hoped that in his new position he could give the fleet a real air arm, but the officer who was assigned his former duties in the Office of Naval Operations was Clarence K. Bronson, a junior grade lieutenant who could not possibly hope to achieve what a captain had failed to accomplish.

Bristol was deeply disturbed when he learned that Bronson was subordinated to Captain Josiah S. McKeen of the Atlantic Fleet's destroyer force, who had no interest in the air arm. The situation further deteriorated when Bronson was killed on November 8 in an airplane by the premature explosion of a bomb.

The farce finally ended when Bristol was detached as Commander of the Air Service on December 12, and the functions of the command—but not the title—were transferred to Rear Admiral Albert Gleaves, Commander, Destroyer Force, Atlantic Fleet.

Commander H.C. Mustin, now in charge at Pensacola, received permission to establish an experimental department for testing new aircraft and their equipment. Naval pilots daily flew their Curtiss N-9 floatplanes with their front-mounted engines while they tested individual bomb racks attached to the bottom of their fuselages with special releasing-bar mechanisms. Meanwhile, other pilots continued to practice with hand-held bombs.

Throughout 1916 aerial bombing techniques showed encouraging improvements, particularly after telescopic sights became available. They were illuminated, thus making night missions practical.

Flight and navigation instruments vital to accurate bombing were also tested. Such instrumentation included the magnetic compass for flying straight courses, the airspeed meter, and an inclinometer (a spirit level in disguise), which indicated proper coordination of the ailerons and rudder in flight. Finally, the group developed a gyroscopic baseline indicator as a flying aid during the final run to the target. This was useful in conjunction with the magnetic compass, which was often erratic in rough air.

The men who developed these new aids to flight performed miracles for their time. Today's turn-and-bank indicator evolved from the baseline indicator of 1916.

Despite these new instruments, most bombardiers of that period couldn't hit the proverbial broad side of a barn. In combat, the best bombing was accomplished by veterans of the wooden stick era, who through some innate sense could lean over their cockpits and drop bombs with reasonable accuracy. Obviously, a better system was needed.

The battle of Verdun in February 1916 cost the Germans and French about 350,000 men each. With the ground warfare achieving no gains for either side, in March the Germans began extensive submarine warfare.

In May, British and German fleets fought a major sea battle off the Jutland Peninsula in the North Sea. The British fleet lost more ships, but won the battle because the German Navy never again ventured out to sea except in submarines.

New German naval airships took to the air during the early summer of 1916. These "super" airships were designed specifically for altitudes up to 24,000 feet. With supercharged engines and oxygen supplied to the crews, they cruised at ninety miles per hour.

These flights were perilous indeed to German crews because of high-velocity winds and extremely cold temperatures. Once the engines failed, they quickly froze because they were water-cooled. It was not at all unusual for airships to drift helplessly because of frozen engines.

These improved airships were constructed so the Germans could launch another bombing offensive during July 1916. There had been a general lull in airship flying since air superiority was lost to the British during the Battle of the Somme. That battle began in July and continued through November. The British had lost 400,000 men, the French another 200,000, and the Germans about 450,000. These enormous losses were incurred with no strategic gains for either side.

The first aircraft production contract was signed by the United States Navy with Glenn Curtiss on August 10, 1916, and covered thirty training planes. The contract stressed the importance of early delivery of N-9 two-seater trainers starting in November and ending in February of 1917.

Despite the bitterness of Bristol, Chambers, and Fiske concerning the slow progress of Naval Aviation, the General Board recommended a great increase in service planes, calling for 564 trainers. Both Towers and Mustin deserve credit for this first long-range program because they fought aggressively for it.

Bristol wanted $13 million, of which $6 million would be set aside for two carriers. He lost this round: the Naval Appropriations Act of August 29, 1916, provided

$3.5 million for aviation in fiscal 1917, but the appeal for carriers was stricken. The bill also provided for the establishment of a Naval Flying Corps and a reserve corps. Despite the fact the bill called for a separate corps, the corps was never created, though the bill did provide openings for 150 officers and 350 enlisted men from the Navy and the Marines.

By the end of 1916, sixty planes were on order. Curtiss had the greatest share with its order for thirty N-9 seaplanes.

Although the United States had still not entered the worldwide conflict, the American military services were preparing for the inevitable. Lieutenant Commander J.C. Edwards proposed construction of a large flying boat bomber capable of carrying 5,000-pound loads. "It should be powered by five steam turbine engines instead of the new Liberty power plant. The large Capronis now in use as the standard heavy bomber of the Allied forces should be replaced," he said.

Despite such views, the General Board went on record as opposing the further development of heavy bombardment aircraft. "In our judgment," they said, "such aircraft should be deleted from the combat roster. Airships of the Zeppelin type are likely to become the standard bombing machine of the World War. Airships able to hover motionless over a target can score far more hits than a moving airplane."

The Navy Secretary, Josephus Daniels, twice had to warn Admiral Fiske to cease making open assertions to newsmen that the Navy must build a fleet of battle planes capable of hitting the enemy where and when the blow was least expected. "Such battle planes should be a major instrument in Naval warfare," he said, claiming his arguments were sound, but his zealous enthusiasm made him many enemies. He contended such aircraft could be built in six months, whereas it required years to build surface squadrons. His insistence prodded the reluctant Naval Bureau of Ordnance to design 200- and 300-pound bombs and 600-pound aerial torpedoes.

Unfortunately, Naval aviation remained hopelessly confused because the Naval Board's 1916 report stated, "Aeronautics does not offer a prospect of becoming the principal means of exercising compelling force against the enemy—except limited to scouting, patrolling and spotting."

By the end of the year the once mighty zeppelins found it almost impossible to continue their raids due to the high toll British night-fighters were exacting. On most missions at least half of these dirigibles were being shot down. Heavier-than-air planes were finally recognized in Europe as more practical because of their superior speed and maneuverability. Unfortunately, many high officers of the United States Navy refused to concede the truth.

With the advent of 1917, it became apparent to most military leaders that opposing land armies in Europe were stalemated in a grim war of attrition. Ground soldiers merely tried to stay alive in the miles of cold, muddy trenches until at last top strategists on both sides turned hopefully to airpower to free them from their dilemma.

2

The Navy in World War I

A wave of optimism swept Allied Europe April 6, 1917, when the United States declared war on the Central Powers. Losses in human life and materiel had been staggering, and there seemed no hope of winning the war solely with ground troops. The free world was enthralled, therefore, by the rumors that the United States would send 100,000 bombing planes to Europe within six months.

Among top Allied military commanders there were some who said, "Large numbers of heavy bombing aircraft can bring about a quick German surrender." Admiral Fiske said, "Aerial bombardment and aerial torpedoes are the key to defeating Germany." Fiske had long supported torpedoes, but the aircraft available were incapable of carrying a torpedo large enough to destroy a warship. Such a torpedo, it was known, would have to weigh at least 800 pounds.

Italian authorities claimed that large warplanes could do at long range what ground and naval guns could only do at short range. The Italians—world leaders in the production of large bombers—urged the United States to purchase Italian bombers as soon as possible and win the war. By this time, the Italian Air Force was flying Caproni bombers over Austria directly across the Adriatic Sea.

But on the day the United States entered the war, the United States Navy had fifty-four aircraft, an airship, and three balloons, none of which was suited for combat, with forty-eight officers and 239 enlisted men with some experience in aviation, but all based on a single air station. Thirty-eight of the officers were pilots. Forty-five of the aircraft were seaplane trainers, six were flying boats, and there were three land planes.

The German submarine war was at its peak when the United States entered the conflict, and containing it was on the minds of all Allied leaders. It was agreed that naval aviation's primary role should be relegated to the antisubmarine campaign.

Lieutenant Colonel William "Billy" Mitchell of the Army Air Service was ordered to Europe as an observer, arriving in France just after the United States' declaration of war.

He immediately went to the front and became the first American Army officer to fly over enemy lines. Mitchell's cables to the War Department calling for the establishment of an aviation branch in the American Military Mission in France were ignored. But he convinced the French General Staff to make such an appeal. President Woodrow Wilson, after he received a telegram from the French premier, Alexandre Ribot, ordered that an appropriation bill be submitted to Congress to pay for such a mission.

After General John J. Pershing arrived in Paris in June 1917, with the advanced echelons of the American Expeditionary Force, Mitchell was selected to create an aviation program for his command. Although Mitchell was selected to run the program as a full colonel, his "air section" three months after the declaration of war still had only a single French Nieuport pursuit plane, with Mitchell as the sole pilot.

In Washington, Secretary of War Newton D. Baker announced that the United States was building 20,000 warplanes, but this was a gross exaggeration. The actual production from July 1917 through July 1918 averaged only 144 aircraft a month.

It had been six years since the first aircraft was acquired by the Navy and its first pilots trained, but the United States Navy still had no strong organization for its aviation force. A few uncertain steps had been taken to develop operational units, but the emphasis was on training. Even the training program was a shambles, recovering from the effects of a six-month training interruption the year before because of a growing number of aircraft accidents.

Although $3.5 million had been approved for Naval Aviation for this fiscal year, the Naval Flying Corps authorized by the Naval Appropriations Act of 1917 had not been established, and Navy orders for new planes were still held in abeyance, although the money for them had been appropriated. Fortunately for American aircraft manufacturers, they were receiving orders from the Allied powers in Europe. Suitable sites for air bases along the East Coast had been recommended, although their construction was still in the planning stages.

Delivery of the first N-9s (Navy version of the famed Army "Jenny") in 1916 had helped to boost training hours, and experiments with shipboard catapults proved their effectiveness. The USS *North Carolina* was so equipped to operate aircraft, and the USS *Huntington* and the USS *Seattle* were undergoing installation of such equipment.

Aviation elements in the Naval Militia and in the National Naval Volunteers had been established prior to America's entry into the war, but with little practical Navy support. But a start had been made in training men to fly airplanes and training other men as mechanics to keep the planes in service. Student groups at universities raised money to purchase equipment and to hire instructors to teach them to fly. Yale University had a particularly good training program. Other young men were learning to fly on their own, hoping to later join these student groups.

Glenn Curtiss had long been active in setting up schools to train pilots, and now other manufacturers did likewise. Rear Admiral Robert E. Peary, who had long fostered the development of aviation, raised money by individual subscription to form the National Coastal Patrol Commission. Its first unit—the Aerial Coastal Patrol No. 1—had men from the first Yale unit.

Although the air station at Pensacola had been activated in January, 1914, its flight test program remained small, with no formal curriculum. Training during the previous three years had been sporadic on an individual basis rather than the more desirable group basis.

Now the Navy's training program was revised to permit assignment of new classes every three months for an 18-month course for both heavier-than-air and lighter-than-air pilots. The organization quickly proved to be impractical, and heavier-than-air pilots were assigned to Pensacola and lighter-than-air to Goodyear's dirigible school at Akron.

On the day war was declared, the Secretary of the Navy established standard flight clothing for the Naval Flying Service. Clothing was specialized to meet the special needs of fliers operating under all weather conditions. It consisted of a tan sheepskin long coat, short coat and trousers, moleskin hood, goggles, black leather gloves, soft leather boots, waders, broughams, and life belts. In September, a winged fowl anchor was adopted as an official device to be worn on the left breast by all qualified naval aviators.

The Joint Technical Board on Aircraft recommended the initial program to produce aircraft for the war effort on May 23, by calling for the production of 300 training planes, 200 service seaplanes, 100 speed scouts, and 100 large seaplanes. The Curtiss N-9 and R-6 were listed as most satisfactory for school and service airplanes, but no others were considered sufficiently developed to permit their selection.

The Navy's DN-1 airship flew on April 20 at Pensacola, but its performance was so unsatisfactory that it made only two more flights that month before it was permanently grounded as unsafe.

The Navy's first successful airship, the B-1, was tested satisfactorily on May 30 by Goodyear pilot R.H. Upson. It made an overnight test flight from Chicago to Akron.

The construction of five prototype models of 8- and 12-cylinder Liberty engines was authorized on June 4 by the Aircraft Production Board and the Joint Technical Board on Aircraft. The design of these engines began on May 29 and was based on conservative engineering practices adapted to mass production techniques long used in the manufacture of automobile engines. Two engineers resolved the new engine's technical problems in a Washington hotel room. J.G. Vincent and E.J. Hall of the Packard Motor Car Company and the Hall-Scott Motor Car Company were responsible for the most noteworthy achievement in aviation during the war. At L.W.F. Engineering Company, College Point, Long Island, a company airplane was redesigned as a test bed for the Liberty engine and made its first flight on August 21, 1917. Although the first engine had eight cylinders, it was decided production models should have twelve to provide more horsepower. Each engine weighed 804 pounds and produced 420 horsepower, and they were of good design. Production lagged because it took so long to "debug" the engine before it could be produced. One of the problems was that automobile executives were running the aircraft production program. They were good at mass production techniques, but they knew little about the production of aircraft engines.

In May, at the urgent request of the French government, seven American officers and 122 men in training at Pensacola were organized into the First Aeronautic Detachment of the United States Navy. With Lieutenant Kenneth Whiting in command, this flying unit arrived in France on June 5, 1917. Only four officers were qualified pilots, and the detachment had no airplanes.

The pilots of Whiting's detachment were as odd a military group as one could imagine, but they had one predominant characteristic—they all wanted to fly. The detachment included mechanics, carpenters, college students, taxi drivers, and farmers, and they could not wait to get into combat. They were partially trained at Pensacola, and completed their training with French instructors. Later they flew with French, British, and Italian squadrons. Most of them never fired an American machine gun or dropped an American bomb—or even saw an American-made plane—until they returned home.

When they arrived in France, General Pershing was in England, still en route to his command of the American Expeditionary Force. The Army's First Division was still back home, and the draft law was just being implemented.

Joe C. Cline had enlisted on April 3 in the United States Navy as a Landsman for Quartermaster (Aviation) after having served four years in the Illinois Naval Militia. At Pensacola, he found to his surprise that there was no cadet status for aviators, and no ground school or flight instruction for large groups. There were just a few regular Naval Aviation officers attached to the station responsible for their training.

After three weeks of drilling and some Navy indoctrination (most of which Cline had received in the Militia at home). Pensacola's commander issued a call for volunteers for duty in a foreign land. Fifty Landsmen for Quartermaster and fifty Landsmen for Machinist Mate were selected. Cline was one of those selected to be trained as a pilot. Machinist Mates were to be trained in maintenance and overhaul duty, but they later ended up as observers, machine gunners, or bombardiers.

After the detachment arrived in France nobody knew what to do with them. Whiting rushed off to Paris to see the American ambassador, the Naval Attaché, the French Minister of Marine, and several others. These officials agreed that the French should train them, and supply the detachment with airplanes and the accessories needed to wage war in the air. Whiting was told that while his men were in training three air stations would be constructed for them.

They trained in the French Caudron G-3—a biplane with warping wings and a two-seat cockpit. None of the group spoke French, and their instructors didn't speak English. After takeoff, the instructor, in the rear seat, would turn control over to his student. If the nose was too high, the instructor would push forward on the student's helmet. If it was too low, he would pull back on his helmet. If the left wing was down, he'd tap on his right shoulder; right wing down, he'd tap on the left. Each flight lasted about twenty minutes.

On the ground, the instructor would pull out a pasteboard card with a line drawn down the middle. One side was written in English and the other in French. The instructor explained the mistakes his student had made while in flight, chewing him out in French, while pointing to the English translation.

Despite these adverse conditions, about two-thirds of the group qualified to solo with less than five hours of dual instruction. Then they went to other bases to complete their pilot training. In less than four months, with an average of thirty-two hours of training, they were qualified pilots.

The American Naval Commander in Chief in Europe, Admiral William S. Sims, was later shocked to learn that the enterprising Whiting planned to set up a base at Dunkerque for offensive operations against German submarine bases. Sims first heard about it from the British, but was forced to agree that, under the circumstances, Whiting's assumption of responsibility was commendable.

With Sims's encouragement, Whiting wrote the admiral a memorandum that explained the functions coastal stations and their seaplanes could provide. Not all his ideas were adopted, but his memorandum set forth the policy that governed the buildup of American naval aviation during that first year.

Sims demanded that Washington officials send both ships and planes to cover the battlefields and oceans. "The sooner planes get overseas," he said, "the sooner the U-boats will be defeated."

John Towers, a veteran of six years of flying, and now a Lieutenant Commander, had taken over the aviation desk in the Office of Naval Operations following Bronson's death the previous December. He quickly achieved what Chambers and Bristol had long fought for—centralized supervision of aviation activities. The Appropriations Act of 1917 also added another $3 million to the funds previously allocated for aviation for a naval air program. It was hoped this money would give American forces a continuous bombing offensive with long-range, heavily armed planes under American command.

On July 23, 1917, fifty men arrived at Cambridge on the campus of the Massachusetts Institute of Technology from First Naval District headquarters in Boston. They were met by Lieutenant Edward H. McKittrick. These men were the first of over 4,000 who would receive their introduction to the naval service at that school and go on to carry out their duties at home and abroad. Many would enter flight training after six weeks of ground school and become naval aviators.

The British had pioneered the concept of using civilian educational institutions for the initial stages of military training. But MIT was the first school of its kind established by the Navy, and it became the principal source of trained men for the growing naval air service.

Meanwhile Yale also trained a unit of reserve pilots for the Navy. After the United States declared war on Germany, some Yale-trained men went overseas to command air stations and fight in combat.

When it appeared likely that existing aircraft plants in the country would not be able to cope with the large orders offered them by the Army and Navy, the Department of the Navy ordered the construction of its own factory. The Naval Aircraft Factory at Philadelphia was ordered to manufacture at least some of the aircraft on order and to design and build wholly new airplanes. It was further assigned the responsibility for accumulating data by which the Navy could be guided in dealing with costs of other contracts by privately owned factories.

Commander F.G. Coburn of the Construction Corps was directed by the Secretary of the Navy in June to recommend a suitable location for such a factory and estimate its cost. He was told that the factory must be capable of producing 1,000 training seaplanes a year or their equivalent.

After visiting several plants—particularly the Curtiss Company plant in Buffalo, which was the only factory in the United States capable of producing aircraft in quantity—he made his recommendations.

On July 27 Secretary of the Navy Josephus Daniels approved the project. It called for the expenditure of $1 million at the Philadelphia Navy Yard, where ample land was available and the location was considered advantageous in terms of labor, material, and transportation. Coburn pointed out in his report that the Delaware River was an ideal testing place for the Navy's seaplanes. It was decided to build a permanent plant, which was completed by November 28, 1917—110 days after groundbreaking—with Coburn as its first manager. Four engineers were hired, but only ten of the technical men who were hired had previous aircraft experience. Thus it was necessary to sign them up as a special employee class exempt from competitive examination. The first female employees went to work there in December as inspectors.

The original proposal for the factory specified that it was for the construction of training planes, but when it was realized that other factories were producing sufficient quantities, the factory began production of Curtiss H-16 twin-engine flying boats. These planes were desperately needed to counter the German submarine menace. The upper wingspan of the big flying boat measured ninety-six feet and its hull was forty-six feet long. It was powered by two Liberty engines, armed with four machine guns, and carried a crew of a pilot, one or two observers, a mechanic, and a wireless operator. The first two Navy-built H-16s were shipped to England in late March 1918.

The previous December, orders had been issued to vastly expand the factory, because it now had orders to build 864 H-16s in addition to the original fifty. This was done. By the end of the war, more than forty acres were occupied by the enlarged plant, representing a total investment of more than $4 million.

By the summer of 1918, the factory was building the F-5L flying boat, which was based on an experimental British prototype. It had greater endurance than the H-16 and was capable of carrying a heavier bomb load. Under its lower wing, the new bomber carried two depth charges. When a submarine was sighted, the pilot released them.

In June, production reached one aircraft per day. By the end of the year, the total output of the Naval Aircraft Factory included 183 twin-engine flying boats, with fifty sets of spare parts. Of this number, the last thirty-three were F-5Ls.

This was a remarkable achievement, coming only six years after the Navy had purchased its first aircraft. It was accomplished by an inexperienced group of workers who were trained to be skilled aircraft producers.

The United States Naval Air Force, Foreign Service, executed thirty attacks against enemy submarines. Ten were considered to have been at least partially successful. Approximately one hundred tons of high explosives were dropped during 22,000 flights over 800,000 nautical miles of submarine-infested ocean. Almost always the damage inflicted by aircraft on submarines was of an indirect nature, because destroyers were summoned to complete the submarine's destruction. Therefore destroyers got the credit, although the aircraft was responsible for making the "kill" possible. Ensign John P. McNamara made the first recorded attack on March 25, 1918, while serving at the Royal Navy Air Station at Portland, England.

The United States Naval Air Station at Ile Tudy, France, had encountered more action than any other because of its location. Two coastal convoys passed through its sector daily, one bound north, the other south. Around Penmarch Point, the water was deep near shore, free of reefs and sand bars, and ideally suited to submarine operations. This sector was marked off into twenty-five-mile squares—subdivided into squares of five miles—so that planes could report positions every half hour and be quickly and accurately located. Communication was maintained with shore bases by radio and pigeons, and with ships by message buoys, phosphorus buoys, Very pistols, and blinker lights.

A section of two planes escorted each convoy. As the sector was too long to be covered entirely by two planes, it was necessary to send out another section to relieve the first when the convoy was roughly halfway through the area. This necessi-

Close-up of F-5L. (Courtesy of U.S. Navy.)

tated the use of eight planes per day just for convoy work. In addition, there was also a section known as the "Alert," which was available to take to the air from daybreak to darkness in response to a report of a submarine. When a convoy was picked up, the planes would first circle over it. While one plane remained near the convoy, the other would fly as far as ten to fifteen miles ahead, zigzagging broadly on both sides. This plane would return, again circle the convoy, and repeat the same maneuver, time and time again. Thus the convoy was well protected from surprise.

On April 23, 1918—a convoy escort of two Donnet-Denhaut seaplanes piloted by Ensign E.R. Smith and R.H. Harrell—joined the southbound convoy of about twenty ships six miles north of Penmarch Point. The weather was foggy so they first flew to the rear of the convoy to look for stragglers. Then they flew a wide circle toward the main body of ships. Shortly afterward they sighted a suspicious wake, apparently made by a submarine at high speed. They immediately dove to the attack and Smith released two bombs. The first landed on the fore part of the wake, and the second ten feet ahead. Peering down, Smith and Harrell noted that their explosions had created a heavy disturbance in the water. This was followed by air bubbles. Harrell believed that Smith's bombs had been so successful that he decided not to drop his. He marked the spot with a phosphorus buoy and circled while Smith flew to a destroyer, the USS *Stewart,* and dropped a message buoy. The *Stewart* was soon at the scene with the French gunboat *Ardente,* and they dropped three depth charges. The pilots, circling overhead, noticed small pieces of wreckage, particles of sea growth, and large quantities of oil bubbling to the surface. They returned to their base and reported that they had attacked a German U-boat and believed they had sunk her.

Ensign Smith and his observer, Chief O.E. Williams, were officially credited by French naval authorities with sinking a submarine. They were awarded the Croix de Guerre with Palm Cluster.

One of the most ambitious of all Navy projects was the plan to establish a strategic bombing force in the Calais-Dunkerque area to attack the German submarine bases at Ostend, Zeebrugge, and Bruges in Belgium. The Northern Bombing Group was created, following recommendations by Lieutenant Robert Lovett, Major A.A. Cunningham (U.S. Marine Corps), and Commander Towers to the General Board. They called for round-the-clock bombardment by both Navy and Marine pilots. They would fly Capronis, Handley-Pages, and the smaller DH-4s.

Despite bitter objections by army brass, the plan was set up. Army officers complained loudly. "The Navy is overstepping the bound by even thinking of flying bombing missions over land," they claimed. The Navy replied, "Destroying enemy submarines is our business, even if the U-boats are in their home pens."

Captain David C. Hanrahan, a decorated naval veteran of the Spanish-American War, was placed in command of the group. It was organized into a night wing of Navy pilots and a day wing of Marines. The night group, although commanded by Robert Lovett—thirty years later an outstanding secretary of defense—had one unhappy experience after another, and achieved little success.

The first contingents arrived overseas during June of 1918 after extensive training at Pensacola and Miami. Navy men were assigned to British squadrons in England for checkout in late-model aircraft, while the Marines were sent to France in July for similar assignments with British and French squadrons for advanced training.

Marine Squadron 9 received its first combat lesson when a flight of eight DeHavilland bombers made an independent raid with the Northern Bombing Group on October 14, 1918, dropping 2,200 pounds of bombs on a railroad junction at Thielt-Rivy in Belgium.

The Army did not give in easily to what they considered were their inland bombing rights. Inasmuch as multiengine bomber production was controlled by the Army, the Navy found itself in a difficult position, because the Army refused to share its new Handley-Pages with the Navy. Top Army officers said indignantly, "Why doesn't the Navy stick to its flying boats and leave inland attacks to the Army?" The problem became so acute that it was referred to General Pershing, Supreme Allied Commander of the American Expeditionary Force in Europe. Pershing, possessing an understanding viewpoint of the controversy, told Admiral Sims, "This is a job for naval aviation."

General Benjamin Foulois, Army Air Services chief, was not receptive to his commander's orders. After some delay, a few Handley-Pages were delivered to Naval Air Station (N.A.S.) Killingholme. They turned out to be poorly constructed and unsafe for flight. Navy pilots refused to fly them. They sent strongly worded complaints to Army headquarters. This action brought only promises to replace the defective parts.

In desperation, the Navy purchased 110 of the slower Capronis, but they were even worse than the Handley-Pages. Of the eighteen bombers delivered in July, wings were warped due to poor crating, and the electrical systems were completely inadequate. Both shortcomings were due primarily to inferior workmanship because of mass production by unskilled workers.

Attempts to fly the Capronis across the Alps from Italy to England met with disaster. Most crash-landed along the route because of engine failures, fires, jammed controls, and erroneous navigation. The first Caproni assigned to a night mission bombed the docks at Ostend on August 15 with 1,200 pounds of bombs.

American-built DeHavilland fighter-bombers now began to arrive in France. This was the first visual evidence of the plan to fill the skies over Germany with American bombers. It impressed no one, least of all the Germans.

Construction of British-designed bombers in the United States was so bad that many had to be torn apart and rebuilt by experienced British and French mechanics.

Results against the submarine bases might have been substantial if only the Americans had been able to match aircraft procurement with the readiness of the shore establishment and the assignment of trained personnel. But the Allied ground offensive brought the war to an end before the air offensive had a chance to show what it could do.

American Marine, Navy, and Army pilots eventually flew rebuilt DHs into combat, and through their personal heroism they chalked up remarkable records.

Congressional committees read about the pace of aircraft production, charging favoritism in the award of some contracts. They were particularly incensed because it was obvious that profits on some airplanes and engines were excessive.

It was an embarrassing period for the aircraft industry, one in which they were not wholly at fault because they had been asked to do an impossible job in too short a time. Even the Military Affairs Subcommittee admitted they did not imply their criticism should be directed at all companies because much had been accomplished. One of the problems was within the government itself. The subcommittee recognized this, recommending that a single man be placed in charge of a department of aviation. The chairman said, "Aviation should be organized similar to our departments of the Army and Navy." Officials of the Army and Navy fought stubbornly and unrealistically against such an independent air force, and were successful in keeping such an organization out of the military system until after World War II.

Despite the debacle in aircraft procurement, some excellent planes were produced. The Thomas-Morse MB-2 was comparable to the French Spad. The Chance Vought VE7 trainer was also outstanding, as was Grover Loening's M-8 monoplane. In addition, Aeromarine and Burgess made some good airplanes for the Navy, in addition to those made in its own factory.

But the committee was adamant in calling aircraft procurement a national scandal. In its report to the full Senate on August 22, 1918, it said in part that a substantial portion of the $649 million appropriated for military aircraft had undoubtedly been wasted. Although the senators were furious, the United States had committed itself to its allies, so another appropriation of almost $900 million was authorized.

The United States had created a sorry spectacle in the eyes of the world by its much-vaunted claims of mass production of airplanes. The truth was that by July 1, 1918, only 67 of 601 American-made but British-designed DH-4 aircraft had reached the battle fronts, but not a single one had seen combat. Congress had approved the production of 8,500 DH-4s, but work had to be stopped until defects in design and workmanship could be resolved. On August 7 a squadron of eighteen American-made DH-4s finally made it over the German lines for the first time.

German Gotha bombers replaced zeppelins in the winter of 1917, and British pursuit squadrons were recalled from the front to thwart this new threat to the home front. After months of devastation, British fighters became successful. The Germans then turned to night bombing, using huge five-engine Riesenflugzeug 501 bombers and Gothas. By the spring of 1918, British defenses were so strong that on one mission only three out of twenty-seven Gothas and 501s made it back to Germany.

Perhaps the most active American Naval Air Station in Europe was the one at Porto Corsini, fifty miles south of Venice, which was taken over by the U.S. Navy from the Italian government on July 24, 1918. The Italians furnished everything for the base except food and clothing, including two-seater M-8 flying boats for bombing operations and M-5 Macchi fighters. The main objective of the group was to bomb the Austrian naval base at Pola, where battleships and cruisers were based, along with German and Austrian submarines. Although Porto Corsini was only sixty-four miles from Pola, the base was an unfortunate choice because landings and takeoffs had to be made on a 100-foot-wide canal against a prevailing wind that blew at right angles to the canal.

During the first mission, on August 21, seven planes (two bombers and five fighters) were dispatched to Pola to drop propaganda leaflets. Shortly after takeoff, a bomber and a fighter had to abort. The other bomber braved heavy antiaircraft fire over Pola while dropping its leaflets. It fled for home when five enemy Albatross fighters appeared.

The Macchi fighters—flown by Ensigns George H. Ludlow, E.H. "Pete" Parker, Dudley A. Vorhees, and Charles H. Hammann—were at 8,000 feet when Ludlow gave the order to attack the Albatrosses, which were hurtling down from 12,000 feet. Ludlow focused his attention on the lead plane, letting off a burst of fire, then swung to engage the plane on his left. When the enemy leader sought to dive away, Parker followed him down. When his right gun jammed, Parker pulled out, firing from his one good gun at another Austrian plane that swept into view. Vorhees no sooner got into action than his guns jammed and he had to leave the dogfight.

The bomber headed home as Ludlow and Hammann took on the two planes of the second Austrian section. Ludlow soon found himself in a fight with three Austrians. He drove one out of the fight and watched it as it started to smoke. In the next instant he was hit hard. He had taken hits on his propeller and engine, and oil streamed behind his plane as it caught fire. Ludlow's plane went into a spin, but he managed to pull out and make a landing five miles off the harbor's entrance.

Hammann looked down and saw Ludlow's plane on the water. He was determined to rescue his comrade despite the risks. The wind was blowing at about twenty miles an hour and the sea was choppy. Hammann realized that a landing under such circumstances would be rough, but now he learned that his own plane had been damaged and he might not be able to take off after rescuing Ludlow. Ludlow's proximity to the harbor—and the possibility that the Austrians would soon comprehend what he intended to do—further complicated his problem. Even worse was the possibility that they might be shot as spies if they were captured. (The Austrians had let it be known that anyone caught engaged in dropping propaganda leaflets would be shot as spies.)

Hammann spiraled down and landed beside Ludlow's crippled plane. He opened the port in the bottom of his hull, kicked holes in the wings to make his Macchi sink faster, and jumped over to Ludlow's plane. He climbed up behind the pilot's seat and sat under the engine, holding to the struts to keep from being swept into the propeller or off into the sea.

The tiny Macchi was built to carry only one person. Hammann hadn't the least idea how to get Ludlow's plane into the air. The bow of the plane, already damaged by machine gun fire, was smashed as the craft gathered speed. Finally the little seaplane got off the water.

Before he left the area Hammann fired his remaining ammunition into his wrecked plane and watched it sink. He refused to give the enemy the satisfaction of recovering it. He began his sixty-mile flight back to Porto Corsini, expecting at any moment to come under attack. Incredibly, the Austrians made no attempt to follow the damaged plane.

At Porto Corsini Hammann made a good landing in the canal, but water poured through the bow and turned the Macchi over. It had brought them home safely, but now it was a complete wreck. The two fliers climbed out, assisted by men in boats that had come to assist them. Ludlow had suffered a bad gash on his forehead and Hammann was badly bruised, but both returned to duty a few days later.

The president of the United States presented Ensign Hammann the Medal of Honor—the first awarded a United States Navy aviator. Less than a year later Hammann met his death in a Macchi plane of the same type he had flown at Porto Corsini.

While flying with the Royal Air Force in a Sopwith Camel, nineteen-year-old Lieutenant David S. Ingalls scored successes unmatched by any other American naval aviator during the war. He was assigned to Squadron 213 in Flanders in August 1918 to fly missions against German airfields. Almost daily his Camel swept across them, wheels almost touching the ground, as he fired his guns and dropped bombs with devastating effect.

On August 11, Ingalls and an RAF officer in Sopwith Camels reached an altitude of 14,000 feet not far from Dixmude when they sighted an enemy Albatross. This two-seater was flying toward the Allied lines at 10,000 feet. The German pilot evidently spotted them just as they saw him because he dove toward Ostend. The Camels attacked, and the leader, firing about 150 rounds in short bursts at 150 yards, followed the enemy plane down to 5,000 feet. Just as the Camel broke off combat the Albatross went into a slow spin, and the two pilots saw it head for the ground out of control.

Two days later, on the night of August 13, Ingalls flew over the German airfield at Varsenaere. He made a low-level attack, spraying 450 rounds of machine gun fire into the facility as the surprised Germans made desperate efforts to shoot him down with their antiaircraft guns. Ingalls swung in a wide circle and again headed for the hangers, letting loose four bombs and "putting out searchlights, scattering Germans, and messing things up generally," as he later reported.

He repeated this maneuver on September 15 at the German airfield at Uytkerke. He made a low-level attack out of the clouds upon the German hangars, firing 400 rounds from his Lewis machine gun into the canvas structure. As he swung up, he released four bombs upon the Fokkers parked on the field below.

On his return flight, Ingalls sighted an enemy two-seater Rumpler west of Ostend at 6,000 feet. Along with Royal Air Force Lieutenant H.C. Smith, also flying a Camel, Ingalls went after it. The enemy pilot turned and dove toward Ostend, but that did not save him. Ingalls and Smith followed him down, firing 400 rounds at close range, sending the German plane crashing in flames just off the beach.

Another time, upon returning to home base, Ingalls dove toward the ground, leveling off at 200 feet, just high enough to escape the antiaircraft fire. But he soon realized he was facing heavy machine gun fire because the Germans were wise to such tricks.

Suddenly, he heard a rat-tat—a noise signaling his engine had been hit. Fuel poured out of the tank below his seat, and clouds of white vapor rose from it. He had flown over a machine gun nest, and the gunners followed him as he crossed the front lines. He didn't dare take evasive action, because he expected his controls would fail at any moment. Carefully, he used the rudder to make little turns toward home. With relief he passed out of range of the machine guns, but he knew he had to land—and quickly. He brought his Camel in slowly over the trees, nursing his engine, and managed to land in an open space in friendly territory.

His airplane was all shot up. A burst of several bullets had perforated the tank under his seat, and all but one strand of wires connected to the elevator was severed. One aileron had been hit at the hinge, and there were holes in the wings.

Ingalls and two RAF fliers went out September 18. They sighted a kite balloon at 3,500 feet in the La Barrière area. They promptly fired at it and watched the observers bail out. As Ingalls pulled away he looked back and saw a fire flare in the bag just before it crumpled in a mass of flames and dropped directly on the balloon sheds, setting them ablaze.

Two days later, Ingalls flew escort for bombers heading for Bruges. When the formation sighted four enemy planes heading toward the DeHavilland bombers at 15,000 feet, the Camels attacked. Ingalls singled out a Fokker that was pursuing a DH-4. He fired at one hundred yards and the Fokker dove vertically, leaving a white smoke trail. It was last seen out of control, very low, near Bruges, and still descending.

Ingalls swung to attack another Fokker at only twenty-five yards range. The German pilot turned his plane on its back, spinning it as it dove for the ground. Ingalls wasn't certain of its destruction so he did not claim it.

Ingalls and four other British Camel pilots flew over Flanders on September 22, seeking German hangars and ammunition trains as their preferred targets. Ingalls dropped four bombs on a German ammunition dump and blew it up, along with a number of wagons loaded with shells. Later he dropped four more bombs on a large hut filled with explosives at Wercken. His next target was the railway station at Thourot, where the Germans had an enormous supply dump. He scored two direct hits on it. On their return to base the fliers encountered a horse transport and immediately dove to the attack. They killed twenty-five Germans and forty-five horses.

Ingalls went out on the 24th and encountered an enemy Rumpler over Nieuport. He and the officer in another plane fired 200 rounds at close range, and the Rumpler crashed in flames.

The British Air Ministry honored Ingalls's gallantry with the Distinguished Flying Cross, saying, "His keenness, courage and utter disregard of danger are exceptional and an example to all. He is one of the finest men Number 213 Squadron ever had." They named him Acting Flight Commander over their own pilots.

The United States Navy presented him with the Distinguish Service Medal for "exceptionally meritorious service in a duty of great responsibility." Ingalls was only nineteen years old when he became an ace (five kills or more), creating a legend in the annals of naval aviation.

Realistic German leaders had seen the end in sight by August 1918, after the failure of their great spring offensive in June, when American ground troops fought their first important battle at Chateau Thierry, as they and the French stopped the German advance. Other factors contributing to the feeling of defeatism in Germany were the failure of the Central Powers to defeat the Allies in the second battle of the Marne in July and August, and the subsequent Allied offensive at Amiens, St. Mihiel, and other key points. The obvious panic of Germany's military leaders contributed to a sense of hopelessness among the German people. Their attitude worsened when Allied leaders began a knockout blow with 896,000 American troops and 135,000 French troops. To the north, the British breached the Hindenburg Line—Germany's last line of defense—with massive drives. During the battles of the Argonne Forest and at Ypres the Germans were forced back, but the cost was high—120,000 casualties. The German military leaders had had enough. They asked for an armistice on October 4. While the ground armies advanced on all fronts, Allied Navy and Army bombing squadrons stepped up their efforts, contributing to the feeling of impending disaster. The formal surrender came on November 11, 1918.

World War I was won primarily on land and at sea. The total contribution of all aircraft participating in this "war to end all wars" had little effect on the final outcome.

Naval flying boat squadrons turned in extraordinary performance records during the war's final ten months. Only three Allied ships were lost to enemy submarines or naval gunfire in the patrol area off the coast of France. In contrast, one year earlier, an average of a ship a day was sunk.

United States Navy patrol aircraft attacked twenty-five German U-boats, sank four, and damaged another twelve during the last months. The Navy Curtiss flying boats, capable of flying the Atlantic, would have been of great assistance, but the war ended before they reached the production stage.

These NC (Navy Curtiss) flying boats were the largest flying boats of their time. They were developed by naval constructors in cooperation with the Curtiss Aeroplane and Motor Corporation. The first, the NC-1, flew in October, 1918, and on November 27 carried a record of fifty-one men into the air. Three of the NCs had four engines, but the other seven were equipped with three engines.

Navy fliers escorted 477 convoys bringing troops and supplies from the United States. In all, they logged 4,314 flights. Numerous enemy surface ships were damaged or sunk as the Allied sea blockade tightened. Most important of all, Navy aircraft of the Allied powers helped to bottle up the powerful German submarine fleet, which had nearly brought disaster to the Allied cause just a year earlier.

Flying boats were also used during the last year of the war to hunt down and destroy the few remaining German airships still flying high-altitude observation missions. Two or possibly three airships were destroyed with aerial bombs and machine guns.

In the closing months, an agreement was reached to put eighteen squadrons in the Adriatic, twelve for day operations and six for night work.

Land planes were to be used now that seaplanes had lost favor as large bombers. Extensive plans were made for operations in 1919, but the armistice halted them.

The small Navy air branch of early 1917 had expanded mightily by the end of the war, with 6,716 officers and 30,693 enlisted men in Navy units and 282 officers and 2,180 enlisted men in the Marine Corps. In nineteen months, forty-three air

Curtiss NC-4 flying boat. (Courtesy of U.S. Navy.)

stations were placed in operation at home and abroad, instead of the single one in service at the start of the war. Aviation personnel numbered 39,871, close to the total of all personnel in the prewar Navy. Of these men, 18,000 officers and enlisted men had been sent abroad with 570 aircraft. A total of 2,107 aircraft were assigned to all stations, as well as fifteen dirigibles and 215 kite and free balloons.

In these days of primitive radio communications, trained pigeons served the Navy with distinction, as 829 birds flew 10,995 missions with Navy pilots. They carried only 230 messages—219 successfully—to their destinations, and at war's end eleven pigeons were declared missing in action.

After the armistice, contracts for more than 1,400 aircraft and aeronautical equipment were cancelled, and manufacture was permitted of only those items that would be less expensive to complete than to discontinue. Public sale of surplus aircraft was initiated, and many ex-Navy fliers purchased them. An F-boat sold for $1,800 and the HS-2— sold originally to the government for $18,480—was sold for $6,160. H-16s were reduced from $33,159 to $11,053. Many aviators used these planes to operate civilian flying schools or inaugurated flying boat passenger service to resort areas along the coasts.

The Allied powers, flushed with the warm glow of victory, for the most part discounted the potentials of aerial bombardment. The Germans thought otherwise, and the Allies were to find out twenty-one years later how well the Germans had learned a vital lesson.

Part II

A Nation Divided Over Airpower

3

Aerial Bombs Sink a Battleship

Naval aviation suffered drastically as a result of demobilization. Bombing squadrons almost disappeared, and the Navy was limited by law to only six seaplane bases in the continental United States. The Navy proceeded, however, with the NC transatlantic flying boat project that was initiated in 1913. Three aircraft were produced and designated as NC-10, NC-3, and NC-4.

The First Marine Air Squadron, Second Marine Brigade, was sent to the Dominican Republic's San Pedro Island in February 1919 to assist Marine ground troops in putting down a native rebellion. Other First Marine pilots were dispatched to neighboring Haiti for action against the Cacos rebels. When the ground forces were pinned down by native jungle fighters on San Pedro Island, headquarters was asked for immediate help. This was impossible because reinforcements were a ten-day march away.

Lieutenant E.H. Sanderson, a Marine with Randolph Field Army Air Service training in dive bombing, strapped a gun barrel to his Curtiss Jenny directly in front of the windshield as a makeshift bombsight, placed a bomb in a gunny sack underneath his wing and outboard of the propeller, and took off for a personal observation.

Over the target area, he spotted signal banners indicating where the troops wanted him to plant his bomb. Pushing the Jenny's nose down, he flew the aircraft carefully until the banner was within his gun barrel sight. He hesitated a fraction of a second, pulled a lanyard to release the bomb, and quickly pulled up. He was much too close to the ground as he pulled back on the stick, with the Jenny creaking in all her joints. She responded, though, and then he heard the bomb hit. Peering down, he saw grimly that his bomb had scored a good hit. At last the Marines in the dark, dank jungle were safe. This proved to be the first time that true dive-bombing was used in combat. It augured well for the future.

The feasibility of using voice radio and telephone relays for air-to-ground communications was demonstrated on March 12, 1919, when Lieutenant Harry Sadenwater, in an airborne flying boat, carried on a conversation with the Secretary of the Navy, who was seated at his desk in the Navy Department some sixty-five miles away.

The next day the Chief of Naval Operations issued a preliminary program for postwar naval airplane development. He said that specialized types—such as fighters, torpedo carriers, and bombers for fleet use—would be needed. On land, he called for the development of single engine, twin-engine, and long-distance patrol and bomber planes. He specified that the Marine corps should acquire a combination of land planes and seaplanes.

An F-5L flying boat—equipped with two 400-horsepower Liberty engines and flown by Lieutenant H.D. Grow—set an unofficial seaplane duration record on April 26 that remained unbroken until 1925. He flew out of Hampton Roads, completing a flight of twenty hours and nineteen minutes over a distance of 1,250 nautical miles. At the time seaplanes were not recognized as a separate class, so the flight was not registered.

Lieutenant Commander Richard E. Byrd—who developed and tested navigational equipment for the forthcoming NC transatlantic flight—requested on April 28 that the Naval Observatory supply bubble levels, which he adapted for attachment to navigational sextants. The bubbles provided an artificial horizon, making it possible to use these instruments for astronomical observations from aircraft.

In the first Atlantic flight from the United States to England, the NC-4, with Lieutenant Commander Albert C. Read in charge, landed safely on May 31, 1919, at Plymouth, England. The other two flying boats were not so fortunate. They landed short of their destination—in the Azores—due to heavy fog. High waves battered both aircraft, and they were unable to take off. The NC-1 sank shortly after her crew was rescued by a freighter. The NC-3, despite thirty-foot waves, taxied 200 miles to Ponta Delgada. Further flight was prevented by extensive structural damage to a wing.

The flight of the NC flying boats proved the extreme strength built into their wooden hulls, but the bobtail hull shape failed to provide adequate buoyancy to keep the tail surfaces clear of the water at anchor or during taxiing. This was a good lesson for future flying boat designers, and they heeded it. The modern double and triple-stepped hulls owe their design to early lessons taught them by the NCs.

This same year the Navy drew up plans for a large "boat" based on data obtained from the NC aircraft. Lack of funds prevented completion of the project, but the radical design called for a 70,000-pound triplane powered by nine Liberty engines housed in three individual pods.

After the war, despite an outward appearance of serenity, all was not well within the Department of the Navy. Naval aviation had a few H-16 flying boat bombers, which participated in fleet maneuvers with a high degree of sufficiency, but they were obsolete. They were replaced by F-5L flying boats. Other than three Thomas-Morse and Sopwith land planes—former fighter planes now used as light bombers—there was no naval bombardment aviation worthy of the name.

Naval aviation had yet to obtain the aircraft carrier that was so desperately needed, and ship operations were confined to the small wooden platform deck constructed over a gun turret on board the USS *Texas*.

The General Board submitted the last of a series of reports on June 23, 1919, to the Secretary of the Navy on a policy for developing a naval air service. The board concluded that aviation had become an essential arm of the fleet, and urged adoption of a broad program for peacetime development that would establish a naval air service "capable of accompanying and operating with the Fleet in all waters of the globe." The secretary made some modifications, but approved the program on July 24. It served to offer direction for the rest of the year for a number of actions. One, authorized on July 1, called for the installation of launching platforms on two main turrets on each of the Navy's eight battleships.

Ten days later the Naval Appropriations Act for fiscal 1920 made several provisions that affected the future of naval aviation. It provided for conversion of the collier *Jupiter* into an aircraft carrier, later named the *Langley,* and for conversion of two merchant ships into seaplane tenders. Only one tender was completed, and it was named the *Wright*. The act also approved construction of one rigid dirigible later designated the ZR-1, or *Shenandoah,* and purchase of another from the British Air Ministry. The ZR-2 broke into two parts and crashed in England on its fourth trial flight. Twenty-eight British nationals and sixteen Americans died in the disaster. The appropriations bill unfortunately limited the number of heavier-than-air stations that could be maintained along the coasts of the continental United States to six.

On August 1, the Aviation Division of the Office of the Chief of Naval Operations was abolished, and its functions were reassigned to other divisions and to the Bureau of Navigation. The Director of Naval Aviation retained his title as head of the Aviation Section of the Planning Division. In effect, Naval Aviation had a director with no authority to direct.

A small number of Marine Corps aviators who took part in interservice training in 1919 now made a profound impact upon the direction that Naval Aviation would take in the future. They were ordered to the Army's Randolph Field in Texas to check out in land planes.

Most Navy and Marine pilots had never flown anything except float planes and flying boats. While at Randolph, they experimented with a new technique called diving, which Army fliers had learned from the British and the French. They believed bombs could be dropped with greater accuracy from dives than from level flight.

The experts—though they could hardly be called that—gathered the expectant Marine pilots in the briefing room the first day and outlined the way diving should be accomplished. "Fly directly over the target," they said, "roll your aircraft into a near-vertical high-speed dive, then release your bomb-load prior to pull-out."

Marine pilots soon realized their aircraft could not withstand the strain of such a rapid pullout. Until new aircraft were designed for such a mission, it was possible only to teach the maneuver as part of their advanced flight-training courses. But they learned important lessons they would put to good use later on. They realized that once the technical barriers were overcome, dive-bombing would be practical in attacking moving targets at sea.

★ ★ ★

During the war, Admiral Fiske fought for the introduction of torpedoes when three Turkish ships were destroyed by such weapons in the Sea of Marmara by planes of the Royal Naval Aircraft Service in the ill-fated Gallipoli campaign. A German torpedo plane sank three British ships in the North Sea in 1917. These two successes, however, were the only ones during four years of war.

The British Admiralty actually placed the world's first aerial torpedo squadron in service aboard the carrier HMS *Argus* in 1918.

After the war, Admiral Fiske kept the question of aerial torpedoes alive. In February of 1919 he was the man responsible for the experimental torpedo launchings at Pensacola. The first successful tests, however, were not made until May at Hampton Roads, using both dummy and live torpedoes dropped from Curtiss R-6s. These tests continued until 1920, when further tests were made with new Martin torpedo bombers, the first plane to be so designated.

In May 1920, the Atlantic Fleet's torpedo plane squadron was formed at Yorktown, Virginia. It evolved from an element of Air Detachment, Atlantic Fleet, organized on February 3, 1919. This marked the first provision for aviation in the organization of the postwar fleet. A similar air detachment had been organized in the Pacific Fleet in September 1919. It also contained an element that emerged in 1920 as a torpedo squadron.

The bitter struggle between battleship admirals and the new aviation-minded junior officers broke into the open in 1920. It reached such intensity that the controversy threatened the very foundations of the Navy's traditional military mission. The problem was how to administer naval aviation within the Department of the Navy.

Commander Jerome C. Hunsaker, a dedicated officer of both the surface fleet and naval aviation, was blunt in his demands for action. He had been aware of the radical plans for abolishing both Army and Navy aviation departments ever since he came home on the *Aquitania* following the war.

"Army General Mitchell personally told me he was coming home to demand a separate Air Force of the United States," he said. "Although many of my fellow officers in the Navy deride the General's outspoken views, I take him seriously."

He was almost alone in his concern within the Navy because practically every influential military officer in Washington underestimated the determination of Brigadier General William "Billy" Mitchell to launch his crusade.

The Navy's part in the coming struggle—which was to last seven years and rock the nation—began innocently enough. Lieutenant Commander H.T. Bartlett suggested a series of aerial bombing experiments using older warships that were destined to be consigned to the scrap heap. He recommended that the USS *Indiana,* a veteran of the Spanish-American War, and captured German warships be used as target vessels. The third of a series of tests began on October 14, 1920, and was conducted at Tangier Sound on Chesapeake Bay under carefully controlled conditions to determine both the accuracy with which bombs could be dropped on stationary targets and the damage caused by near misses and direct hits.

The first tests were underwater concussion tests against the *Indiana*'s hull to simulate near misses by aerial bombs. They caused the ship to sink slowly, and she was towed to shallow water and run aground. Large 1,800-pound bombs were set off on deck, inflicting heavy damage to turrets and superstructures. These tests were only preliminary, but the consensus among air enthusiasts was that a warship could be put out of action or sunk. The next step was to use bombers.

After Mitchell returned to the United States he unleashed an attack that pulled no punches. "Was it not true," he charged, "that Congress had appropriated one billion dollars for military aircraft production during World War I?"

His words made headlines and the newspapers eagerly used them. He told the American people, "Only 196 DeHavilland light bombers and a few Liberty engines were all the nation had to show for such a huge expenditure. Not only that," he said significantly, "most of these bombers were poorly made and unairworthy."

There were many in government who tried to shut him up. They quickly learned the caliber of the individual with whom they were dealing. "Was it not true," he said, "that America failed to deliver combat aircraft worthy of the name to her own Army and Navy fliers risking their lives in Europe?" Mitchell was mad, and he had every right to be. No one could deny that, for the most part, he spoke the truth. A juicy national scandal promoted avidly by the nation's press began in earnest.

Navy Admiral Fiske added his voice to the controversy. "If the Division of Aeronautics had not been effectively abolished when aviation-minded Mark Bristol was purged from that office in 1916, the United States Navy could have started a flotilla of bombing machines and torpedo planes across the ocean the day war was declared against Germany April 6, 1917!"

Mitchell challenged the right of the War Department to keep down the aviators. He lashed out on all sides and the Navy felt the sting of his harsh words. "The Navy is no longer the best means of protecting the nation from sea attack." He requested that the Army be allowed to participate in bombing tests proposed by the Navy. He pressed anew for the establishment of a separate air force. He reiterated that airpower alone won the war. This statement reduced the general's credibility, however, because it was not true.

Mitchell added to the general confusion by recommending that the Navy construct a fleet of aircraft carriers to transport its bombing planes at sea. He qualified this statement later. "Aircraft carriers would be useless against first class air powers." At the same time, he said, "Aerial torpedoes are useless against moving targets at sea."

During a postwar Board of Inquiry to investigate charges of misconduct and fraud in connection with wartime procurement, the Army general erroneously claimed the Navy's General Board agreed with his views.

The Navy–Mitchell controversy burst out afresh when the Army registered a complaint with the Secretary of War that it could not obtain suitable target ships from the Navy to conduct its own unbiased bombing tests.

Mitchell was not alone in his concern about airpower. Admiral W.F. Fullam spoke prophetically when he said publicly, "With the advent of airpower and the submarine, I can see a time when battleships and cruisers will be driven from the seas. In the future, seapower will be dependent upon control of the air. A large fleet of airplanes properly armed can hold off an attacking fleet. Seapower is dependent upon airpower!"

Navy brass were appalled by such words from one of their own. This was heresy! The press called the following uproar "The Dance of the Bureaucrats," as each group maneuvered to maintain their status in the defense system. Politicians watched public opinion polls because Navy and Army funds soon had to be allotted.

Conservative members of the Army and Navy had tried to soothe seething emotions among those they considered "hotheads" by publishing a policy statement on

January 8, 1920, that defined the functions of Army, Navy, and Marine aircraft as guidelines for procurement, training, and expansion of operating bases. It set forth the conditions under which air operations would be coordinated in coastal defense; it spelled out the means by which duplication of effort could be avoided; and it provided for the free exchange of technical information.

But General Mitchell persisted in his demands that the Navy share some of its captured German warships for use as target ships. The Navy reluctantly agreed that Army pilots could make a series of tests. A set of operational rules was established and Navy Captain A.W. Johnson, who had earlier been placed in charge of the Navy tests, was made responsible for the combined tests, which were scheduled over a four-year period. He was a good choice because he was a firm believer in aviation, and had formerly been in charge of naval aviation in the Fleet. When Mitchell learned that he must serve under a "mere" Navy captain, he exploded with a terrible wrath. "Just another trick to be expected from the Navy," he irately told intimates.

Just when things seemed to be moving along smoothly, loud debate began about the bombing rules. Secretary of the Navy Josephus Daniels brought ridicule upon himself by saying, "Battleships can't be sunk by bomber aircraft. I'll stand bareheaded on the bridge of the target vessel during an attack." Fortunately for him, he wasn't taken seriously.

Behind the scenes the Army sought to discredit the Navy's war potential in the newspapers. They leaked stories that no modern battleship had ever been able to sink another battleship during combat. They prompted the press to demand that the Army be given a chance to prove its bombers could do what a capital ship could not do.

Specifically, Army aviators claimed that one large bomb, properly placed, could knock a warship's exposed water condenser pipes off the bottom of its hull by underwater shock waves. When this happened, they claimed, the engines would be useless. They were ready to prove that their large bombs could crush hull plates by the tremendous forces unleashed by underwater shock waves. They based their assertions on a series of underwater bombing tests conducted in Chesapeake Bay using TNT.

When the time came for the first series of combined tests in 1921, naval aviators moved in to the NAS *Norfolk* with their F5L flying boats and Marine DeHavilland land planes. The Army established test headquarters at Langley Field, Virginia, for their Handley-Page and Caproni heavy bombers. They would carry the big bombs, but Mitchell's team also had fifty other airplanes, including DeHavillands, Martin MB-1s, and small pursuit aircraft.

Air crews practiced day and night from both low and high altitudes. Army pilots, flying to and from the practice area just above the waves, were surprised to learn for the first time that their aircraft had an unusual lift developed by the winds because of the cushion-like effect of air compressed between the planes' wings and the surface of the water.

These tests were not just a publicity stunt. Not only was an important controversy resolved, but the intensive efforts on both sides brought vitally needed improvements in combat techniques and equipment. The Navy concentrated on the development of arresting gear, already planning to operate bombing planes from the carrier *Langley*.

Despite renewed efforts by high-ranking officers to bring about a new spirit of interdepartmental coordination within the Navy in the face of Army threats, the feud between the battleship admirals and naval aviators continued.

Captain Chambers urged some sort of reorganization. He was not alone in voicing concern for reestablishment of more normal internal relations. It was evident to a growing number of officers that Naval Aviation would perish unless changes in departmental organization were made quickly. They knew there had to be an end to the continual bickering between aviators and surface officers.

In an atmosphere of unusual harmony, a separate Bureau of Aeronautics was recommended by Captain William A. Moffett after he was named Director of Naval Aviation on March 7. Congress approved the bureau on July 12 and charged it with responsibility for naval aeronautics. Moffett proved to have a genius for public relations, and his ability as a superb negotiator influenced the promotion of airpower to a profound degree.

There were some in Naval Aviation who felt General Mitchell should be given the credit. "Hadn't he built a fire under the Navy exactly when it needed it most?" they said. Mitchell, however, kept up his unrelenting attacks to destroy Naval Aviation.

The fight between the Army and the Navy took on a new slant now as both services fought for the greater share of military appropriations for aeronautical developments. When a board of inquiry was established to determine the future direction of United States air policy, the meetings were often vitriolic.

Admiral S. Sims, President of the Naval War College, and naval inventor Admiral Bradley Fiske, shared most of Mitchell's views, except for his proposal for a united air service. Sims, in particular, had fought for modernization of the United States Navy since the turn of the century. During sessions of the House Naval Affairs Committee early in 1921, Sims said that air stations along the coasts of the nation would provide the best system of defense against invasion. "The fleet with the greatest number of planes will command the air," he said.

He was asked if battleships should be discarded. "I would prefer right now to put money into airplane carriers and the development of airplanes. The airplane carrier will become a more powerful capital ship than the battleship. It is my belief the future will show that the fleet with twenty carriers, instead of sixteen battleships and four carriers, will inevitably knock the other fleet out."

At last, the military appropriations bill for the fiscal year 1922 decreed that the Army Air Service would henceforth control all aerial operations from and over land. It specified that the Navy should control all aerial operations attached to the Fleet. Mitchell interpreted these guidelines in his own fashion, and promptly issued orders for the Army to take over thirteen land-based Naval Air Stations.

The uproar over this action was the worst observers had ever seen. The War Department delayed the move temporarily, then no action was taken in the following years. To quell the "hotheads" on both sides, the War Department ordered bombing tests against German target ships to proceed immediately. Again, arguments started about who was to bomb what, with the Army demanding exclusive bombing exercises.

When the Navy anchored the German submarine U-117 off Hampton Roads, Virginia, letting the Army bomb it only after Navy and Marine pilots had made their runs, the whole affair threatened to get completely out of hand. When Navy airplanes sank the submarine with twelve bombs before Army aviators had a chance to drop their bombs, tempers got out of control.

In subsequent tests during July 1921, Army bomber pilots easily sank the German destroyer G-102.

Events moved swiftly, and on July 18 the German light cruiser *Frankfurt* was sunk in two hours by both Navy and Army bomber crews. One 600-pound bomb was credited with actually dealing the mortal blow, landing alongside and caving in a large section of the steel hull by concussion.

At last the big moment arrived, and tests were scheduled against the German battleship *Ostfriesland*. It was the most modern warship of its time. Would these tests finally resolve the arguments over whether bombing planes could sink a battleship? The world would soon know, but it was years before the argument was settled to most people's satisfaction.

Days of elaborate planning preceded the bombing of the *Ostfriesland*. They were days of charges and countercharges by both sides. General Mitchell strongly protested the decision by the Navy to anchor the ship in deep water off the Virginia Capes to eliminate damage by shock waves bouncing off the ocean floor. "The Navy will jeopardize Army lives because this distance is at the extreme range of Army aircraft," he said in protest. The Navy refused to back down, but established a line of rescue ships from Thimble Shoals in Chesapeake Bay seaward to the target area.

Blimps were assigned to photograph and observe the bombing and also to provide radio wind direction and velocity for the bombardiers.

Captain Johnson assigned himself to the control ship *Shawmutt* to direct the operation. Mitchell had a special radio-equipped DH-4B light bomber from which he planned to direct the Army's operations.

All parties agreed to a system of visual signals. A flag draped on the *Shawmutt*'s forecastle meant cease bombing. Heavy smoke from the stacks signaled emergency—cease bombing immediately.

The plan called for an inspection of the ship between attacks by a naval team to assess progressive damage. One order was explicit: "Stop bombing if a 1,000-pound bomb scores a direct hit so the inspection party can make an on-the-spot damage estimate."

July 20, 1921, was the start of a series of tests which had aroused the interest of the world. At first, salvos from deck guns of ships in the Fleet inflicted severe damage to superstructure and guns, but failed to damage the hull. The *Ostfriesland* was a sturdy ship, and had been built on Admiral von Tirpitz's orders to make her as unsinkable as possible. Four thicknesses of metal protected her massive hull, and she was divided into watertight compartments. There were no openings whatsoever in the hull, so she was considered leakproof. When she was launched, the Germans deliberately lied about her true gross weight of 27,000 tons.

The Atlantic Fleet of eight battleships, cruisers, destroyers, and auxiliary craft had hundreds of government officials, military attachés of foreign governments, aeronautical experts, and approximately fifty newspapermen on board the morning of July 20 for the start of the long-awaited bombing tests against a capital ship. Bombs ranging in size from 230 to 1,000 pounds were scheduled to be dropped by both Navy and Army fliers. The two services conducted their separate operations with different goals. The Navy was interested principally in learning what effect bombs would have on a battleship. Mitchell was interested only in proving that bombers could sink a battleship.

Mitchell's team faced their first test confidently, although they showed the strain of weeks of training. They waited impatiently at Langley Field for word from the command ship.

One delay followed another because of bad weather in the target area. Mitchell impatiently dispatched a plane to learn the cause of the delay. Before the plane returned, he reacted furiously to an order from Admiral Hilary Jones, in charge of the battleship force, for all to return to port. In typically impetuous fashion, Mitchell told his pilots, "The Navy High Command is using a bad weather excuse to insinuate Army aircraft aren't operable in strong winds."

Jones shortly reversed himself, not because of the blast from Mitchell, but because the storm abated. At 11:35, the admiral gave the signal to proceed with the program. During the next hour and seventeen minutes, Marine and Navy fliers dropped fifty-two small bombs on the *Ostfriesland,* scoring six direct hits.

After examinations by the board of observers, Army fliers awaited word from Captain Johnson on the *Shawmutt.* It was not immediately forthcoming because, Johnson said later, he did not want to keep the airplanes waiting overhead. He was thunderstruck, therefore, when he received word that Mitchell had dispatched his Army bombers without orders.

The bombers remained in the skies for fifty-seven minutes before the order for attack was given at 3:42. Lieutenant Clayton L. Bissell deployed the land planes in column formation at an altitude of 1,500 feet. They proceeded to drop five 600-pound bombs, scoring two direct hits.

The *Ostfriesland* listed astern, but she had suffered no fatal injuries. It appeared to many observers that Mitchell would be proved wrong. One Navy expert said, "The ship could not be sunk by bombing." Among high government officials, it was obvious that little impression had been made. General Pershing and Secretary of War John W. Weeks did not even remain for the following tests.

The next day might provide the proof of Mitchell's contentions, because the 20th Squadron at Langley planned to use 1,000- and 2,000-pound bombs.

Nature was kind to the operation, with clear skies and only a light wind. Mitchell, after a personal inspection of each of his bombers, radioed the control ship at 7:00 A.M.: "Our bombers are on their way." This infuriated the Navy, particularly when he said bluntly, "I want no interference by naval aircraft."

The *Ostfriesland* appeared ahead of the bomber formation, her gray sides marked into sections by vertical white lines, plainly showing the years of neglect since she had been captured.

After one 1,000-pound bomb scored a direct hit on the ship, Captain Johnson said, "Run up the cease-fire flag." To his consternation, the signal was ignored and Army pilots continued to make their runs. The captain strode back and forth on the bridge of his ship livid with rage. "Display the emergency signal!" he ordered. It also was ignored, and five more big bombs were dropped from above even though the *Shawmutt* was within a thousand feet of the *Ostfriesland.*

Johnson flashed a message over Navy radio directly to General Mitchell's plane. "Cease fire! Cease fire! Standby for boarding!" Mitchell radioed back, "Bombing will continue until further signal from Army bomber number 24...repeat...bombing to continue until further signal from bomber 24!"

This led to further vituperation, but Captain Johnson ordered the *Shawmutt* to move away from the target area immediately. He later charged, "My signals were deliberately disregarded!" Lieutenant Bissell, leading the formation, testified later there was no intentional violation of the orders.

Mitchell finally ordered his bombers back to their bases despite the fact they had nine 1,000-pound bombs left. In a gesture of protest, they went within a half mile of the Atlantic Fleet and exploded their bombs in the sea, much to the consternation of observers.

Meanwhile, the inspection teams found that two direct hits had gone through the upper decks, but the protective deck was unharmed. Although much damage was caused, the ship was judged seaworthy.

Close bursts of heavy bombers cause havoc aboard the *Ostfriesland.*
(Courtesy of U.S. Air Force.)

Mitchell, who had been champing at the bit at Langley, awaited final word to proceed with the bombing. At last the order came at 10:30 to send his aircraft to the target area, but with only three of the large bombs.

This was the final straw. "I have an agreement with the Navy that my men will be allowed two direct hits on deck," he roared. "This latest order makes it impossible of achievement."

Mitchell wired the control ship. "Martin bomber and Handley-Page formation proceeding to target with 2,000-pound bombs. Unless two direct hits are made, attacks will continue until the Army has secured them."

No further word was received from the Navy, so a few minutes after noon Captain W.R. Lawson led his "V" formation to the area. A Handley-Page developed trouble and failed to make the rendezvous. The others moved into column formation and saluted Mitchell as they passed his airplane in review.

The bombers remained in column formation at an altitude of 2,500 feet and headed in for their drops. The first big bomb was released—the airplane lifting slightly in the air—and headed toward the ship. It exploded 100 feet off the starboard bow, causing no apparent damage.

Bombs tumbled from the sky in rapid succession, followed by the tremendous "water hammer." The second landed 300 feet off the bow, but the third hit the side armor, glanced off, and exploded alongside with such force the warship tilted slightly in the water.

To Mitchell's discerning eyes, the third bomb off her port beam would have been close enough to knock her out of action in wartime. He believed that if her magazines had been full, such as they would have been in wartime, the "water hammer" would have caused a destructive explosion.

Mitchell noted with satisfaction that his bombers were getting the range, as bombs four and five dropped on the port side near the mainmast and twenty-five feet off the port side. The ship shuddered with the force of these bombs as the massive battleship reared upward by water displacement pressures far in excess of the weight of the battleship. Mitchell was in a state of exultation as water poured entirely over the ship as she rocked back and forth.

The *Ostfriesland* seemed in bad shape, and spectators gazed in awe as she settled down and rode the waves gently. The sixth bomb landed near the starboard side and lifted her stern high in the air. When she fell back, there was an obvious list to port. "She's sinking!" came the agonized cry from the control ship. Most Navy officers refused to concede. They had seen warships stay afloat in wartime even after cruel wounds had been inflicted.

While the officers of the two services argued with one another, the *Ostfriesland* rolled again to port, her bow rising higher in the air. Thousands of spectators crowded the decks of the Atlantic Fleet, staring incredulously as the indomitable dreadnought prepared to make her final plunge. The ship filled rapidly with water, and her bow was high enough for all to see the large hole in her hull.

Above this scene of drama on the ocean's surface, Mitchell and his jubilant airmen watched eagerly as the ship heeled to port, exposing countless holes in her hull. Then the bow lifted even higher and her stern quickly became awash. She momentarily resisted the watery death awaiting her, but once her masts and funnels were underwater, the end came quickly and the waves closed over the once proud bow.

Some captains and admirals openly wept as a Handley-Page swept triumphantly over the watery grave of the huge ship, dropping the seventh and last big bomb. It exploded within the swirling mass of bubbles arising from the sunken ship. The tremendous "water hammer" seemed to sound the death knell of the battleship as a major weapon of war.

Army General Charles T. Menoher turned to Secretary of the Navy Edwin Denby. "I guess the Navy will get its airplane carriers now." Rear Admiral William A. Moffett, new chief of Naval Aviation, was the only participant to speak out publicly. "We must put planes on battleships and get aircraft carriers quickly." "Day of Battleship Ended!" cried the nation's press as General Mitchell continued his fight for a single United States Air Force.

Although President Woodrow Wilson had issued orders to Secretary of War Baker that, "All military officers should refrain from releasing personal opinions to the press and before investigating committees without specific permission from the War Department," Mitchell had ignored them. After President Warren G. Harding had assumed office in 1921, Mitchell assumed the new president was sympathetic to a unified air force. He was appalled when Harding said, "Aviation is inseparable from either the Army or Navy."

The controversy raged for several years and Congress, exploring rumors of aircraft scandals, established the Lampert Committee on March 24, 1924, to probe the entire field of airpower and its relationship to national defense. Mitchell was one of the witnesses. "If we were plunged into war tomorrow it would take us two years to meet on a

A mighty warship succumbs to bombs. (Courtesy of U.S. Air Force.)

par with England or Japan," he said. He struck hard at the theme that airpower now was dominant in war. Admiral Moffett, in his new position, also testified that at least three-fourths of the Navy's planes were either obsolete or unfit for use. "The safety of ships in the next war will depend," he said, "to a great measure on aviation."

Mitchell's outspoken criticism of the military establishment cost him the top job in the Army Air Service, which was given to General Mason Patrick, who had been Mitchell's boss during World War I in Europe.

Mitchell took his case to Europe, where his former comrades understood his views about airpower. Between December 1923 and July 1924 he had conducted an extensive tour of the Pacific and the Far East. There he wired his findings back to the War Department that the Hawaiian Islands' air defenses were dangerously vulnerable to an attack by Japan.

In February 1925, Secretary of War John W. Weeks warned Mitchell that he would be fired unless he ceased his public complaints about the state of America's defenses. He returned Mitchell to his permanent rank of colonel and assigned him to a base as a corps aviation officer.

When the Navy's airship *Shenandoah* crashed on September 3 in a storm, Mitchell denounced the "incompetence of the national defense by the Navy and War Departments." He said he spoke as a patriotic citizen, claiming he could no longer stand by and watch the disgusting performances "…at the expense of the lives of our people, and the delusion of the American people."

The Lampert Committee released its findings in December. It announced that it favored the establishment of a department of national defense, with adequate representation in high military councils. It voiced concern for the state of the aircraft industry and recommended that government plants be stopped from competing with private industry. It said it favored spending of $10 million annually for new flying equipment. The committee's report was so general in nature that it satisfied no one.

Mitchell continued to speak out. He was determined to crucify government aviation officials and the War and Navy Departments, whom, he charged, "were responsible for incompetency, criminal negligence, and almost treasonable administration of national defense." He expected to be court-martialed, and he was not surprised when the War Department issued orders for him to stand trial for his "unmilitary behavior and gross insubordination."

In December, near the end of his lengthy trial, the Morrow Board, with questionable taste, released its findings on the subject of airpower. It and the Lampert Committee had conducted separate investigations. The Board discounted many of Mitchell's charges without mentioning him, blithely informing the nation it was in no danger from air attack by foreign enemies, and that the nation's airpower compared favorably with other nations. It said a unified department of national defense would be both uneconomical and inefficient. This incredibly naive report served to lull the nation into a false sense of security.

Will Rogers, America's favorite humorist, reflected the views of many Americans when he said, "[I] see by the newspaper this morning that government officials say there is no danger from Europe from airplanes. When we nearly lose the next war, as we probably will, we can lay it onto one thing and that will be the jealousy of the Army and Navy toward aviation. They have belittled [airplanes] since it started and will keep on doing it till they have something dropped on them from one."

Mitchell was found guilty of violating the 96th Article of War. He was sentenced to five years suspension of rank and loss of command and pay. President Coolidge restored Mitchell to half pay but he promptly resigned from the Army on February 1, 1926, to continue his fight directly to the American people through books, articles, and speeches. Mitchell's stormy seventeen-year Army career had begun in 1909 when he was taught to fly by Orville Wright himself.

Mitchell later was invited to testify before Congress, where his popularity was still high. For centuries, he told congressmen, wars had been fought by armies each trying to get at the interior, sensitive areas of the other side's sources of support for war operations—the will of the people, the manufacture of munitions, and the economic structure of a state. The airplane, he said, could fly over struggling armies and strike at the interior structure supporting the enemy nation. "If this kind of war is waged against the right targets," he said, "and with an adequate number of aircraft, the results could culminate in a war-winning strategy."

In the past, it had been the Army's and the Navy's view that airpower could only play a tactical role in direct support of surface forces. Mitchell's doctrine of strategic airpower envisioned air forces that could use their unique ability of flying over battlefields to strike directly at an enemy's cities and their industrial establishments, thereby destroying their ability and will to fight.

In his earlier report to the War Department in 1924, Mitchell had predicted a Japanese attack on Pearl Harbor. Now he argued that Japan could be defeated in any future conflict only if the United States undertook the development of a long-range bomber with a range of 5,000 miles and a service ceiling of 35,000 feet. This was a radical idea in 1926, but it resulted in the B-29 Superfortress of World War II.

Unintentionally, Mitchell had forced the Navy to make more realistic appraisals of its own aviation activities, hastening integration of Naval Aviation into the U.S. Fleet.

During the years the airpower issue dominated the minds of military men, the civilian emphasis was on disarmament. The Washington Disarmament Treaty was signed on February 6, 1922, by representatives of the British Empire, France, Italy, Japan, and the United States. It established a tonnage ratio of 5-5-3 for capital ships of Great Britain, the United States, and Japan, and a lesser figure for France and Italy. The same ratio was set for aircraft carrier tonnage, with overall limits of 135,000 for Great Britain and the United States and 81,000 for Japan. The treaty also limited any new carrier to 27,000 tons, with a provision that if total carrier tonnage was not exceeded thereby, nations could build two carriers of not more than 33,000 tons each or obtain them by converting existing or partially constructed ships that otherwise would be scrapped under terms of this treaty.

Secretly, Japan made plans to ignore the treaty's terms and began to build up its Navy beyond these approved limits. Negotiators were far too naive in their assumption that all nations would abide by a treaty just because their representatives signed it.

4

A Pattern for the Future

The Navy's first carrier, a flush-deck, electrically powered former collier, was commissioned the USS *Langley* on March 20, 1922. Its deck measured 534 feet by 64 feet, roughly the size of the "baby" flattops of World War II.

Admiral Moffett told a large gathering, "The air fleet of an enemy will never get within striking distance of our coasts while our aircraft carriers are able to carry the preponderance of airpower to sea."

Donald W. Douglas, who had moved to California following the war to start his own aircraft company, drew up plans for a new torpedo plane called the DT-1. After the first airplane was built, he personally took the drawings to Washington to present them to the Bureau of Aeronautics.

"We like your plans," Mr. Douglas was told. "Build three of them for us. Can you produce them for $40,000 each?"

"We can," he said eagerly.

The planes were completed later in 1922 and flown to Naval Air Station, San Diego. Originally they were fitted with twin floats, but wheels were later installed to permit an interchangeable landing gear for quick conversion from floats to wheels. They proved eminently satisfactory, because their 400-horsepower Liberty engines permitted them to climb to 7,000 feet in ten minutes and attain a top speed of 100 miles an hour. Each carried a 1,700-pound torpedo under the belly on a special rack.

Torpedo and horizontal bombing became the established offensive missions for naval aircraft. The Atlantic Fleet's shore-based Torpedo and Bombing Plane Squadron One made Navy history on September 27, 1922, when its PT-1s and PT-2s attacked the *Arkansas,* which, in formation with other battleships, steamed at full speed seventy miles out of Norfolk. Even though the *Arkansas* attempted to dodge the darting airplanes, seven nonlethal hits were made. Admiral Moffett, reflecting the jubilant mood of his fliers, said, "This proves torpedoes can be launched from aircraft and made to run straight!"

Since a large airplane was required to carry torpedoes or heavy bombs, it was decided to combine the two missions along with a third mission—scouting. A num-

The Navy's first carrier, the USS *Langley,* is depicted by artist R.G. Smith with the DT aircraft. (Courtesy of McDonnell Douglas.)

ber of "three-in-one" type aircraft such as the PTs, DTs, and SCs were built, some of them becoming the first to fly from a carrier.

The rigid airship USS *Los Angeles* (ZR-3) was ordered on June 26 from the Zeppelin Airship Company in Germany. This zeppelin, part of war reparations, was obtained as a nonmilitary aircraft under the terms approved by the Conference of Ambassadors on December 16, 1921.

Aviation was employed during a four-day exercise starting on February 18, 1923, to work out a defense of the Panama Canal. Blue Fleet and Army coastal and air units defending the canal were assisted by the operations of eighteen patrol planes of Scouting Plane Squadron One based on the tenders *Wright,* the *Sandpiper,* and the *Teal.* The lack of carriers and planes for the attacking Black Fleet was made up by designating two battleships as simulated carriers. On the approach, the *Oklahoma* launched a seaplane by catapult to scout ahead of the force. Early the following morning a single plane representing an air group took off from Naranyas Cays and flew in undetected, and without either air opposition or antiaircraft fire, theoretically destroyed Gatun Spillway with ten miniature bombs.

On April 15 the Naval Research Laboratory reported that equipment for radio control of aircraft had been demonstrated in an F5L and found satisfactory up to a range of ten miles. It also stated that radio control of an airplane during landing and takeoff was feasible.

U.S. Navy aircraft won first and second places in the international seaplane race on September 28, 1923, for the Schneider Cup at Cowes, England. The winning plane established a new world record for seaplanes, with a speed of 169.89 miles per hour for 200 kilometers. Lieutenant David Rittenhouse, the new record holder, marked up 177.38 miles per hour for the race, and Lieutenant Rutledge Irvine placed second with 173.46 miles per hour. Both were flying CR-3s equipped with Curtiss D-12 engines.

Navy planes swept the Pulitzer Trophy Race at St. Louis on October 6, taking the first four places all at faster speeds than the winning time of the previous year. Both first and second places bettered the world's speed record, with the winner, Lieutenant A.J. Williams in an R2C, setting new records for 100 and 200 kilometers at 243.812 and 243.673 miles per hour, respectively.

Since the war, Navy fliers consistently established new world records not only for speed but also for endurance and payload. Some of the reason for their recent successes was the fact that the Bureau of Aeronautics, on February 21, 1923, severely restricted the repair and reuse of engines over two years old. Now the Navy was able to dispose of its World War I engines and equip aircraft with new and better engines. At last free of the millstone of stocks of obsolete engines, the Navy was able to sponsor the development of new and improved aircraft engines to meet its requirements.

Major R.E. Rowell of the Marine Corps had taken advanced flight training at the Army's Kelly Field in Texas, receiving instructions in dive-bombing. Some Army aviators derided this form of attack. When he took command of Marine Squadron VO-1M, Naval Air Station, San Diego in August of 1924, Rowell started to train this aggressive group of fliers as a dive-bombing outfit. "Warships can be destroyed by dive bombing," he said emphatically. "Moving targets at sea will find it difficult to withstand such an attack."

Captain J.M. Reeves, Commander of Aircraft Battle Force, believed him. Undaunted by wordy assaults from surface officers, he assembled Pacific Battle Fleet aircraft at San Diego to try out new tactics. Copies of what became known as "Reeves's Thousand-and-one-Questions" were circulated to fliers in a thought-provoking endeavor to find something of value in naval air warfare.

It was a period of experimentation for these young and eager men. They made dives from altitudes of up to 10,000 feet in their Curtiss F6Cs, often coming in from several directions to confuse an enemy. Later, they made dive-bombing tests against target ships at sea, using cameras to score "hits."

Lieutenant Commander O.B. Hardison, who led similar dive-bombing experiments with the Atlantic Battle Fleet, told his superiors, "Bombers can dive so fast from high altitude that deck gunners won't see them until it is too late." West Coast pilots agreed. "If you dive from out of the sun or clouds, or from different directions, bombers will so confuse shipboard gunners their fire will be ineffective." This was only a start, but these early tests proved invaluable to Naval Aviation. Dive-bombing became a standard part of Fleet operations.

A special board, to consider recent developments in Navy aviation, and headed by the new Chief of Naval Operations Admiral E.W. Eberle, submitted its report to the Secretary of the Navy on January 17, 1925. In it Eberle recommended a policy for the development of the Navy in its various branches, but the board devoted most of its discussion to the importance of the battleship, although its recommendations focused on aviation. The board recommended that carriers be built up to treaty limits, that the *Lexington* and the *Saratoga* be completed expeditiously, that a new 23,000-ton carrier be laid down, and that a progressive aircraft building program be established to ensure a complete complement of modern planes for the Fleet. The board also recommended

The USS *Saratoga* rides at anchor in San Francisco Bay in 1930.
(Courtesy of U.S. Navy.)

expansion of aviation offerings at the Naval Academy, assignment of all qualified academy graduates to aviator or observer training after two years of sea duty, and the establishment of a definite policy to govern the assignment of officers to aviation.

A Fleet "problem," scheduled to start on March 2, became the first to incorporate aircraft carrier operations, now that the *Langley* had joined the Fleet. For the next ten days the air activity of the *Langley* was limited to scouting in advance of the Black Fleet. Its performance convinced the Commander in Chief, Admiral R.E. Coontz, to recommend that completion of the *Lexington* and the *Saratoga* be speeded up as much as possible. He also recommended that steps be taken to ensure development of planes of greater durability, dependability, and radius, and that catapult and recovery gear be further improved. He added that the present equipment permitted the catapulting of planes from battleships and cruisers on a routine basis. A new type of catapult—using gunpowder instead of the mechanical system—had been developed in 1922, and was now used on all capital ships. The feasibility of using flush-deck catapults to launch land planes was demonstrated on April 2. A DT-2, piloted by Lieutenant Commander C.P. Mason, was launched from the *Langley* while it was moored to the dock in San Diego.

The rigid dirigible *Shenandoah* was torn apart in a severe line squall before daylight on September 3, 1925, over Byesville, Ohio. The control car and aft sections of the hull fell directly to the ground, while the forward section with seven men aboard free-ballooned for an hour before it landed all of them safely twelve miles from the crash site. In all there were twenty-nine survivors, but fourteen were killed.

Competitive trials by Consolidated, Curtiss, and Huff Deland aircraft companies were now held to acquire a new airplane as land, sea gunnery, and training planes. Consolidated won the competition on December 18 at the Naval Air Station at Anacostia, leading to the delivery of the NY series of training planes, which continued in service into the 1930s.

The Secretary of the Navy directed on April 21, 1926, that starting with the class of 1926 all graduates of the Naval Academy would be given a course of twenty-five hours of flight instruction during their first year of sea duty, and that, for the purpose of providing this instruction, flight schools would be established at the Naval Air Stations at Hampton Roads and San Diego.

Lieutenant Commander Richard E. Byrd as navigator, and Floyd Bennett at the controls of a Fokker trimotor called the *Josephine Ford,* completed the first flight over the North Pole on May 9, 1926. They returned to base at Kings Bay, Spitzbergen, completing the round trip in fifteen and a half hours.

On the 24th, an act of Congress implemented recommendations made by the Morrow Board governing the future of Naval Aviation. It specified that command of aviation stations, schools, and tactical flight units be assigned to Navy aviators. In a far-reaching decision, command of aircraft carriers and tenders was ordered to be assigned to aviators or naval observers. The act further specified that the office of the Assistant Secretary of the Navy be created to foster naval aeronautics. Most importantly, Congress ordered that a five-year aircraft program be set up, under which the planes on hand would be increased to 1,000 useful planes.

A milestone was reached on July 28, 1926, when the submarine S-1 surfaced and launched a Cox-Klemin XS-2 seaplane flown by Lieutenant D.C. Allen. It also recovered the aircraft and submerged, completing the first cycle of operations in a series of tests investigating the feasibility of basing aircraft on submarines.

In a display of tactics developed by VF-2, Lieutenant Commander F.D. Wagner led his F6C-2 Curtiss fighters on October 22, 1926, in a simulated attack on the heavy ships of the Pacific Fleet as it sortied from San Pedro. In almost vertical dives from 12,000 feet—at the precise time the fleet had been warned—the squadron achieved complete surprise. Fleet and ship commanders were so impressed with the effectiveness of this spectacular approach that there was unanimous agreement that such an attack would succeed over any defense. This was the first Fleet demonstration of dive-bombing. Although the tactic had been perfected by the squadron in an independently initiated project, the obvious nature of the solution to the problem of effective bomb delivery was evident. VF-5 on the East Coast had simultaneously developed a similar tactic.

Rear Admiral James M. Reeves, Commander of Aircraft Squadrons, Battle Fleet, reported December 13 on the results of this dive-bombing exercise (officially it was known as light bombing) conducted in the formal Fleet gunnery competition. One Marine and two Navy fighter squadrons participated. The Marine and Navy fighters made forty-five-degree dives from 2,500 feet, and at an altitude of 400 feet they dropped twenty-five-pound fragmentation bombs. Observation squadrons dropped from 1,000 feet. Pilots of VF-2, commanded by Lieutenant Commander F.D. Wagner and flying F6Cs and FB-5s, scored nineteen hits with forty-five bombs on a target one hundred feet by forty-five feet. This tactic was developed to disable or demolish flight decks, destroy aircraft on enemy carriers, attack exposed personnel on ship or shore, and attack light surface craft and submarines.

The practicality of dive-bombing tactics was proved on July 20, 1927, during the fight by Marines against General Augusto Sandino's rebel forces in Nicaragua. Major Ross E. Rowell led a flight of five DHs in a strafing and dive-bombing attack against bandit forces surrounding a garrison of American Marines at Ocotal. Curiosity of the rebels was aroused by these persistent attacks, and much to the amazement of the American pilots they hurried into a clearing every time the planes appeared overhead. Exploding bombs created havoc, causing at least 300 casualties, but it took the rebels some time to learn these "big birds" were destructive weapons of war.

Once, during a particularly tight ground operation, Marine aviators forced Sandino's banditos back into the hills. Pilots flew five to seven hours a day, stopping only to refuel and reload, until success marked their efforts. Although instances of diving attacks had occurred during World War I and Marine Corps pilots had used the same technique in the Dominican Republic and Haiti in 1919, this attack was made according to a doctrine that was developed in training and represented the first organized dive-bombing attack in combat.

Marine aviators proved adept at this type of warfare. They were the only aviation personnel to see combat service between the Great Wars.

Dive-bombing was perfected to a fine art during occupation duty in China, but Marine Captain J.T. Moore put on an unscheduled performance he was not anxious to repeat. Marine pilots, dulled by the monotony of neutrality patrols, looked forward to anything that would provide exciting action. Noticing morale was slipping, Moore called his pilots together: "Let's give the people of Tientsin a demonstration they'll never forget." Faces lit up and the pilots gave a shout of approval.

Assembling his planes high above an enthusiastic throng on the ground, Moore talked to his pilots on the radio: "Follow me." Thousands of Chinese watched intently as the airplanes hurtled earthward. Not a sound could be heard in that vast throng except an occasional sucking in of breath. Moore made his pullout sharply, and his head jerked back in horror as the wings of his plane ripped off with a sharp explosive crack. Moore dove over the side and quickly opened his parachute. It blossomed above him just as he landed in a mud-filled ditch.

Momentarily stunned by the shock of the landing, he staggered to his feet, dripping mud from head to toe. The Chinese vacated the scene when pieces of Moore's plane rained down upon them, then they raced back to him. Slowly, painfully, he crawled out of the ditch as the Chinese cheered wildly. He looked at them with amazement, and barely able to stand, slumped helplessly to the ground. "Do it again!" they yelled.

A major advance in the transition from wooden to metal aircraft structures resulted from the Naval Aircraft Factory's report on May 23, 1927, that the corrosion of aluminum by saltwater—hitherto a serious obstacle to the use of aluminum alloys on naval aircraft—could be decreased by the application of anodic coatings. Through the use of electrolytic action, aluminum could be coated with a protective film.

Dive-bombing came under official study May 27, 1927, when the Chief of Naval Operations ordered the Commander in Chief, Battle Fleet, to conduct tests to evaluate its effectiveness against moving targets. VF-5 was ordered to participate in the tests in the hope that the need for special aircraft and units could be developed for the future. They did so in the late summer and early fall of 1927. The tests proved so successful that new equipment was placed on order and dive-bombing received official sanction as a standard method of attack.

Newer versions of the DT series made it possible for pilots to adapt the planes to level bombing and as a scout bomber, as well as their original use as a torpedo bomber.

Still another triple-threat plane came along when the Douglas twin-engine T2D-1 became one of the earliest combination land-and-float planes. It had folding wings for ease of storage on carriers, and could be equipped with either floats or wheels.

When the new 450-horsepower Wright Cyclone engine was mounted in the Navy's new DT-6, it marked an important continuation of the air-cooled radial engine era, which lasted into the jet age.

Up to this time Fleet exercises had used water-based torpedo planes with efficiency, but accuracy was limited because of the delicate torpedo construction. Designers eventually constructed a torpedo that could withstand the shock when dropped from a speeding airplane.

Most important of all, through the determined efforts of Admiral Moffett, two carriers—the USS *Saratoga* and the USS *Lexington*—were commissioned on December 14, 1927, and January 5, 1928. They had been converted from two heavy cruiser hulls that had been scrapped under terms of 1922's Washington Disarmament Treaty. Weighing 33,000 tons each, their flight decks were 800 feet long and 100 feet wide, with control islands far to starboard. Marc A. Mitscher, a heroic airman of later years, made the first landing on the *Saratoga*.

Political opposition and technical problems continued to retard the program planned for Naval Aviation. Operational techniques for using aircraft at sea were expanded, however, and experimental launches were made by a DT-2 from the deck of the *Langley* using an early mechanical flywheel catapult. When the T2D was placed into service it was first flown from the *Langley* using a conventional wheel-type landing gear and arresting hook. Pilots suffered numerous cases of "instrument face" as they tossed violently in their cockpits when the outmoded arresting gear brought them to a savage stop.

By 1928 the Navy had 829 aircraft of all types—close to the 1,000 specified for the end of its five-year plan. Two hundred and eighty-five of these aircraft were operated from carriers.

The Douglas torpedo bomber had outlived its usefulness by this time, so the Navy turned to Martin T4Ms and the Great Lakes TGs. They were similar aircraft in many respects, and became the standard torpedo airplanes in the Navy.

During their return voyage from Fleet maneuvers in Hawaiian waters in September, pilots of VB-2B, a light bombing squadron based at North Island, San Diego, practiced dive-bombing tactics in the hope they could represent the Navy in the National Air Races in Los Angeles.

With their Boeing F2Bs, they coordinated their flights from the deck of the *Saratoga* with the carrier's antiaircraft crews. Arriving over their target ships, the F2Bs executed a reverse turn by the first section, followed by crossovers by the other two sections. They soon learned that this maneuver permitted the lead section to dive on a target from ahead while the other sections dove from the port and starboard at speeds of up to 250 miles per hour.

These razzle-dazzle tactics proved necessary because Navy pilots knew that antiaircraft batteries could zero in on a long line of dive-bombers with comparative ease. Attacks from three directions, however, proved to be almost insurmountable obstacles for the *Saratoga*'s fire controllers, because the time interval was too short to make accurate computations and change fuse settings.

In the past, practice missions used only twenty-five-pound dummy bombs. Ship admirals decided that VB-2B pilots should use 500 or 1,000-pound bombs.

During a briefing, one admiral said, "I've been listening to these 'hotshot' pilots brag for months. It's time they showed us what they can really do." His white moustache and red face quivered with indignation. "The problem seems simple to me," he said pompously. "All they have to do is dive straight down with a 2,000-pound torpedo. They can't miss."

Most pilots kept their thoughts to themselves. One, however, expressed the feelings of all. "I'm not about to strap myself to a torpedo in a vertical dive," he said. Then, with a trace of humor, "I'd like to sling the admiral under my plane and take him for a vertical dunking into the ocean!"

Extensive bombing tests were ordered against moving targets at sea by the Chief of Naval Operations. No bombs would be dropped, because assigned targets were ships of the Atlantic and Pacific Fleets.

Lieutenant Commander F.D. Wagner led his VP-2 North Island squadron in a series of simulated diving attacks against ships off San Pedro, California. After the ships were informed of the precise time that each attack would be made, the Curtiss F6C-2 fighter-bombers dove from 12,000 feet. Despite the advance warning, the ships were caught completely by surprise. The squadron commander gleefully told his pilots when they landed at Long Beach, "We were on the ground before the ships' crews manned their battle stations!"

VB-2B earned the right to represent the Navy at the National Air Races at Mines Field—now Los Angeles International Airport—in September. On opening day, fifteen sleek F2Bs droned into position at 15,000 feet. Squadron Leader Lieutenant D.W. Tomlinson signaled for them to peel off. Each plane hurtled earthward toward the whitewashed target circle in front of the reviewing stands. Moments later, the bombers pulled out in front of the roaring crowd as their bombs slammed into the circle. When the demonstration was repeated the following year at the National Air Races in Cleveland, Ohio, one interested observer made notes for future use. Major Ernst Udet of the German Air Force was so impressed by this new technique that he took the idea back to Germany.

Aerial bombardment now entered three definite stages and set the pattern for the future. Horizontal bombing and dive-bombing were accomplished from high altitudes, while torpedo bombing became a "just above the water" operation.

Dive-bombing was a natural for the Navy because aircraft could fly out of anti-aircraft range and the dive itself—almost straight down at high speeds—offered the best elements of a surprise attack. This technique proved successful even against warships as they twisted and turned at top speeds on the ocean's surface.

More powerful bombers were desirable, so in 1928 the Bureau of Aeronautics issued formal specifications for a new series of aircraft capable of carrying bombs up to 1,000 pounds.

Martin Aircraft Company received a development contract for a special dive-bomber known as XT5M-l. A similar bomber was built by the Naval Aircraft Factory and designed as the XT2N-1. They both were open-cockpit biplanes of conventional design, but they had special extra-strength bomb rack installations.

Curtiss F8Cs appeared in 1928 to replace the old DH-4Bs of World War I. At this time the name "Helldiver" became synonymous with Marine and Navy dive bombers. The F8Cs carried two 116-pound bombs under each wing and a 500-pounder under the fuselage. There was no bomb-ejecting mechanism, so bombs were released in a shallow dive to clear the propeller. These were rugged airplanes that could make vertical dives at nearly full throttle and used telescopic sights mounted ahead of the cockpit.

Planes from the carrier *Langley* proved the vulnerability of the Panama Canal to air attack in 1929, and within the Navy a growing awareness of the primary importance of naval aviation became evident, promoted in large part by Moffett's untiring efforts.

This was also the year that Moffett fostered increasing technical improvements within the Bureau of Aeronautics. It marked the use of wheel brakes and tail wheels for carrier aircraft, which greatly increased pilot safety.

Norden Mark II gyroscopic bombsights were tested this year at the Naval Ordnance Proving Ground. Both Norden and Sperry bombsights evolved from early devices dating back to World War I. They proved so reliable that two bombs out of three landed within twenty-five feet of the aiming point on the first try. Carl Norden's bombsight became the first effective high-altitude sight used by the Navy, and later models were adopted by the Army Air Corps.

As the 1920s drew to a close, the Navy's air arm was steadily increasing in size and strength, while the Army and its Air Corps seemed to remain largely dormant. Now three carriers were in full operation, patrol squadrons were performing scouting functions, and aircraft were regularly assigned to battleships and cruisers. Together these elements played important roles in the annual Fleet war games.

Impressive technical progress had characterized the previous decade. Despite the limited availability of funds, the radial air-cooled engine was developed as an efficient and reliable source of power. Better instruments came into use and an accurate bombsight was developed. Aircraft were now equipped with oleo struts, while folding wings facilitated carrier operations. Each year Navy aircraft flew faster, higher, and longer, setting many world records.

Tactics were improved and dive-bombing was established, despite earlier misgivings by many pilots. Marine Corps expeditionary troops learned through experience the value of air support. The techniques of torpedo attack, scouting, spotting for gunfire, and operating from advance bases became routine. The skill of Navy pilots permitted new uses for aircraft in polar exploration and photographic surveys. It was now evident to most people that the Navy had solved its basic and unique problem of integrating aviation into the operations of its surface ships.

The period was not without controversy. Angry statements by proponents of airpower, and by their detractors, made lively reading in the nation's newspapers. There were charges of duplication, inefficiency, prejudice, and jealousy. Airpower's role—and its application to the nation's coastal defenses—was debated hotly. Even the further need of a Navy was questioned. Naval aviators frequently voiced their complaints about career limitations and lack of command positions. The aircraft industry railed against small defense contracts, and at government procurement policies and competitive practices. Most firms were at or near the bankruptcy stage.

Aircraft engineers realized in 1928 that their industry had clung too long to outmoded ideas, and that metal should be used for greater strength and durability on wings and fuselages, replacing fabric and wood. (The Stout ST-1 in 1922 was the first all-metal airplane, but it had inadequate longitudinal stability.) The entire industry—around the world—was moving in the direction of greater use of light-gauge aluminum alloys, not only for structural tubing for the fuselage, but also for internally stressed wings. These alloys came into use after the Aluminum Company of America developed techniques to manufacture the odd shapes needed for airframe construction. Ford and foreign manufacturers had been experimenting with all-metal airplanes that used corrugated metal skin surfaces, and now the rest of the industry prepared to follow their lead.

Commander Richard E. Byrd, as navigator, and pilot Floyd Bennett, at the controls of a trimotor Fokker, made the first flight over the South Pole on November 28, 1929. After circling the Pole, the *Floyd Bennett* returned to its base at Little America on McMurdo Sound. The historic flight took almost nineteen hours, including a fuel stop on its return flight.

Fleet exercises in 1930 used the carrier group for the first time as a complete tactical unit, a trend that had a vast significance in the years ahead.

The United States now felt the impact of the worst depression in its history. Naval aviation suffered along with everything else. The early years of the 1930s were to mark a new era in naval aviation, laying the foundations of air and sea power for World War II. Due to limited funds and congressional apathy, this foundation proved to be a slim margin for the tragic days ahead.

A naval treaty was signed in London on April 22, 1930, by the signatories of the Washington Naval Treaty of 1922, which carried forward the general limitations of that earlier agreement and provided for further reductions in naval armament. Under terms applicable to Naval Aviation, the definition of an aircraft carrier was broadened to include ships of any tonnage designed primarily for aircraft operations. It was agreed that installation of a landing-on or flying-off platform on a warship used primarily for other purposes would not identify it as an aircraft carrier. It was also agreed that no capital ship in existence on April 1, 1930, would be fitted with such a platform or deck.

Secretary of State Henry L. Stimson spoke by radio from London to the United States at the conference's conclusion. He said it had given him "more confidence in my belief that the peaceful methods of diplomacy can eventually take the place of war." The depth of his misjudgment would soon become obvious.

On November 28, 1930, the Chief of Naval Operations, Admiral W.V. Pratt, issued a new Navy air policy, to become effective on April 1 of the following year. It reorganized aviation and established it as an integral part of the Fleet to operate with it under the direct command of the Commander in Chief, United States Fleet. This policy stressed the importance of Fleet mobility and the need for offensive action in protecting the United States against invasion from overseas. It assigned to Naval Aviation the primary responsibility for development of the Fleet's offensive power at sea and at advanced shore bases. Participation in coastal defense was reduced to a secondary task. The policy also specified that air stations in strategic naval operating areas henceforth would be assigned to, and operated by, the Fleet. The shore command was given responsibility only for shore stations, which performed training, testing, and aircraft repairs.

Tests were completed at the Naval Proving Grounds in January of 1931 on new displacement gear for dive-bombers. The danger of a bomb colliding with its releasing airplane had haunted pilots for years. This danger was eliminated by new gear that swung the bomb away from the airplane, preventing any danger of contact after its release.

On January 9 the Chief of Naval Operations, Admiral Pratt, and Army Chief of Staff General Douglas MacArthur reached agreement about operations of their air branches. Their respective air forces had had a long history of interservice controversy about the division of responsibilities for coastal defenses. The two service chiefs agreed that the functions of the two air forces would remain closely associated with those of their parent services. Naval Aviation was defined as an element of the Fleet, to move with it and carry out its primary mission. The Army Air Corps was listed as a land-based air arm, to be employed as an essential part of the Army in performing its general mission, which was defined as defense of the coastal regions of the United States and its overseas possessions.

When the beloved Knute Rockne, Notre Dame's celebrated football coach, was killed in the crash of a commercial airliner on March 31, 1931, there was a tremendous hue and cry in the world's press about the inadequacy of commercial transports, engendering questions of safety about all aircraft, including those flown by the armed forces. There was almost universal condemnation of airlines and plane manufacturers, and accusations that commercial aircraft were obsolete and unsafe. The public outrage was so great that the crash of the Fokker trimotor, which suffered a major structural failure because of its largely wood construction, dealt an almost mortal blow to the infant airline industry. Although the framework of the Fokker's fuselage was made of tubular aluminum, the rest of the plane was constructed of wood. Casein glue (a phosphoprotein precipitated from milk), was used to bond the wing's wooden spars and outer "skin," but after water penetrated the wing's interior over a long period of time the glue produced fungus growths that eventually weakened the structure and brought about a total collapse. The Civil Aeronautics Authority insisted on so many modifications to the Fokker that further operations of the trimotor were impossible.

A contract for the XFF-1 two-seat fighter—the first naval aircraft to incorporate retractable landing gear—was awarded on April 2, 1931, to Grumman. Tests had shown that such gear would substantially improve aerodynamic cleanness and increase performance.

Another basic improvement had been tested for several years. The use of variable-pitch propellers had demonstrated shorter takeoff runs and an increase in speed. Rear Admiral Moffett, who had been following the testing and evaluation of such propellers, acted on September 10, 1931, ordering that the tests at Anacostia be expedited. A variable-pitch propeller installed on the Curtiss F6C-4 demonstrated a twenty percent improvement in efficiency.

Advocates of airpower had a new champion in the election of Franklin D. Roosevelt as president in the fall of 1932. Mrs. Herbert Hoover, who had christened the USS *Akron* on August 8, 1931—the first of two huge dirigibles built for the Navy by the Goodyear Company—had had nothing but praise for it. It was designed by Karl Arnstein, who was responsible for most of Germany's wartime zeppelins. The first rigid airship, the *Akron* was 785 feet long, and with its sister ship the *Macon* it was expected to form the backbone of the Navy's air fleet. An innovation was incorporated that excited the imagination of the public—it was equipped to launch and recover aircraft.

Americans were particularly shocked when the *Akron* crashed into the sea on April 3, 1933, killing seventy-three people. Among the dead was Admiral Moffett. The energetic aviation leader had served as chief of the Bureau of Aeronautics for twelve years, and had fought stubbornly for a lighter-than-air program, so his death was a blow not only to airship enthusiasts but to all naval aviators.

When the German airship *Hindenburg* was lost on May 6, serious doubts began to be expressed about the safety of these huge airships.

Ernest J. King, a hard-driving officer with courage and conviction, took over Moffett's job a month later. Moffett had told King in 1926 that the U.S. Congress was expected to legislate that all aviation units must be commanded by naval aviators or observers. As a result, Moffett said, carriers and naval air stations would need senior-rated officers. Moffett assured King that, despite his age, if he learned to fly he would be given command of an aircraft carrier.

King had no doubt that the Navy's future was in aviation, but he was fifty years old and not anxious to learn what he considered a young man's profession. But he had applied earlier for command of a light cruiser, and he quickly realized he had little chance of getting such an assignment because there were so many senior captains ahead of him. He decided to accept Moffett's suggestion.

Moffett put King in command of a seaplane tender, the USS *Wright*. Her former commander, Captain John V. Babcock, admitted to King that his ship was rather loosely run—which King learned was an understatement—so his first task was to establish discipline while the *Wright* was being refitted.

King decided his own pilots should teach him to fly, and his progress was so rapid that he asked Moffett to classify him as a student aviator. Moffett agreed, but he insisted that King take formal flight training at Pensacola.

King was assigned as a student in 1927, and he was treated like all the others even though he was a captain. His relations with other students were strained at first, but that all changed after a few nightly poker games where the liquor flowed freely. King had long had a serious drinking problem, and he had periodically quit. When he found that everyone else was drinking after hours, he enthusiastically joined them. His previous flight training proved beneficial, and he was able to solo soon after he arrived at Pensacola. He was not a natural pilot, and he was fully aware of his limitations. But he developed habits of safety in training and performed only those flights necessary to qualify as a pilot. He received his wings on May 27, 1927, after flying the minimum of 200 hours.

Admiral Moffett later recommended that the Bureau of Navigation give King command of the USS *Lexington*. This was the Navy's best aviation command, but bureau officials rejected Moffett's recommendation—quite possibly because of King's reputation as a heavy drinker despite his admitted brilliance as an officer.

Moffett still believed in King, and he made him his assistant. The inevitable happened between two such strong personalities, and their clashes interfered with the operation of the Bureau of Aeronautics. When King requested a change of duty nine months later, Moffett was glad to get rid of him. He sent him to command the Naval Air Station at Norfolk, Virginia.

Moffett had promised King that he would give him command of the *Saratoga* in the summer of 1930. King objected, wanting instead the command of the *Lexington,* which he considered the finest ship in the Navy. This time King was successful, and he took command of the "Lex" in June. He was surprised to find that he had inherited another "loose" ship, and immediately set about tightening discipline. By the time four squadrons of sixty-five planes arrived in February of 1931 King had established tight control of the carrier's men by getting rid of officers and men who didn't shape up to his ideas of a disciplined ship. Pilots reacted angrily when King insisted they inspect their aircraft before and after each flight. Such a practice eventually became a standard operating procedure throughout the Navy's carriers.

After Moffett died in 1933, King lobbied strongly for the job as chief of the Bureau of Aeronautics. He went directly to Admiral Pratt, the chief of naval operations, but Pratt refused to consider his request. Secretary of the Navy Claude A. Swanson admired King, and recommended him to President Roosevelt. Such support paid off: King got the job and was promoted to rear admiral.

Although the General Board recommended five 23,000-ton aircraft carriers and five 13,800-ton smaller carriers during the mid-1920s, only the USS *Ranger* was authorized. Its keel had been laid on September 26, 1931, and it was the first ship to be designed as a carrier. The others had used converted hulls.

The much-maligned National Industrial Recovery Act was now of great assistance in the maintenance of an effective Fleet air arm. Roosevelt directed on June 16, 1933, that $238 million in NIRA funds be used to engineer and construct two new carriers and other ships for the Navy. As a result the USS *Yorktown,* laid down on May 21, and the USS *Enterprise,* on July 16, 1934, became available for World War II.

In Congress, $7.5 million in funds from the National Industrial Recovery Act, passed June 16, 1933, authorized the Navy to add planes to its previous authorization for 1,000 planes and maintain them with modern equipment.

Dr. A. Hoyt Taylor, head of the Radio Division of the Naval Research Laboratory, on March 14, 1934, authorized a project for development of pulse radar (as it was later called) to detect ships and aircraft. The basic concept, which had been proposed by Leo C. Young, involved special sending, receiving, and display equipment, all mounted in close proximity. This radar equipment would send out pulses of radio energy of a few microseconds in duration separated by time intervals that were tens of thousands of times stronger than the duration of a pulse. A target was indicated when an echo was received. The time it took for the echo to travel to the target and back indicated its distance, while the bearing was identified by the directional sending or receiving antennas. The Naval Research Laboratory had been exploring continuous radio waves for nearly four years, but the pulse technique promised greater utility because it provided range and bearing as well as detection. Most importantly, the new apparatus could be installed on a single ship. The pulse technique resolved from new techniques in the radio industry, including the cathode ray tube, higher-powered transmitting tubes, and special receiving tubes.

In Congress, the Vinson-Trammel bill was passed on March 27 and approved by the president, authorizing the Navy to add 650 planes to its previous authorization for 1,000 planes. This permitted the Navy to bring its air strength up to the amount approved by the London and Washington treaties of 1921 and 1930, respectively, but, most important of all, it specified no exact limit. The act authorized construction of a number of ships, including one aircraft carrier of about 15,000 tons. It also provided that not less than ten percent of authorized aircraft and engines be manufactured in government plants. Under the bill's authorization, the *Wasp* was laid down in 1936.

Although some individuals in the Bureau of Aeronautics clung stubbornly to the obsolete biplane configuration, it had to be replaced. Despite their advantageous slow landing speed characteristics aboard carriers, biplanes inhibited an increase in performance. The bureau circulated specifications for a new dive-bomber in 1934. Officials stated that the new airplane should be for dive-bombing and scouting only, and must possess substantial range and high payload. Specifically, they said the new airplane had to include a bomb displacement or swing-down gear to toss bombs away from the airplane and free of the propeller. Dive brakes to reduce speed in steep dives were considered absolutely essential.

The Vought XSBU-1 proved to have the highest all-around performance. Unfortunately its dive brake system—actually only a reversible propeller—was not satisfactory. The airplane was ordered into production even though its operations were severely restricted because of its constant propeller problems.

The Martin entry was an improved version of their earlier T5M, later redesignated the BM-1. It proved to be the best dive-bomber, capable of pullouts with a 1,000-pound bomb attached.

The Douglas-Northrop XBT-1 was flown for the first time in July 1935. It showed promise even at first, and its designers' hopes were realized far beyond their fondest dreams when it led later to the famous SBD Dauntless.

The first XTBD-1 Devastator—destined to revolutionize torpedo warfare—was delivered to the Navy for testing in 1935. An all-metal, low-wing monoplane, it carried a crew of three: pilot, bombardier, and rear gunner.

Naval aviation came of age when the first ship designed from the keel up as an aircraft carrier, the USS *Ranger*, was assigned to the Pacific Fleet. It was urgently needed because the aging *Langley* had been reduced to limited duty.

The first all-Marine flight squadrons went to sea at this time aboard the *Lexington* and the *Saratoga*. They had Boeing F4B-4 fighter-bombers of a new design capable of carrying several-hundred-pound bombs under their wings.

The TBD Devastator proved so reliable in Fleet operations that the torpedo plane and the dive-bomber now moved into more equal balance for sea warfare. One hundred and thirty TBDs were ordered, with delivery to squadrons in 1937.

The airship *Akron* was lost in a storm over the Atlantic in 1933, killing all seventy-seven men on board. Two years later its sister ship, the *Macon*, crashed off Point Sur, California, with two fatalities. A tailfin, damaged in a storm, caused the *Macon* to crash. It broke loose and punctured the airship's skin, rupturing three helium cells, and the ship plunged into the ocean. All but two of the eighty-three men on board survived the crash and were picked up by lifeboats. This marked the end of production of the huge dirigibles. Now the airship *Los Angeles* was the only rigid airship in the Navy, and its days were numbered. With the loss of the two airships, the Navy finally ended its fascination with such ships. They would have been extremely vulnerable in wartime, and of little use.

A BT-1 in flight. (Courtesy of McDonnell Douglas.)

The Aviation Cadet Act, passed on April 15, 1935, created the grade of aviation cadet in the Naval and Marine Corps Reserves, and set up a new program for pilot training in which otherwise qualified college students between the ages of eighteen and twenty-eight could be eligible for one year of flight instruction, pay benefits, uniform gratuities, and insurance. After serving three additional years on active duty, cadets would be commissioned as ensigns or second lieutenants. The act provided for a bonus of $1,500, and each officer would be returned to inactive duty as a member of the Naval or Marine Reserve.

By 1935 Adolf Hitler pressed development of the German Air Force, despite his public protestations of peace. Ernst Udet, who had been enthralled by the possibilities of dive-bombing while attending the National Air Races in Cleveland, Ohio, in 1929, had imparted his thoughts to some World War I military friends. Once he got the ear of Air Marshal Hermann Goering after Hitler came to power, development of this new tactic was pushed secretly in Germany. Hitler himself ordered construction of the Luftwaffe's first dive-bomber, the JU-47. It subsequently evolved into the deadly JU-87 Stuka series.

The dreaded Stuka, soon to strike terror among civilians and soldiers all over Europe, was the first dive-bomber to be placed in mass production. Eventually 5,000 were built. They were created despite Allied inspection teams endeavoring to preserve terms of the 1918 armistice. Disguised as fast, light transport and airmail planes, the prototype Stukas were produced in neutral countries and financed by German and Russian funds.

An SBD drops its bomb just before pull-out. (Courtesy of McDonnell Douglas.)

The Stuka first saw action in the Spanish Civil War in 1937. Their pilots developed a surprise attack that was devastating. The Stuka, while aiming at one target, would suddenly change direction in a dive and head for another. Unusual air brakes, permitting it to twist and turn in a dive, made it all but impossible to hit with ground fire.

During a period of technological advances and growing world tension, Rear Admiral Arthur B. Cook replaced King on June 12, 1936, as Chief of the Bureau of Aeronautics. Formerly a destroyer officer—and like Admiral Halsey a latecomer to the aviation field—Cook brought a tremendous enthusiasm to his new job. This was typical of this energetic individual, who was not afraid to speak his mind on a controversial subject.

A contract was awarded to Consolidated on July 23, 1936, for the XPB2Y-1 four-engine flying boat. It had been selected for development as a result of a design competition held late the previous year, and in later configurations it became the Navy's only four-engine flying boat.

On September 15, the USS *Langley,* the Navy's first aircraft carrier, was detached from Battle Force and assigned to Aircraft Base Force. There it was assigned duty as a seaplane tender. After a brief period of operation, she went into the yard for conversion, emerging early in 1937 with the forward part of her flight deck removed.

Refinements in aircraft design continued under Admiral Cook at an accelerated pace as the Navy found it necessary to start instrument courses for its pilots who were flying the newer airplanes.

After Douglas Aircraft Company's BT-2s entered production and fifty-five were built, they were redesignated as SBD-1 Dauntlesses, a name that would later earn enduring fame in the history of naval aviation.

5

A Two Ocean Navy

In France and England military aviation suffered from many of the circumstances American aviation had to contend with, including bureaucratic control that was blinded to the necessity of keeping abreast of important technical changes and plans for future wars. As a result, crews lacked proper equipment and training, particularly in locating and destroying targets.

In 1935, Admiral King had sought the three-star billet as Commander, Aircraft, Battle Force. As a vice admiral he would be in charge of the Navy's four carriers and their squadrons. His reputation for drinking and his abrasive nature lost him that job, and instead he was named to the rear admiral's post of Commander, Aircraft, Base Force in charge of seaplane patrol squadrons.

Despite his disappointment, characteristically he became a tough taskmaster, forcing pilots to fly under all kinds of conditions. His organization was renamed Aircraft, Scouting Force, in 1938. King's toughness became legendary. He openly cursed those whom he found to be weak and ineffectual—shocking behavior for an officer in the staid Navy of the 1930s—but he proved that seaplanes could be the eyes and ears of the Fleet.

King did become Commander, Aircraft, Battle Force, in January of 1938. His force included the carriers *Lexington,* the *Saratoga,* and the *Ranger,* and later the newly commissioned *Yorktown* and *Enterprise* in the spring of 1939. Despite his brusque, almost insulting manner, he trained some of the best captains in naval aviation. Several of them proved outstanding in the years ahead.

Impatient with bureaucratic maneuverings, King requested that a scouting force—with light cruisers, destroyers, and patrol planes—be added to his carriers to provide a stronger striking force. His request was forwarded to the Bureau of Navigation, where it was received with outrage because his suggestion would combine two vice admiral commands into one under his command. King explained that such a command would release the thirty-three-knot carriers from duty with twenty-one-knot battleships and provide a fast-moving, far-ranging strike force for independent action.

Admiral Adolphus Andrews, who headed the bureau, was a longtime battleship advocate. Although he grudgingly admitted that King's proposal had some merit, he claimed that carriers were too vulnerable to operate without battleships for protection. King was not surprised when his request was turned down and Andrews was given command of Aircraft, Scouting Force as a vice admiral.

But these years were a valuable training period for King that benefited the U.S. Navy in the years to come. His ideas for the employment of aircraft carriers, and the tactics developed by him and his staff, were later adopted as standard operating procedures.

Great Britain, France, and Italy agreed at the Munich Conference September 30, 1938, to permit Hitler's partition of Czechoslovakia. Prime Minister Neville Chamberlain of Great Britain said this action would assure "peace in our time."

A few weeks later, a doubtful United States strengthened the Atlantic squadron. Unfortunately, there were no carriers to adequately assure a strong defense line in the Atlantic because they were needed even more in the Pacific.

Although dive-bombing was an accepted part of warfare, tactics differed among Navy squadrons as dual-purpose missions became common during operational maneuvers. Dive-bombers often flew scouting missions, with external fuel tanks replacing the usual heavy bombs.

Long distance flights made it necessary to use radio direction-finding equipment and automatic pilots to relieve pilots on wearisome missions. Technical improvements also provided new bomb racks and releasing equipment, plus illuminated bombsights. Airplanes gained greater range and efficiency once 1,000-horsepower radial engines were installed.

Early pioneers of dive-bombing felt a sense of vindication, because not all nations were convinced it was an important means of attack. Only a few remembered that it was the U.S. Marines who had started the practice in 1919.

By January 1939, another world war was imminent, with the time and place dictated by Hitler's Nazi Germany. After the U.S. Congress accepted President Roosevelt's recommendations for rearmament and an increase in naval strength above the replacements permitted by the 1934 Vinson-Trammel Act, it also approved legislation guaranteeing the nation's neutrality.

Admiral William D. Leahy, Chief of Naval Operations, formed the Atlantic Squadron in 1939 to protect the United States' inherent rights to the sea-lanes. Rear Admiral A.W. Johnson—the same man who had been in charge of the 1921 bombing tests—was given seven cruisers and seven destroyers to cover his vast responsibilities, but his command had only eighty airplanes.

A contract was signed in 1939 for the first experimental Curtiss SB2C dive-bomber. The Helldiver, an all-metal, mid-wing aircraft, had serious problems, but it later served with distinction in the Fleet.

That same year a contract was signed with Grumman for another experimental torpedo plane, later known as the TBF Avenger. After many difficulties with the early models it served well in the latter days of World War II.

Also in 1939 production of SBD-1s stabilized, with an order for Marine dive-bombing squadrons. New Douglas SBD-2s and SBD-3s with self-sealing fuel tanks were right behind them.

Admiral King was now convinced he was qualified to be chief of naval operations or commander of the U.S. Fleet. He wrote a letter accompanying his fitness report seeking either job, and pointing to his personnel file for justification. It was filled with letters of commendation and the fact that he had served in all branches of the Navy.

The Aviation Cadet Act of 1935 was revised on June 13, 1939, to provide for the immediate commissioning as ensigns or second lieutenants of all cadets on active service and the future commissioning of others upon completion of their flight training. The law also extended the service limitation to seven years after completion of training, of which the first four would be required, and provided for promotion to the next higher grade on the basis of examination after three years of service.

In the summer of 1939, Chief of Naval Operations Admiral William D. Leahy informed President Roosevelt that he wanted to retire. King had sought the job, but his name was not on the list submitted to the president. When Rear Admiral Harold H. Stark was nominated, King believed his career was at an end. At the time there were only three admirals of flag rank with aviation experience, and he was one of them. The others were Rear Admiral William F. Halsey and Charles A. Blakely, but King was resigned to the fact that the top Navy job would be given to one of the seventy-four officers of flag rank who belonged to the surface Navy. But President Roosevelt had heard of King's excessive drinking problem, and he did not personally know him, which was almost a prerequisite for advancing to the top post.

King was sixty years old in July 1939 when he was appointed as a member of the General Board. He had no illusions that his Navy career was coming to an end because this was where elderly admirals were placed until they reached the age of retirement at 64.

Despite a nonaggression pact signed in 1934, Germany invaded Poland on September 1, 1939, and two days later Great Britain and France declared war on Germany. Thus the war that had seemed inevitable for more than a year began with strong thrusts of German panzer units against the Polish nation. Russian troops invaded Poland from the east on September 17, and on September 28 a German–Russian agreement divided Poland between the two countries as resistance ended.

President Roosevelt on September 5 proclaimed the United States would remain neutral in the European war, and directed that the Navy organize a Neutrality Patrol. In complying, the Chief of Naval Operations ordered the commander of the Atlantic Squadron to establish air and ship reconnaissance of the sea approaches to the United States and the West Indies for the purpose of reporting and tracking any belligerent air, surface, or underwater units in the area.

Three days later the president announced the existence of a limited national emergency and directed that measures be taken for strengthening the nation's defenses within the limits of peacetime authorizations.

Despite highly vocal isolationists at home and a stubborn Congress, Roosevelt described this new act as a "strengthening of national defense for proper observances safeguarding and enforcing the neutrality of the United States."

Naval Aviation really came into its own when Rear Admiral John Towers—a Curtiss school graduate of 1911 and the Navy's most qualified pilot—became head of the Bureau of Aeronautics. Three months after war broke out, he demanded an increase in trained men. Fortunately, the Naval Reserve Act of 1939 followed his recommendations and set a minimum strength of 6,000 in the Naval Reserve. Towers vigorously promoted ways to destroy submarines from the air and, with the assistance of Captain John S. McCain, initiated plans to use aircraft during landing operations.

For months the war remained on a "sitzkrieg" basis, but Germany ended the stalemate when it invaded Denmark and Norway on April 9, 1940. Then, on May 10, the mighty Luftwaffe bombed airfields in the Low Countries, and invaded the Netherlands.

Free nations everywhere mourned the capitulation of King Leopold of Belgium a week later. They gained little condolence from the rescue of 335,000 out of the 400,000 Allied soldiers at Dunkerque by British civilian and naval craft between May 26 and June 3. Italy entered the war on June 10 and invaded France. The Germans entered undefended Paris on June 14 and France and Germany signed an armistice at Compiègne eight days later. In a crisis mood, Roosevelt signed the "Two Ocean" Navy bill, authorizing the greatest military expansion in the country's history, and providing for 1,325,000 tons of new combat ships. On June 14, the Naval Expansionary Act authorized an increase in carrier tonnage to 79,500 tons over the limits imposed in 1938, and increased aircraft strength to 4,500. Congress raised the ceiling to 10,000 airplanes the following year.

Referring to Mussolini's war declaration against France, Roosevelt, in a rare exhibition of public contempt, told Congress, "The hand that held the dagger has stuck it into the back of its neighbor!"

In exchange for fifty four-stack destroyers, Great Britain by formal agreement ceded to the United States for a period of ninety-nine years sites for naval and air bases in the Bahamas, Jamaica, St. Lucia, Trinidad, Antigua, and British Guiana. It also extended similar rights freely, and without consideration, for bases in Bermuda and Newfoundland. Acquisition of these sites advanced the United States' sea frontiers several hundred miles and provided bases from which ships and aircraft could cover strategically important sea approaches to the nation's coasts and to the Panama Canal.

A reorganization of the Fleet on November 1, 1940, changed the administrative organization of aviation by dividing the forces between two oceans, which was the beginning of the independent development of forces to respond to new strategic requirements. In the Atlantic, aviation was transferred from Scouting Force to Patrol Force, which was formed in place of the Atlantic Squadron as a Fleet command parallel to Scouting Force, and set up under Commander, Aircraft Patrol Force and Commander, Patrol Wings Force. In the Pacific, Patrol Wings remained attached to Scouting Force under the combined command of Patrol Wings and Aircraft Scouting Force.

Atlantic and Pacific Fleets were established on February 1, 1941, to complete the division begun the previous November, thereby changing the titles of aviation commands in the Atlantic Fleet to Aircraft, Atlantic Flight and Patrol Wings, Atlantic Fleet. No change was made in the Pacific Fleet's aviation organization.

Admiral Stark, ever since he became chief of naval operations, had tried to get Admiral King an active command because he was convinced that his talents were being wasted on the General Board. In December 1940, King was given command of the Atlantic Squadron. Even though the position called for a rear admiral in command, obviating a promotion for King, he accepted the assignment with gratitude. Any job was preferable to the one he had.

Then King's command was designated the Atlantic Fleet. It was one of three fleets, the others being the Pacific and Asiatic Fleets. The title Commander in Chief, U.S. Fleet, which had called for a full admiral, was abolished.

King had no illusions. He anticipated that he would soon be replaced. He wrote Admiral Stark that there was too much business as usual and that most ships needed to be made combat-ready. He urged that steps be taken immediately to make them so. Secretary of the Navy Frank Knox agreed with King, but nothing was done to implement his suggestions. King reached his sixty-third birthday on November 23, 1941—a year short of the Navy's mandatory retirement age.

On March 11, 1941, the president was empowered by an act of Congress to provide goods and services to those nations whose existence was deemed vital to the defense of the United States. Thus was initiated a Lend-Lease program under which large quantities of munitions and implements of war were delivered to the nation's Allies.

Embattled England faced a supply crisis that year. When the American merchant ship *Robin Moor* was torpedoed and sunk by a German submarine in the South Atlantic, Roosevelt came to an important decision. In a radio address to the nation six days later, on May 27, he called for an unlimited national emergency that would entail expansion of the naval patrol in the Atlantic, and require that all of its military and civilian defenses be placed in a state of readiness to repel all acts or threats of aggression directed toward any part of the Western Hemisphere.

The character of the war changed drastically when Germany invaded Russia on June 22, 1941. Roosevelt viewed the invasion as a chance to limit America's involvement by supplying Lend-Lease to the Russians as well as the British. He had promised the American people that he would not send their sons to foreign wars.

While Americans sat soberly by their radios, the familiar calm, cultured voice gave them the word. "Our patrols are helping now to insure delivery of the needed supplies to Britain. All additional measures necessary to deliver the goods will be taken."

The president did not go into details, but part of his program became evident. First, an American Naval Task Force landed Marines at Reykjavik, Iceland on July 9, taking over occupation from the British, with the reluctant consent of the Icelanders. This provided a necessary base to police the patrol lines between Europe and the United States.

Meanwhile, neutrality violations became increasingly serious. A German submarine made its first attack on an American warship when it fired torpedoes at the USS *Greer* in early September. The ship was on a mail run to Iceland. The torpedoes fortunately missed, and the *Greer* trailed the submarine for three hours, acting as a spotter for a British patrol plane. Three days later the merchant ship *Steel Seafarer* was sunk by German aircraft in the Red Sea.

Roosevelt spoke to the nation on September 11 and compared Hitler to a snake, "When you see a rattlesnake poised to strike you do not wait until he has struck you before you crush him...from now on if German or Italian war vessels enter the waters, the protection of which is necessary for American defense, they do so at their own peril!"

Despite his "shoot-on-sight" order, the USS *Kearney* became the second victim of Nazi torpedoes during the early morning hours of October 17 while it sought to rescue the crews of three ships in a British convoy that had been attacked by a wolf pack of submarines. It was a new destroyer, built only the year before. It managed to limp into port carrying the bodies of eleven crewmen killed in the attack.

A 16,000-ton oil tanker, the USS *Salinas,* was struck on October 30, but it was not fatally injured.

Until now Fleet ships had been damaged but not sunk. The following day, as public indignation rose to a fever pitch, the USS *Reuben James* sank—with a loss of 197 lives—after breaking in two following a German torpedo attack. The old four-stacker was the first American Navy ship to be sunk by enemy action since 1918.

Indignation mounted and demands for action became more insistent. Two weeks later Congress removed the restrictive clauses from the 1939 Neutrality Act. A month earlier Secretary of the Navy Frank Knox and Chief of Naval Operations Admiral Harold E. Stark had appeared before the House Foreign Affairs Committee to urge repeal of the whole bill.

Roosevelt told the world, "Merchantmen will be armed and permitted to enter harbors of belligerent nations."

The war controversy had been bitter, but the torpedoing of the *Kearney* and the *Salinas* and the sinking of the *Reuben James* overcame most of the opposition.

As Americans awoke each morning, chilled by the hysterical ranting of Adolf Hitler over their radios, and disgusted by the submissive "Sieg Heils" of the Nazis, they felt helpless in the face of one disaster after another. In Russia, German armies were moving on Moscow, and it seemed the Russian capital would soon fall.

The Truman Committee, meeting in 1941, was not convinced of the potential superiority of the Navy's dive-bombers. "Dive-bombers should not be procured in numbers by the Navy," they reported. "All funds for this purpose should be cut." Navy officials, realizing their dive-bombers had not been proven in combat, but knowing they might soon be engaged in a fight for survival in the Pacific, voiced vigorous disagreement. "The SBDs are in being, combat ready," they replied, "while newer airplanes are still on the drafting boards."

Navy and Marine aviation had grown tremendously by the start of December 1941, with 5,233 aircraft of all types, including trainers, and 5,900 pilots. There were five large and one small aircraft carriers, five patrol wings, and a few advance air bases.

Little did the American people realize at the time the ultimate price the nation would pay for its lack of foresight.

Part III

World War II

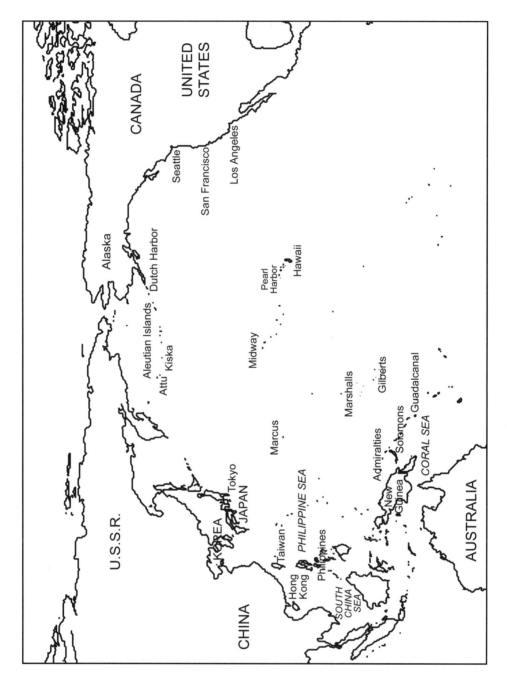

The Pacific theater. (Drawn by Stewart Slack.)

6

Intercept and Destroy

The Imperial Japanese Navy—although smaller than the U.S. Navy—had a distinct advantage in that its ships were concentrated in the Pacific Ocean, while the United States had responsibilities in the Atlantic as well. Of the American Navy's 216 major ships, 114 remained on duty in the Atlantic because of the German submarine menace to Allied shipping.

Japan also used to the fullest extent the potent element of surprise, preparing for war while technically remaining at peace. The 180-odd ships of her navy could therefore be poised to strike in overwhelming numbers in a particular area of the Pacific most advantageous to her strategic plans.

Six American carriers were in operational readiness when war came on December 7, 1941, and the newly commissioned *Hornet* quickly terminated her shakedown cruise the day the Japanese attacked Pearl Harbor. In the Atlantic were the flattops the *Wasp,* the *Yorktown,* and the *Ranger.* In the Pacific, the *Saratoga* was laid up for repairs at San Diego and the *Enterprise* under Admiral Halsey was 200 miles west of Pearl Harbor on its way back after delivering twelve Marine fighter pilots and their Grumman Wildcats to Wake Island. The carrier *Lexington,* with three heavy cruisers and five destroyers under Rear Admiral John H. Newton, was 425 miles southeast of Midway to deliver Major C.J. Chappell Jr.'s VMSB-231 scout-bomber squadron—which also included eighteen SB2U Vindicators—to the almost defenseless island. Halsey, upon hearing of the attack on Pearl Harbor, headed back toward Pearl Harbor at full speed.

The Japanese—increasingly arrogant as successes mounted for Hitler and Mussolini—continued talks with State Department officials in Washington while Japan's mighty warships headed for their historic rendezvous.

Admiral Stark had sent a war warning to Admiral Husband E. Kimmel, Commander in Chief, Pacific Fleet, on November 27, telling him to expect an aggressive move by Japan within the next few days. Hawaii was not specifically mentioned in the warning, but the Philippines, Thailand, the Kra Isthmus, and Borneo were named as possible targets of Japanese attacks. Stark's warning had ordered him to prepare to "exercise deployment preparatory to carrying out the existing war plan." It was not more explicit because the United States at that point had only broken Japan's diplomatic code and not its military code.

From transmissions in Japanese diplomatic code, American officials had learned that Tokyo had requested that its Honolulu consul provide periodic reports about American ship positions at Pearl Harbor and any movement within the harbor. This kind of information was sought by the Japanese from their consuls throughout the world, and by itself it was not significant. Admiral Kimmel and General Walter Short were not advised of this intelligence because it was restricted to top people in Washington, and Hawaii was considered too full of spies and Japanese nationals.

Neither Kimmel nor Lieutenant General Short, commander of all Army forces in Hawaii, believed that Stark's message indicated there was a Japanese threat to the Hawaiian Islands. Actually there had been no air raid drill since November 12. For protection against sabotage, Short ordered all Army aircraft parked wingtip-to-wingtip, making them particularly vulnerable to air attacks.

Although the Navy had sent flying boats on patrol, these were essentially training flights, because no system had been set up to seek out unfriendly ships or submarines.

Army Chief of Staff George C. Marshall was shocked to receive a decoded message that the government had intercepted in Washington on December 7. It was the wire from Tokyo to the admiral of the Japanese Fleet to take action against elements of the U.S. Fleet, although there was still no mention of Pearl Harbor. Marshall quickly wrote a warning to generals and admirals in the Pacific area. "Expect a full-scale attack by Japan at any moment. Take all necessary precautions." Unfortunately, the Army circuits to Hawaii were out of commission and the man who had been ordered to dispatch the message immediately sent it by Western Union, even though direct Navy channels were available. The delivery boy did not arrive at Fort Shafter Army Headquarters until four hours after the attack. By then it was 3:00 Sunday afternoon, December 7.

Earlier that day an Army technician practicing with a radarscope to improve his skill detected a mysterious blip on his screen. After another and yet another showed up, he put in a hasty call to his lieutenant. "Forget it," he was told.

A little later, early risers idly noticed the sunlight flashing from the wings of airplanes circling lazily over Ford Island. One observer counted twenty-five. A Navy man, who thought they looked different than familiar American types, squinted into the bright morning sun. He had no binoculars so he could not be sure, but he was not concerned. The small group watched as another squadron flew into the area from seaward.

They gaped in astonishment as a dive-bomber from the high group shrieked earthward, heading directly for Ford Island Naval Air Station. Onlookers were so shocked as the first bombs hit that they stood petrified as the bombers attacked in screaming dives.

It was some time before the military establishment realized the extent of the attack by carriers of the Imperial Japanese Navy under Vice Admiral Chuichi Nagumo. His task force of thirty-one ships consisted of six carriers, two battleships, two heavy cruisers, one light cruiser, nine destroyers, and supporting ships.

The sun's first rays were just above the horizon on the morning of December 7 when planes of the *Enterprise* roared off the carrier and headed for Ford Island, 200 miles away. Radio monitors on the ship, listening with a bored air to the voices of the pilots from the carrier's planes, jerked to abrupt attention when they heard, "Don't shoot! This is an American plane!" The room became a beehive of activity, and the lackadaisical altitude that had prevailed for hours evaporated in an instant. "That's Ensign Manuel Gonzales," one of them said. "I'd

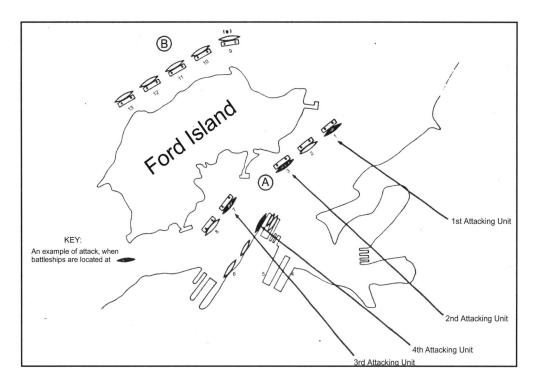

High-level bombing attack plan against battleships moored singly in Battleship Row. Reproduced from Fuchida's original chart. (See bibliography.)

know his voice anywhere." The naval reservist's radio went dead, and the monitors looked at one another in astonishment, never suspecting the tragedy occurring so many miles away.

The *Enterprise*'s planes were caught by surprise when they arrived over Pearl Harbor, and before they knew what was happening, eight other pilots and their gunners were shot down by the attackers.

Japanese pilots, who had been briefed thoroughly for what Roosevelt later called, "a day that will live in infamy," searched for their primary targets. "Destroy the carriers!" they had been told. They had been informed that the carriers were normally berthed on the northwest side of Ford Island. Where were they? Nagumo had learned from Tokyo the night before that the carriers had departed Pearl Harbor, but he remained hopeful that some might have returned.

The leader, thinking he had spotted the *Saratoga,* signaled for attack. As he roared down, he failed to realize that he was sighting on the old target ship *Utah*. It looked somewhat like a carrier, stripped down and timber-covered, but the *Saratoga* was 2,000 miles away in San Diego.

Another group of Japanese planes headed for the flying boat base at Kaneohe. The Japanese High Command in Tokyo had issued strict orders that airplanes on the ground or on water should be given top priority.

The Naval Air Station at Ford Island paused momentarily in astonishment after the first bomb hit. The commander of Patrol Wing 2 broadcast, "Air raid, Pearl Harbor! This is no drill! This is no drill!"

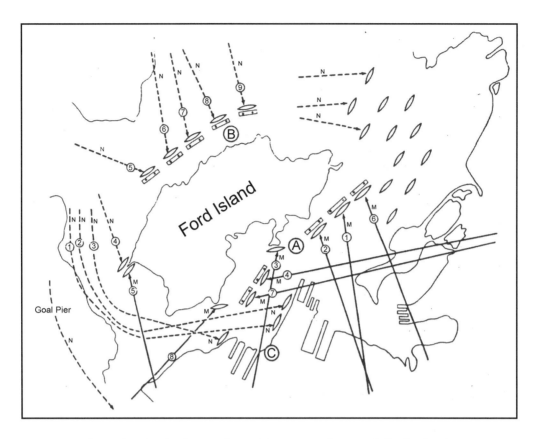

Torpedo attack plan against ships moored around Ford Island.
Reproduced from Fuchida's original chart. (See bibliography.)

The Japanese used horizontal torpedo planes and dive-bombers in devastating attacks against military installations at the Army base at Hickam. Lieutenant Cecil Durbin, a B-17 pilot based there, got up early that morning and was walking across the field to the tennis courts to play a few games. When the bombs started to drop, he looked at the diving planes in shocked amazement. He started to run and, seeing a heavy vehicle up ahead, dove under it. When the bombs stopped falling, he noticed with shock that he had sought sanctuary beneath a gasoline truck.

The Japanese had planned well. Torpedo planes sought out the heavy ships. Their torpedoes had been adapted especially for shallow water. Wooden fins added buoyancy, and short-run exploders made them deadly weapons in the harbor. Despite the incredible destruction, the Japanese lost only twenty-nine aircraft during the attack.

There was only a short respite after the initial wave before dive-bombers screamed down against "battleship row." The Japanese, taking advantage of complete air superiority, made second and third runs with their high-level bombers to assure perfect aim as they dropped their bombs on the battleships.

At 9:45 A.M., after all attacks ceased, a heavy pall of smoke hung over the stricken area. Huge explosions, hurling smoke and flames into the cloudy sky, seemed incongruous to those who had been familiar with the magical softness of the tropical air.

Mighty warships succumb to the fierce Japanese attack at Pearl Harbor.
(Courtesy of U.S. Navy.)

Exact assessment of the damage would take time, but at Kaneohe Bay all thirty-four planes were out of commission. Some were salvaged later, and three patrol planes on dawn alert were saved.

Rear Admiral P.N.L. Bellinger, senior officer of the Naval Air Station at Pearl Harbor, arrived while the attack was under way. He was appalled by the bombed hangars and twisted wreckage of his airplanes. "Get the planes in the air!" he ordered. It was too late. After the first Japanese wave of bombers, only three American aircraft were capable of flight.

The *Enterprise,* returning from Wake Island, sent its Scouting 6 airplanes ahead to land at Ewa airfield. Those that survived the first attack found their own gunners on the ground to be as much of a menace as the Japanese. After thirteen dive-bombers taxied up to the blasted operations building, they were sent out immediately. Lieutenant G.D. Dickinson destroyed one Japanese plane over Pearl Harbor, but found his own plane riddled so he had to bail out.

There had been 301 Navy planes of all types in the Oahu area before the attack, including fifty-four noncombat types. Of this number, ninety-nine were still in storage as replacements. When the last Japanese plane headed back to its carrier, only fifty-two planes were able to take to the air, and sixteen of these were noncombat types. The air station at Kaneohe was a shambles and the destruction at Ford Island Naval Air Station was just as complete.

The captain of the *Lexington* placed his ship in maximum readiness, with all battle stations manned, and hatches and doors secured for combat operations. Command headquarters at Pearl Harbor sent out an emergency dispatch to all ships at sea: "Intercept and destroy enemy! Believed retreating on a course between Pearl and Jaluit. Intercept and destroy!" It was a futile gesture, though understandable at the time under the stress of great emotion. The American Navy had suffered its worst defeat in history.

Japanese broadcasts exultantly proclaimed a great victory, saying that Pearl Harbor had been wiped off the map. They even claimed that the *Lexington* had been sunk. (Actually it had been en route to Midway to deliver planes but had turned back to Pearl Harbor after the attack.)

It had taken only an hour and fifty minutes, but 253 Japanese planes had brought disaster to the U.S. Pacific Fleet. If they had continued their attacks that day, and in the following days, not one battleship would have risen from its watery grave.

Admiral Isoroku Yamamoto, commander of the Combined Fleet and the First Fleet, had urged Nagumo to continue the attacks, but the cautious admiral headed for home instead. The heavyset, gray-haired Nagumo was a veteran seaman, but he was not an airman. A battleship veteran, he had considered carriers too vulnerable to use in an attack on Pearl Harbor, and had opposed the attack ever since its recommendation by Admiral Yamamoto. He was a torpedo expert who was convinced that sea battles could only be won by battleships. Most American and British admirals of the time agreed with him. Prior to the attack, he had believed he would lose half of his carriers.

Yamamoto was disgusted by Nagumo's lack of aggressiveness. A brilliant naval strategist, Yamamoto—like Admiral King—had a drinking problem, which he learned early to control. He had been given command of all Japanese naval units, although he stressed that the Imperial Fleet faced grave risks in any war with the Allies. He told intimates, "If it is necessary to fight, in the first six months to a year of war against the United States and England I will run wild. I will show you an uninterrupted succession of victories. But I must also tell you that if the war were prolonged for two or three years I have no confidence in our ultimate victory."

Once the Japanese government committed itself to war, Yamamoto realized that the Imperial Fleet could achieve initial success only by delivering a crippling blow to the U.S. Pacific Fleet based at Pearl Harbor.

When he revealed his plans in May 1940, high navy and government officials objected, saying that an attack at such a great distance from the fleet's home base was tantamount to "putting one's head in the lion's mouth." Yamamoto persisted, sending his carriers to Kagoshima to prepare for the attack. The volcanic island of Sakurajima in Kagoshima Bay resembled Pearl Harbor and was used to train Japanese pilots.

Nagumo's race for home waters permitted the U.S. Navy to recover from its devastating defeat. In one of the most remarkable salvage operations the world has ever seen, most of the ships were saved to fight another day.

The *Arizona*, horribly burned and blasted in two, was a total wreck. The *Oklahoma* had capsized, her twisted superstructure on the bottom. The *West Virginia* had settled into shallow water, the *California* was sunk, and the *Nevada* had to be run aground to prevent her sinking. The old *Utah* would cruise no more. She was down, bottoms up, a complete loss. Among other battleships, the *Pennsylvania,* the *Maryland,* and the *Tennessee* were damaged but were repaired in the months to come. Many other ships also were damaged but repairable.

In later months the Japanese were to find to their chagrin that ships they thought had been sunk still had a lease on life that proved to be deadly to their own ships.

Publicly, the Navy admitted the loss of the *Arizona,* the *Utah,* destroyers *Cassin,* the *Downes,* and the *Shaw,* and the minelayer *Oglala.* The *Oklahoma,* which could have been repaired, was scrapped so its materials could be diverted to more worthwhile ships.

The U.S. Congress declared war on Japan December 8 following the president's condemnation of the attack on Pearl Harbor, calling it "a day that will live in infamy." Three days later Germany and Italy declared war on the United States, and the United States reciprocated.

Aircraft from the *Enterprise* attacked and sank the Japanese submarine I-70 on December 10 in waters north of the Hawaiian Islands. This was one of the submarines that had been used to scout the Hawaiian area in connection with the Pearl Harbor attack. It was the first Japanese combat ship sunk by aircraft in this new war.

After the Pearl Harbor disaster, President Roosevelt realized that the Navy had grown too protective and too conservative. It needed men with new ideas, he told his cabinet. Secretary of the Navy Knox told the president that King was his personal choice to be Commander in Chief, United States Fleet, although Roosevelt had abolished the position the year before.

The president had some initial doubts, but he knew that King was an officer of vast knowledge and a commanding presence, despite some personal shortcomings. The more he considered others for the position, the more he realized that no other admiral had the ability and toughness to provide the Navy and the nation with the necessary leadership. He didn't know King well, and he had heard that the admiral was cold and often ruthless in his dealings with subordinates. He reached a decision. King, he told Knox, was ideal for the job of taking command of the U.S. Fleet.

When King was advised of the president's decision, he replied, "When war breaks out, they look for the sons of bitches." One of his first actions was to change the acronym for Commander in Chief, United States Fleet—*Cincus.* He ordered that it be changed immediately to *Cominch,* claiming *Cincus* sounded too much like "sink us."

The Navy had laid down the keels for sixteen ships larger than a cruiser during 1941, and now top priority was given to new aircraft carriers. Orders were issued in Washington to start construction of some of the mightiest battleships the world had ever seen. Meanwhile, those sunk or damaged at Pearl Harbor were redesigned with completely new superstructures and became better warships than they had been in the past.

The Japanese had struck a telling blow, and it was not to be the last. A timetable, established years before, initiated strikes that day on such widely scattered points as Hong Kong, Singapore, Thailand, and Kota Bharu on the Thai-Malayan border.

The attacks were simultaneous on a perimeter extending from Pearl Harbor to the Gulf of Siam. Task forces and scores of troop transports roamed the western side of the Pacific. Tens of thousands of trained soldiers—many hardened veterans of the China war—helped to push Japan's frontiers ever farther from the home islands, hoping eventually to encompass all of East and South Asia. "Japan must with patience wait for the time of confusion in Europe to gain its objectives," Viscount Tani had said fifty years before.

The Japanese premier in 1915, Count Shigenobu Okuma, had inspired Japan's militarists for years with his ringing cry, "Japan will meet Europe on the plains of Asia in the Twentieth Century and wrest from her the mastery of the world." Japan

Fleet Admiral Ernest J. King. (Courtesy of U.S. Navy.)

was on the march after these years of patiently waiting for the opportune moment. She struck at Midway Island, 1,100 miles from Hawaii, and Wake Island, another thousand miles away. The Philippines, only 1,500 miles from Tokyo, was needed for expansion southward, and awaited her turn. That night on Midway, a bright moon silhouetted the military installations—new construction put up hastily while Japan's peace envoys dickered in Washington.

The Marines had been waiting since early morning for the Japanese to arrive after the first announcement of the attack on Pearl Harbor. They remained at their battle stations, a small force completely inadequate to ward off an invasion. One Marine, glancing anxiously at the full moon, whose rays outlined the island, said grimly, "I'll bet the Japs can see us for miles."

Two Japanese warships—a destroyer and a cruiser—went into action later, and Midway's shore batteries answered back. The Japanese had expected to catch the Marines off guard, but when surprise failed they quickly concluded the engagement. They wanted Midway—they needed it badly for their westward expansion plans—but they figured it would be too costly a venture. Little did they realize Midway's defenses were at their weakest in the first few days after Pearl Harbor.

Wake Island, with less than 400 Marines and Navy personnel, stood almost defenseless. Its coral atoll, roughly shaped in the form of a "U," had a Pan American Airways Station and a hotel on one part of the island. Civilians crowded it, hastily preparing military installations. To the few pilots and planes, this former bird sanctuary had been a quiet haven, made dangerous only by the constant menace of the pirate and gooney birds, which made landings and takeoffs hazardous.

Twelve Grumman fighters of Marine Fighter Squadron 211 under Major Paul Putnam had flown in from the *Enterprise* on December 4. They were the first military planes to protect the islands other than an occasional Navy flying boat.

It was December 8 on the other side of the International Date Line when Major James Devereux was forwarded the dispatch about the attack on Pearl Harbor sent to the island's commander.

Twenty-four Japanese bombers hit the airfield flying in "V" formations later that day. Their fragmentation bombs created havoc among the new installations. Only four Marine fighters survived the onslaught undamaged. At times Major Putnam's squadron had difficulty keeping even one fighter in the air. Day after day bombers appeared, dropping their bombs almost unmolested except for an occasional fighter that had been patched up for duty.

A Japanese invasion was routed on the 14th. Major Devereux smiled grimly when he read a wire from Pearl Harbor asking what he needed. He scrawled a message that he needed planes. He did not—as was reported at the time—wire "Send us more Japs." That message was dreamed up by a newsman and was not from Devereux, who had enough problems without more Japanese.

Seventeen SB2U-3 Vindicators of VMSB-231, led by a PBY (Catalina flying boat) of Patrol Wing 1, arrived at Midway Island on December 17 from Oahu, completing the longest mass flight by single-engine aircraft then on record in nine hours and forty-five minutes. This was the same squadron that was en route to Midway on December 7 aboard the *Lexington* when reports of the attack on Pearl Harbor forced it to turn back short of its goal.

The battle for Wake Island ended tragically on December 22 when Japanese land, air, and sea forces overwhelmed all resistance. A small American task force was on its way to relieve the beleaguered garrison on the same day, but it was recalled, as further resistance was futile.

Within a fortnight the carrier *Saratoga* was torpedoed. As the naval command at Pearl Harbor wondered where the ships would be found to fight a delaying action, the carrier had to be sent to dry dock for extensive repairs.

Four hundred Navy personnel and 155 Marines were sacrificed on Guam, a strategic island the Japanese had bombed repeatedly. Resistance ended December 22.

The PBY-5 Catalina was used for patrol missions. (Courtesy of General Dynamics.)

It was 3:00 A.M. on December 8 in the Philippines when Admiral Thomas G. Hart was awakened by the ringing of his bedside telephone and informed of the bombing of Pearl Harbor. After he called his staff together, he told them soberly, "We can expect an attack momentarily. The Japanese must have the Philippines if they wish to move southward. We're right in their path."

At dawn on December 8 Japanese dive-bombers appeared off the southern island of Mindanao, where the Navy had dispersed its flying boats of Patwing (Patrol Wing) 10. Two PBYs were sunk but the tender *Preston* evaded all attacks.

Now that the war had come, Admiral Hart reviewed available forces at his command. They were pitifully inadequate. He had his former flagship the *Houston,* two light cruisers, and thirteen overage destroyers. He was better off in submarines, with twenty-nine at his disposal. The future looked grim but he maintained an air of calm dignity when he called an emergency staff meeting. "I'm not surprised Japan has decided to go to war," he said. "I've expected it since June. What does astonish me is that they struck first at Pearl Harbor."

Army aviation lost most of its strength before noon when a heavy force of Japanese fighters, bombers, and dive-bombers swept across Army airfields near Manila. Two-thirds of the Army's fighters and one half of its bomber strength were destroyed. Despite the warning about the attack on Pearl Harbor, the Army's aircraft were caught on the ground—lined up in rows just like those at Pearl Harbor—and were easily destroyed by the attacking Japanese planes. Such unpreparedness was un-forgivable—and traceable right to General Douglas MacArthur, whose command in the Philippines had been wholly out of touch with reality.

Without fighter protection, Patwing 10's slow PBY Catalinas found patrols not only hazardous but almost suicidal. What was needed was a strong carrier force.

Admiral Hart received a new shock on December 10, when he learned that the new British battleship HMS *Prince of Wales* and the old battle cruiser HMS *Repulse* had been sunk west of the Ananbas Islands off Malaya. Incredibly, it was learned later, the Japanese lost only three planes. This was not the only bad news—Cavite Naval Yard was almost completely destroyed by air attack. To cap the news of the day, Japanese assault forces landed on Luzon at three different places.

Admiral Hart called in Patwing's commander for a brief conference. He noticed Captain F.A. Wagner's haggard face and he felt an understanding sympathy for the brave captain. His two squadrons—VP-102, under Lieutenant Commander E.T. Neale, and VP-101, under Lieutenant Commander J.V. Peterson—had performed miracles with their clumsy PBY Catalinas, even making attacks of a type the air-planes were never designed to perform. These twin-engine aircraft dated back to 1928, and had long been outmoded.

"Disperse your planes at outlying bases, Captain," the admiral said. "Attack only when necessary. It's only a matter of time until the Asiatic Fleet has to retire and we'll need you to search and scout for us."

Losses quickly reduced Patwing 10 to seventeen Catalinas—and only eleven could fly because the others were riddled with bullet holes. Then another PBY was lost at Lake Lanao in an accident.

Admiral Hart was forced to move the seaplane tenders *Langley,* the *Preston,* and the *Heron* to the Netherlands East Indies, and Wagner went along. Of the few planes left at Manila, four were repaired for service.

After the war was only three days old the Japanese had almost destroyed the backbone of the United States Pacific Fleet at Pearl Harbor, and in separate actions thousands of miles apart had sunk two of Great Britain's finest warships. The Japa-nese naval air arm had achieved brilliant victories while operating over 400 miles from their bases.

The American Navy now had only three carriers and a few cruisers, destroyers, and submarines in the Pacific that were available to contest Japan's conquest of the valuable chain of islands to the south. The raw materials that Japan so desperately needed to pursue her expansionist plans in the Far East seemed within reach for the taking. Yamamoto believed that Allied garrisons, unable to be supplied or rein-forced, would soon fall to his victorious forces.

Prime Minister Winston Churchill of Great Britain and President Franklin Roosevelt met with their staffs in Washington two weeks after the Pearl Harbor attack. At the Arcadia Conference both leaders reaffirmed their previous decisions to defeat Germany first, while maintaining a holding action in the Pacific against Japan until a revived Pacific Fleet could take the offensive.

They agreed that a unified command was essential in the Southwest Pacific, to include American, British, Dutch, and Australian forces. Admiral King endorsed General Marshall's suggestion that British General Sir Archibald Wavell be named

Supreme Commander. He was considered the most experienced high-ranking officer in the theater.

Roosevelt insisted that the U.S. Navy go on the offensive, instead of just reacting to each Japanese move. King resisted such a plea, saying that air cover must be supplied before the Pacific Fleet could take the offensive. He was disturbed that Roosevelt did not seem to understand how heavy airpower losses had been in the Pacific. King spoke forcefully about the need to keep the lines of communication open between Hawaii and Samoa, citing it as the number one priority. "All other projects must give to this," he emphasized.

Roosevelt continued to insist upon action, if only against small Japanese task groups. Although the president had approved of Wavell's appointment as Supreme Commander, he privately told King that he was concerned about who would give Wavell orders to go on the offensive. When his concern was presented to Admiral Sir Dudley Pound, Great Britain's Sea Lord, Pound suggested that a Chief of Staff Committee, including American and British service chiefs, be empowered to determine Allied strategy and give the orders to theater commanders after such strategy was approved by the president and the prime minister.

King disliked the suggestion—proposing instead a Southwestern Pacific Council including representatives from all Allied Countries—but General Marshall said he preferred the British suggestion and King did not push his own idea. Quite possibly the idea was not really his own, but one suggested by the president, because it ran counter to his own belief in a strong central command.

The matter was turned over to Churchill and Roosevelt, who continued to support a separate command function. King and the president met for lunch, and the admiral said that he thought the Americans and the British—who were paying for and doing most of the fighting—should determine strategy. He finally persuaded the president to agree with his views. The Combined Chiefs of Staff were formed, with headquarters in Washington. It determined the course of the war from then on. The body proved successful because all members sought to honestly resolve their differences. At first it included Admiral King, head of the United States Fleet; Admiral Harold R. Stark, chief of naval operations; Army Chief of Staff Marshall; and General Henry H. "Hap" Arnold, commanding general of the newly designated United States Army Air Forces—who were all part of the U.S. Joint Chiefs of Staff. On the British side were General Sir John G. Dill, chief of the Imperial General Staff; Admiral Sir Dudley Pound; and Air Marshal Sir Charles Portal of the Royal Air Force. Later, Field Marshal Alan Brooke replaced Dill, and Admiral Stark was dropped after he was replaced as chief of naval operations by Admiral King. Roosevelt's naval adviser, Admiral William D. Leahy, served as unofficial chairman of the U.S. Joint Chiefs.

Admiral Chester W. Nimitz, who had served before the war as chief of the Bureau of Navigation, was now in command of the Pacific Fleet. The 56-year-old Nimitz was a wise choice, although he and King often disagreed about men under Nimitz's command and the conduct of the war. Despite his age, Nimitz was a vigorous man with a pink complexion and blond hair that was now turning white. He was an ideal buffer between his subordinates and the caustic King.

After Nimitz returned to Pearl Harbor, he sought his combat commanders' views about how the war should be prosecuted. Each time one of them returned to Pearl Harbor Nimitz always found time to interview him. When the officer entered his office, Nimitz would stand, greet him genuinely, and answer any of his questions.

Fleet Admiral Chester W. Nimitz. An oil painting by Adrian Lamb.
(Courtesy of U.S. Navy.)

Nimitz was a firm taskmaster, removing any officer who got out of line. He was always willing to give a man a chance, but meanwhile he promoted others who showed promise and did not hesitate to replace those who did not measure up to his strict standards. A calm, prudent man, Nimitz had a tremendous capacity for organization, but he was also an inspiring leader. His appointment soon restored an air of confidence throughout the Pacific Fleet.

Roosevelt had demanded after the Pearl Harbor attack that Admiral Husband E. Kimmel and Lieutenant General Walter Short be relieved of their duties and brought home for investigation of their conduct. Roosevelt wanted them court-martialed and forcibly retired, but Secretary of War Henry Stimson protested that they had merely reflected the apathy of the entire country and that they should not be severely punished.

Roosevelt at first agreed, but in February 1942 he told Stimson that the temper of the American people required their court-martial. General Marshall joined Stimson in protesting such action because he believed it was totally unfair. The two men pointed out that General MacArthur, whose actions were even worse—leaving the Philippines undefended even after he knew the Japanese had attacked Pearl Harbor—had been kept in command and even awarded a Medal of Honor. Although Short and Kimmel were forced to retire, wiser heads prevailed about their courts-martial and they never had to face one. Kimmel had actually sought a court-martial, believing it would clear his name.

The British had been secretly developing radar before the war, and shared its secrets with U.S. scientists. Officials of the U.S. Navy realized early the significance of radar for use with the Fleet, but early models were too bulky for effective use, especially in aircraft. The announcement on December 17, 1941, that the Naval Research Laboratory had completed successful tests of radar in a PBY using a duplexing antenna switch was greeted with some relief. The duplexing switch made it possible to use a single antenna for both transmission of the radar pulse and reception of its echo. It eliminated the necessity of cumbersome antennae and increased the radar system's reliability. This new equipment contributed substantially to the reliability and effectiveness of airborne radar sets.

In the Pacific, Japanese invasion forces followed the island chains ever southward, but they were not confined to an island-hopping strategy. They also reached inland along the Asiatic coast. Their mighty land, sea, and air forces often made wide end runs to occupy territory hitherto deemed safe. It soon became evident to America's military leaders that Japan had the initiative and would not relinquish it without overwhelming opposition.

Before Nimitz left Washington, he and King reached agreement on the strategy to be followed in the next few months. They had no illusions about stopping the Japanese drives, but they hoped that a revitalized Pacific Fleet would slow Japan's eastward movement, which was imperiling the sea-lanes between the United States and Australia.

Most of the Pacific Fleet's effective ships were formed into two task forces, and Vice Admiral Halsey and Rear Admiral Frank J. Fletcher, respectively, were placed in charge. Halsey was given the carrier *Enterprise,* along with three cruisers and six destroyers. Fletcher's flag flew on the *Yorktown,* with two cruisers and four destroyers making up the rest of his slim force. Their first job was to protect loads of troops en route to threatened Samoa.

"What has happened to the United States Navy?" Throughout the free world people asked the question, suspecting the worst. Many were inclined to believe the wild boasts on Japanese radio stations, and the Japanese people tended to believe their propaganda that the U.S. Navy had been "destroyed to pieces."

On January 9, 1942, Halsey studied the latest orders from Nimitz, "You are to raid the Southern Marshall and Northern Gilbert areas as soon as possible. The 'lifeline' must be kept open."

It was 3:00 A.M. the morning of February 1 when Halsey's task force appeared thirty-six miles off Wotje. Airplane engines broke into a roar, drowning out the sounds of the churning waves of the *Enterprise*'s wake and the soft whistle of the trade winds in her superstructure.

The "Big E" turned into the wind, knifing through the water at thirty knots. Fighter planes swarmed off the deck first, then the scout bombers, and lastly the torpedo planes. The thirty-six scout bombers were each loaded with a 500-pound bomb and two 100-pounders. En route, the torpedo bombers separated to make their attacks on Kwajalein Island while the others headed for Roi on the lagoon's northern end.

Lastly, as the *Enterprise* serenely rode the waves, fighter planes charged off her deck, each with a single hundred-pound bomb, and headed for Taroa-Maleolap. One spun crazily near the end of the deck and plunged into the sea before the pilot could free himself. The other five winged their way to the target.

None of the pilots was sure what to expect, because their target maps were only photographic enlargements of clippings from old charts. This was to have an unfortunate effect on the group headed for Roi. In the misty dawn they failed to spot the place, alerting the Japanese and giving them twelve minutes' warning of the attack.

Cruisers and destroyers opened fire while fighter planes strafed the runways on Kwajalein, riddling many planes before they could get off the ground. A few dive-bombers, in tight formation, found the harbor filled with ships. Air Group Commander Howard L. "Brig" Young immediately detached eighteen planes of Bombing 6 headed for Roi and sent them to the opposite end of the long atoll.

Lieutenant Commander William Hollingsworth led them in. He checked his altimeter before signaling to nose over. He glanced at the formation. It looked good. Practiced eyes scanned the fuel gauge and checked the carburetor mixture. Without consciously thinking of what he was doing, he made sure the engine's supercharger was set properly. His years of training were paying off. A glance showed him that the flaps on the engine cowling were closed. Then, drawing his goggles over his eyes, he opened the telescopic aiming sight, pushed back the cockpit canopy with a strong sweep of his right arm, locked it in position, and made final tab adjustments to the rudder and stabilizer. In seconds he checked to see if his diving flaps were open to reduce his diving speed, and retarded the throttle so the propeller turned at only 1,900 rpm.

Hollingsworth nosed over, picking out a large building far below him, and jockeyed stick and rudder to align his SBD (Dauntless dive-bomber) with the target. The wind roared about his head as the speed of the plane increased. He was conscious of flak but, plunging downward, man and plane became one smoothly working machine. He glued his eye to the telescope, taking a quick peep at the altimeter

Fleet Admiral William F. Halsey. (Courtesy of U.S. Navy.)

every chance he could. He was aware of how easy it was to become too fascinated by the job at hand and fly the airplane straight into the ground.

Hollingsworth stared intently into his telescope, watching the bubble level on the bottom and the concentric circles framing the Japanese hangar. The bubble was to the left. He was skidding. He corrected quickly. With a last look at the altimeter showing he was near the release point, he noticed with satisfaction that the bubble was centered. He hesitated a split second to note any side movement. The dive was good! He quickly released his bombs and pulled out.

Without conscious effort his hands closed the diving flaps, pushed the throttle forward, then reset the rudder and horizontal stabilizer for level flight. It had taken less than forty-five seconds. Once the excitement of attack was over he felt as if every organ in the midsection of his body had been savagely pummeled. A black and blue spot slowly grew on his right hand. As he watched it for a second, he wondered when he had slammed his hand into something. There had been no feeling of pain.

Hollingsworth looked down. He saw the second division come in fast and low, with the third behind it. He swallowed nervously as a scout bomber was riddled with flak and plunged into the water. Then a Japanese fighter exploded in the air. He saw the bombs drop as the attackers wheeled out to sea with Japanese fighters hot in pursuit. Two more bombers flamed brightly and plummeted down, but two Japanese planes went with them.

Below them the barracks, gun emplacements, and hangars were on fire. Then Hollingsworth felt his Dauntless rock slightly in the air, and looking down he saw a huge smoke cloud rise quickly into the air. "Ammo dumps," he said to himself. He smiled as one of his pilots yelled on the radio, "Yippee! Right on the button!"

It appeared that every building on the base had been leveled. Billows of smoke rose from several sections of the island, and then he watched with a thrill as a flight of Devastators skimmed the water to launch their torpedoes at warships and target vessels at Kwajalein, the atoll's harbor island.

Air Group Commander Young could see a dozen fires on the island at 8:07. Then an awe-inspiring explosion thrust a bulbous orange blast that billowed out, mushroom-like, high in the air. He made a note that it must have been a gasoline storage tank. He listened eagerly to the radio chatter, "Get out of my way, Joe, that big baby is mine!"

Kwajalein had a brief respite lasting fifty minutes before the third wave from the *Enterprise* arrived. Lieutenant Commander Lance E. Massey led nine torpedo bombers toward the harbor. The fire from the ground was intense but, flying at only 700 feet, they held to their course.

"That CL is making a sneak. Go get him!" Massey called, as he watched a Japanese cruiser head for the harbor's entrance. Three planes took after the wounded ship. The others dropped their torpedoes and climbed away in a hurry.

The Japanese, meanwhile, were so confused that their gunners fired into their own shore batteries and their ships riding at anchor.

Back on the *Enterprise,* Halsey and his staff listened with satisfaction as the radio brought the words from the pilots over the targets. It was a jumble of conversation, but it afforded them an immense satisfaction. They chuckled when they heard, "I got him! Oh mamma! What a sock!"

It had been a day of victory—a limited one because only one transport was sunk and nine other ships were badly damaged—but shore installations had been hit heavily and the atoll commander was killed. More than the damage done, however, was the tremendous lift to morale as the United States, reeling back from defeat, was attacking the enemy in a vulnerable position.

During this action Lieutenant James Gray led his five fighters to Taroa. This arrow-shaped atoll near Kwajalein was a base for Japan's long-range heavy bombers. When Gray led his flight over the island he was astonished to find two long runways with airplanes parked between them. He led the attack, raking facilities and planes. He was joined by four scout observation planes from the USS *Chester.*

Lacking incendiary bullets, their attack was not effective, but the *Chester* and her destroyers moved in close to put salvo after salvo into the island's defenses. The shelling was so heavy even the men on the ships found themselves bracing against it on each firing. The blasts seemed to rob the air from their lungs as they pinned the men against the superstructure.

Hollingsworth, after a quick cup of coffee upon his return from Kwajalein, led his bombers in an attack out of the sun on this new base, which had been established to strike at the American–Australian lifeline. This time they caught twenty-five planes on the ground. Seven of them, including twin-engine bombers, were burned. Then fighters approached from the *Enterprise* and swept the field.

As Hollingsworth led his planes back to the *Enterprise* he noticed nine more bombers headed for Taroa. He knew they would find opposition heavy because they had disturbed a hornet's nest. A bomber was lost in this final action, but three Japanese planes were shot down.

Now it was Wotje's turn. It was the most fully developed island in the Japanese chain of defenses, with a deepwater anchorage and large shore installations. This was a job for surface ships, and only six fighters strafed the harbor and base before salvos from the big ships went into action.

Scout and torpedo bombers returned to Wotje later, but there was nothing worthwhile left to bomb. The cruisers and destroyers had done their job well.

In the whole operation, the *Enterprise* lost five airplanes—a bomber and four scouts—plus the fighter that careened out of control on takeoff. A total of thirty-three planes were damaged and six officers and five enlisted men were killed.

Halsey, after studying the combat reports, gathered his air staff around him, "Our planes are slower than the Japs and less maneuverable," he said. "It is a tribute to your men that they shot down ten of the enemy and destroyed twenty-four planes on the ground."

The Japanese were out for revenge. They called in bombers from their western bases and struck savagely at the "Big E." Fighter pilots, exhausted by the day's operations, had to take to the air again in defense of their ship.

After the cruisers and destroyers rejoined the carrier, Halsey ordered the formation to turn homeward. Fletcher's task force, which had been assigned the southernmost islands of the Marshall group, and Makin in the adjacent Gilbert Islands, found little of military value to bomb. Bad weather was their greatest enemy.

Lieutenant Commander W.O. Burch of the *Yorktown*'s Scouting 5 led his squadron on Makin. He got a hit on a seaplane tender with planes on board and his bomb exploded with a tremendous yellow flame and set the ship on fire. He led his planes next in an attack on four-engine bombers. Burch headed for one, spraying it with machine-gun fire until it exploded violently in front of him. Just then a second bomber blew up as his wingmen let loose. After strafing attacks on the seaplane tender again they headed back for the *Yorktown*.

The raids on the Marshalls and Gilberts had raised morale to a height unknown since the war started. This could not have happened at a more opportune time, because there was nothing but bad news everywhere else in the Pacific. American forces were waging a desperate rearguard action in the Philippines, and the Allied Fleet in the Netherlands East Indies was close to destruction. It was refreshing for members of the Pacific Fleet to be on the offensive, even though this was no solace to the defenders of Manila Bay and the Java Sea.

The two carrier task forces had made a worthy start in taking the war to the enemy by damaging an enemy naval base and an important bomber base. Halsey was the first to acknowledge they had learned much of value for future operations, "If nothing else," he said, "these raids are a valuable exercise in developing task forces for a new kind of war at sea, task forces built around a floating airfield."

This was a start, a feeble one at best, but a start on the long road back toward victory. Ships, planes, and men were tragically short. The planes, in particular, were wholly inadequate for the big job of destroying Japanese aviation and ships across the broad reaches of the Pacific. They needed self-sealing fuel tanks, more armor, and much more firepower. The years of economy prior to the war would have to be paid for in blood.

7

The Times Demand Bold Action

After a few days of rest, Halsey was impatient to get back to sea. He was excited, therefore, when he received orders on February 11 to attack the Japanese invaders at Wake. His *Enterprise* carrier group would go to Wake while the *Yorktown* group would head for Eniwetok. At the last minute the *Yorktown* assignment was cancelled. They had a more important convoy assignment.

The men of the *Enterprise* remembered Wake with a vengeance, because they had been only a day's cruise away with supplies and reinforcements when the island surrendered. The raid on February 24 was a success, causing tremendous destruction.

Halsey headed back to Pearl but, en route, received new orders from the Commander in Chief, Pacific. Marcus Island, only a thousand miles from Tokyo, was to be attacked. Only five miles in circumference, this wedge-shaped island had an important airfield. Its weather reporting station also was of great value to the Japanese fleets in forecasting their weather conditions in advance for their strikes throughout the Pacific.

The audacity of the raid against Marcus on March 4 outraged the Japanese. Although damage to installations was not extensive, it made them reconsider their strategic plans and, possibly, their decision to turn westward in future attacks was based on their fear of what might happen if they did not strengthen their outer perimeter defenses.

As February drew to a close, the American–British–Dutch–Australian Supreme Command came to an end in Japanese waters, although some ships remained to fight without hope of victory. When a large convoy was spotted coming down from the north on February 27, Admiral Karel Doorman of the Royal Netherlands Navy and his remaining ships went into action.

American fighters, meanwhile, were on their way from Australia. They had been placed aboard the tender *Langley*, and the thirty-two P-40s were ready to fight. The British *Seawitch* also had twenty-seven planes en route to the combat zone, but they were crated in her hold. Fifty-nine fighters could not do much at this stage, but it was four times as many as could be found in the islands.

Commander Robert McConnell on the *Langley* wanted to remain at sea until he could make a night entry into Tjilatjap. In this way, he hoped to escape detection by Japanese patrol planes. He glanced at his orders, then looked up at his staff. "We've got to go in by day." He tapped the orders significantly. "The need is urgent."

Japanese aircraft found them the morning of February 27. During the first two attacks, the *Langley* evaded falling bombs. The third wave hit her severely and set her parked aircraft on fire. "Push those planes over the side," the captain ordered. But it was too late. The fire gained such headway that the ship had to be abandoned. It was sunk later by the destroyer *Whipple*. It was a tragic ending for the Navy's first carrier. Originally it was a collier, then it was converted to a flattop, then, when it was too old for combat duty as a carrier, it ended up as a tender.

The *Seawitch* was more successful. Arriving at Tjilatjap on the 20th, her crated airplanes were hastily put ashore.

Admiral Doorman's small fleet failed to stop the Japanese invasion, so the precious P-40s were destroyed by the Dutch before the Japanese landed.

There had been countless displays of desperate energy and courage, but these two admirable qualities were not sufficient to change the situation. What was needed was a large Allied surface fleet, backed by fast attack carriers, but these would come later. At the moment, all the Allied forces could do was retreat and rebuild their shattered commands.

Patrol planes reported Japanese invasion fleets on their way to Java on the 25th of February. When Admiral C.E.L. Helfrich, Royal Netherlands Navy, received orders to scatter the enemy's convoys, he could only shrug helplessly and do his best. The Allied strike force under Admiral Doorman was ordered to attack. It had been in the Java Sea for several days looking for an opportunity to engage the enemy. His command included only five of the original thirteen American destroyers and no airpower to protect his small fleet.

The Battle of the Java Sea was as violent as any surface battle ever fought. When it ended, only the American cruiser *Houston* and the Australian light cruiser *Perth* survived out of a fleet of eighteen warships. Shortly thereafter, after stopping to refuel, these ships were also destroyed, leaving not one survivor of the Battle of the Java Sea.

Allied commanders were stunned by the disaster. They immediately recalled all Allied warships to Australian ports, leaving only American submarines to harass the enemy.

The Combined Chiefs of Staff met in Washington to review the situation. Admiral King recommended that Great Britain assume responsibility for the further defense of Asia by forming a China–Burma–India (CBI) Theater. He said that the American defense of the Philippines undoubtedly would end in failure, and he recommended that the armed forces from Australia and New Zealand be assigned to Admiral Nimitz. After a conference on February 22 between King, Roosevelt, presidential aide Harry Hopkins, and General Marshall, General MacArthur was ordered to withdraw from the Philippines and escape to Australia. It was a sober meeting, with all realizing that conditions in the Pacific could easily end in disaster unless the lifeline between the United States and Australia and New Zealand was kept open.

King met with the American Joint Chiefs of Staff on March 2 and read them a memorandum. "The general scheme or concept of operations is not only to protect the lines of communication with Australia, but in so doing, to set up 'strong points' from which a step-by-step general advance can be made through the New Hebrides,

Solomons, and the Bismarck Archipelago." He explained that after each strong point was seized, Army troops should garrison it, thereby freeing Marines for further conquests in an island-hopping campaign. Despite the "Germany first" decision by the Combined Chiefs, King told them that he wanted to go on the offensive. He said the Allies "were getting licked in the Pacific."

General Marshall disputed King's priorities, telling the Joint Chiefs that Germany was a far greater threat to the Allied powers than Japan. He stressed that a policy of containment in the Pacific should be followed, as already approved by the Combined Chiefs. Marshall said that once Germany was defeated, the United States could go all out to defeat Japan.

King disagreed, saying the United States had no choice but to go on the offensive in the Pacific. He asked Marshall for only two, possibly three Army divisions for garrison duty. The Army Chief of Staff reminded him that they were not available, and would not be for some time. The meeting of the Joint Chiefs ended with no decision about increasing American operations in the Pacific.

After the fall of Singapore on February 15, Prime Minister Churchill wrote President Roosevelt, "We have suffered the greatest disaster in our history." He told the president that he expected even worse news. "It is not easy to assign limits to Japanese aggression." Roosevelt replied on March 9, agreeing with the Prime Minister that the situation was very grave and that the United States was still in retreat.

The war news was dismal on all Allied fronts. German armies were advancing on the Russian front and moving toward the Caucasus. The British fleet in the Mediterranean was reduced to a squadron of cruisers and destroyers by German submarines, Italian midget submarines, and the German Stuka dive-bombers. In the Atlantic Ocean and the Caribbean, German submarines were sinking Allied merchant ships faster than they could be built. In North Africa, the British army had almost been destroyed by General Erwin Rommel. In the Far East, the Japanese were closing in on the last American, British, and Dutch strongholds.

Admiral King's relations with Admiral Stark, Chief of Naval Operations, had been deteriorating ever since he became Commander of the United States Fleet after Pearl Harbor. King talked to the president about the frustrations caused by their overlapping responsibilities. He insisted that the situation must be clarified. "We'll take care of that," Roosevelt told him.

Stark was relieved and assigned to London as Commander of the United States Naval Forces in Europe, and King was appointed Chief of Naval Operations on March 12. Now he had the Navy's two top jobs—the first time the two positions had been filled by one man. There was only one man between King and the president—Navy Secretary Knox.

King now told Knox that he had to have full authority to get the Navy on the offensive, and insisted on the removal of all officers who proved inadequate to their responsibilities. He said he wanted to surround himself with men whom he trusted. Then he asked Knox for authority over all the Navy's bureaus. They had been independent for more than a century. This would have required congressional action, and Knox and the president refused to give King that authority. Roosevelt promised King, however, that he would replace any bureau chief of whom King disapproved. Within the Navy, King became all-powerful after Roosevelt made him directly responsible to the president only. Now even Knox, by law King's superior and long-time supporter, came off second best in the competition with the aggressive King.

President Roosevelt had insisted on early action in Europe, but when it became clear that an American and British invasion of the continent in 1942 was out of the question, he turned more to King, who kept insisting on an offensive in the Far East. Churchill, who had fought against an invasion of Europe in 1942, now went along with Roosevelt for increased operations in the Pacific. Except for operations in India, Burma, and surrounding areas, the British prime minister turned over all operations to the American command.

After MacArthur set up his command in Australia, there were demands by Congress and some elements of the American people that he should be appointed Supreme Commander in the Pacific. General Marshall backed MacArthur, but King did not. He insisted that the U.S. Navy would never become subordinate to an army commander. He argued that the Navy had been preparing to fight a Pacific war for twenty years, and that MacArthur did not understand seapower.

Fortunately, the president was privately opposed to appointing MacArthur Supreme Commander, but he wanted MacArthur to remain in the Pacific because he was fearful that the Republicans might nominate him as his opponent in the 1944 election. With the exception of General Marshall, who always seemed to be in awe of and almost subservient to his former commander, no one in the top military establishment wanted MacArthur as Supreme Commander and most certainly not—as some had proposed—as head of a proposed defense department. So MacArthur remained in the Pacific, where he was forced to share command responsibilities with Nimitz. The Joint Chiefs agreed on March 30 that Nimitz and MacArthur would head independent commands and receive orders from their respective service chiefs. They would both report to the Joint Chiefs. Under the agreement, MacArthur would be responsible for the defense of Australia and its approaches through New Guinea and the Netherlands East Indies and to the east, and for defense of the territory between the Solomon Islands and the New Hebrides. Nimitz was assigned responsibility for all other operations. Later, the Philippines were added to MacArthur's command.

General Marshall was disturbed by this compromise, because it violated his theory of the unity of command, and he knew it would end in constant bickering. King, on the other hand, was pleased that his persistence had won, and that his plan for operations in the Pacific had become the basis for the entire Pacific strategy. Although MacArthur had achieved more authority, the divided command was a strong blow to his ego. King's insistence that the United States take the offensive in the Pacific was just what the president wanted to hear. He gave King his full backing, and it never wavered.

Admiral Chester W. Nimitz was named to command all Navy and Marine organizations, plus Army ground and air units operating over water, on March 26. The Commander in Chief of the Pacific Fleet then proceeded to offer MacArthur's command only minimal assistance. Some submarines and amphibious vehicles were temporarily assigned to MacArthur, but Nimitz used the vast majority of his forces to carry out King's grand plan for defeating the Japanese. He was given specific orders to prepare for major amphibious assaults in the Solomons, New Guinea, and the Bismarck Archipelago, even though his primary responsibility was to maintain the lifeline with the United States' allies down under. King told Nimitz, "Hit the Japs at every opportunity."

The Japanese had gained so many footholds on the islands of the South Pacific that the existence of Australia as a free nation was threatened for the first time in its history. Government officials repeatedly appealed to the United States for more effective naval action to keep her lifeline open.

Admiral Nimitz at Pearl Harbor was finally able to organize a task force around the carrier *Lexington,* with Vice Admiral Wilson Brown in charge. While Halsey was raiding the Marshall and Gilbert Islands on February 1, Brown patrolled the area with his small force.

The Joint Chiefs of Staff in Washington, after carefully reviewing available information, agreed that the Japanese would probably concentrate their forces in the New Guinea–New Britain area. From there they would have excellent harbors to launch operations southward and eastward. Admiral King, in particular, was concerned they might even move toward the Panama Canal Zone. He also did rot rule out another strike at Pearl Harbor. "The Japanese might bypass Australia," he warned the Joint Chiefs. "If communications are cut between this country and the United States, Australia would quickly become impotent in the Pacific War."

When Admiral Brown's force arrived in the area, it was assigned to Vice Admiral Herbert F. Leary, Commander of Forces ANZAC.

The Army, meanwhile, gathered whatever aviation squadrons were available and put them under a unified command.

Brown received permission to attack Rabaul as the first step. This magnificent harbor on the northeastern tip of New Britain was an important assembly base. Brown knew there were great risks in the operation because his task force would face opposition by Japanese warships in waters that were poorly charted. He had no illusions of halting the Japanese with his slim force, but he hoped his actions would force the Japanese to change their plans—and whatever ships he could sink would subtract from their total strength.

A joint operation, the plan called for the *Lexington*'s planes to strike Rabaul from the sea, while heavy U.S. Army bombers under the Australian command approached from the south. Brown hoped if the attack were successful he could order his surface ships to move in and bombard the Japanese anchorage.

The task force headed north into the Solomons, then past Bougainville. Watching intently for signs of Japanese air patrols or lurking submarines, the carrier, four cruisers, and ten destroyers proceeded cautiously through the narrow strait between New Britain and New Ireland.

Brown peered anxiously each day into the sun where Japanese bombers often hid, waiting for the right moment to catch the ships by surprise in a devastating attack. The Japanese were spotted on February 20. The fighter patrol attacked a four-engine bomber and it was shot down. A second bomber flamed and plummeted into the sea, but the third bomber headed away under full power.

Brown watched the fleeing bomber with dismay. He knew this meant trouble, because the Japanese would be alerted. Peering off in the direction of Rabaul, he saw tiny specks high above the horizon. The fighter patrol went into action, splashing eight of the nine Japanese twin-engine bombers.

Exultation aboard the ships was short-lived, because another wave of nine bombers came down out of the sun. Admiral Brown's eyes searched the sky in vain for the combat patrol. Then a lone Navy fighter tore into the formation with blazing guns, and while the admiral watched, five Japanese bombers succumbed to this tenacious fighter, who swung quickly from one bomber to the next until the four other bomb-

Edward H. "Butch" O'Hare in the cockpit of the F4F Wildcat fighter in which he brought down five Japanese planes. (Courtesy of U.S. Navy.)

ers, trailing smoke, headed for Rabaul. The rest of the combat patrol hastened to assist, downing three of the fleeing bombers.

Brown called air operations. "Find out who fought off that last Jap wave of bombers, and send him up." Lieutenant Edward "Butch" O'Hare stood respectfully before the admiral a short while later, embarrassed by being singled out. Brown shook his hand warmly, "Lieutenant, the *Lexington* owes her life to you." O'Hare's face flushed and he stammered a brief thanks. Brown quickly ordered that O'Hare be recommended for the Medal of Honor. In due course, it was approved, and O'Hare found himself in the office of the president of the United States, where the nation's highest award was hung around his neck.

The strike against Rabaul was a failure as far as extensive damage to the harbor and installations was concerned. The *Lexington* lost only two planes and one pilot, Ensign John Woodrow Wilson. Japanese air losses were severe, including two four-engine flying boats and sixteen twin-engine bombers.

The Rabaul attack did not faze the Japanese, however, and their advance forces overran New Britain and New Ireland while landings were made on Bougainville. Japanese bombers were now free to bomb all ports of New Guinea, making the ports untenable for any kind of Allied operations.

The *Yorktown,* Rear Admiral Fletcher's flagship, arrived to augment Brown's force. Now he had two carriers, eight heavy cruisers, and fourteen destroyers. In consultation with his staff, Brown said, "The times demand bold action. We must risk our reserves even though we face far superior forces. We plan to attack Lae and Salamaua and hope to surprise the Japs who are reported to be landing troops there. Here is the plan. Our two carriers will launch from the Australian side of the island, overfly the Owen Stanley Mountains, and attack the Japanese from the rear." There were expressions of surprise at such a bold strike with limited forces. "March 10 is the day," he said in dismissal.

Conferences were held daily to plan the delicate operation, where timing was all-important. Admiral Brown, shortly before the attack, was appalled to learn that four cruisers and four destroyers would be detached from his task force to convoy troop ships from the United States. Fully appreciative of their need to protect vital troop ships at sea, he was also aware that their loss at this critical time could have disastrous consequences for his own operations.

Little seemed to be known about the interior of New Guinea, and the seas around the island were almost uncharted. Brown eventually learned of a pass in the 15,000-foot mountains that was only 7,500 feet high. His intelligence officer said, "It's an air link between Salamaua and the Gulf of Papua. We've talked with pilots who have flown it." Brown nodded. "I'll send Commander Ault ahead to locate the pass and fly figure eights around it. That way he can guide the rest through the mountains."

Lexington's Air Group Commander William B. Ault was first off. Later, the *Lexington* sent eighteen SBD-3s of Scouting 2, twelve SBD-3s of Bombing 2, thirteen TBD-1s of Torpedo 5, and ten F4F-3s of Fighting 42. It was a strong force, carrying forty-eight tons of bombs and thirteen torpedoes.

Ault found the pass and made wide, sweeping figure eights around it. For a time he thought the following planes might have lost him, then he noticed planes heading directly toward him. He counted them eagerly—103 in formation. They poured through the mountain slot and dove steeply at the target. They were thrilled to see two cruisers and four destroyers in the Lee-Salamaua bight, standing by for protection while five transports and two cargo ships unloaded.

The Japanese were caught by surprise when the *Lexington* group struck first. The first bombs were dropped even before the enemy manned their guns. Wave after wave of bombers roared through the slot while the fighters sought vainly for fighter opposition. A single float biplane rose in a futile challenge but was destroyed.

The raid was a success, but Ensign Joseph Philip Johnson and his gunner, J.B. Jewell, were killed. A converted light cruiser, a large minesweeper, and a cargo ship were sunk, and other ships were damaged. The airfield, buildings, and antiaircraft defenses were smothered with bombs. As they headed back to their carriers, they received a report that a large task force was only twenty-five miles away, but they were over the mountains before its carrier planes could attack them.

Ault was first to return. He flew up the carrier's stern at a slow speed in landing configuration, watching the landing signal officer. When the landing signal officer (LSO) signaled he was in position Ault pulled back on the throttle and his SBD slammed down on the deck. He felt the arresting gear seize the plane and bring it to an abrupt stop while the tires squealed and the sturdy plane took its punishment with accustomed ease.

Ault hopped out of his plane and watched his men return. Although he had seen it hundreds of times, he still got a thrill watching the landing signal officer direct the pilots to safe landings. The LSO reminded him of an orchestra leader as his paddles messaged the approaching pilot precise instructions. He watched a heavy TBD Devastator come in. The LSO's paddles warned the pilot that he was too low. The engine broke into a louder roar as the pilot lifted the TBD's nose higher while reaching for the deck. The approach now looked good to Ault, as the LSO signaled the pilot to cut his engine and land.

The Japanese were too firmly entrenched for this raid to alter their plans appreciably, but it did delay their plans against Port Moresby.

Japan's confidence in her home defenses was jarred abruptly on April 18, 1942, when Army bombers dropped bombs on the Tokyo area. Although the Japanese did not know it at the time, the U.S. Navy provided its newest carrier, the *Hornet,* as a floating air base for sixteen B-25 medium Air Force bombers for a strike at their homeland. In a spectacular move proposed by Captain Francis S. Low, on King's staff, to bring the war to the Japanese people, Lieutenant Colonel James Doolittle had hand-picked a group of Air Force pilots and crews of the 17th Air Group to train for a specific and unusual mission. Planes the size of the B-25 had never taken off in such a limited space as a carrier deck. After weeks and weeks of training, particularly in extremely short takeoffs, Doolittle was ready.

It was a joint Army–Navy operation, and the *Hornet* promised to take Doolittle's raiders to within a few hundred miles of Japan's coast to launch the attack upon Tokyo. The *Hornet* left Alameda, California on April 2 and rendezvoused with the *Enterprise* and other ships of Task Force 16 under Admiral Halsey north of the Hawaiian Islands. (The *Enterprise* was needed because the *Hornet's* deck was filled with B-25s, leaving it defenseless.)

Since U.S. forces on the Bataan Peninsula had surrendered April 9, the United States needed a morale booster, and Doolittle and his crews promised to give it to them.

The Halsey task force reached a point 700 miles east of Japan on April 18 on a direct line with Tokyo Bay. This was still a long way from the anticipated strike distance, but after a Japanese patrol ship spotted them, Halsey knew their existence would be broadcast immediately to Tokyo. The raid had to be cancelled or the plans altered. After a conference, Doolittle and Halsey decided to launch at once. Doolittle's raids would have to land at bases in China, many at the extreme limit of their fuel capacity, but it had to be done. The vital task force could not be sacrificed for one bomber raid.

After the planes were launched, Halsey ordered the task force to reverse course and head east at high speed. Halsey broke radio silence. "To Colonel Doolittle and his gallant command, good luck and God bless you—Halsey."

The raid brought new hope to a despair-ridden America, but the bombing itself caused only limited damage. The infuriated Japanese, who had loudly proclaimed that no American bombers would ever appear over their main islands, had some difficulty explaining the event to their own people.

China's Generalissimo Chiang Kai-shek had opposed the Doolittle raid. Knowing that the planes would have to land at his Nationalist bases, he feared retaliation. His worst fears were realized when fifteen crews had to bail out when their planes ran out of fuel. This was due to the fact that they had been forced to take off at a greater distance from Japan than planned. Some landed in Russian territory.

On its way to Japan, an Army B-25 takes off smoothly from the unfamiliar confines of a carrier deck. (Courtesy of U.S. Navy.)

Eight crewmen were captured behind enemy lines in occupied China and tortured, and three were executed.

Two crewmen drowned after ditching. The rest of the fliers reached unoccupied China with the assistance of the people in East China. The Japanese Army retaliated by destroying cities and villages that had assisted the fliers, and up to a quarter of a million Chinese civilians were killed in retaliation. (Between 1931 and 1945 at least 20 million Chinese died at the hands of the Japanese.) The raid may have given a boost to American civilian morale, but the price was not worth it.

Admiral Yamamoto had to apologize to the Emperor for having allowed the bombing, and the Japanese decision to attack Midway and temporarily reduce its drive to the Southwest Pacific may have been linked to fears that the homeland was vulnerable to attack.

Vice Admiral S. Inouye had his headquarters for his 4th Fleet at Rabaul. He was vastly pleased with Japan's Greater East Asia War. He put down his latest intelligence reports and gazed intently at the crowded harbor. It seemed impossible the armed forces of Japan had acquired twelve and a half million square miles of new territory in only five months of war. It was apparent to him that the Americans would find it impossible to stop Japan with their depleted forces. He ordered further advances, particularly to gain mastery of the air over the Coral Sea. His plans included the occupation of the Fiji Islands and Samoa to cut the lifeline between Australia and the United States.

Inouye was not aware, however, that Rear Admiral Frank J. Fletcher's forces, totaling eleven ships south of the Solomons, would shortly be augmented by another task force under Rear Admiral Aubrey Fitch, who had replaced Admiral Wilson Brown in early April. They joined forces at 6:30 A.M. on May 1 while the Japanese were occupying Florida Island in the Solomons and making landings at Tulagi.

Brown, a surface admiral, had been removed by King because he had made only one raid in fifty-four days, saying that raiding fortified Japanese bases was too risky for carriers. Nimitz had accepted Brown's excuse, but King accused Brown of lack of aggressiveness and demanded his removal. Fitch was an experienced carrier man, whereas Brown was not a pilot and had very little background in carrier operations. He was given command of the new amphibious force at San Diego, where his excellent organizing abilities were put to more effective use.

Admiral Fletcher received a decoded report on April 17 (the United States had now broken the Japanese military code) stating that Japanese troopships and the light carrier *Shoho* had left Rabaul and would reach an unknown objective on May 3. Messages also indicated that two large unidentified carriers would leave Truk in the Caroline Islands and head for the Coral Sea. Later it was learned that these ships were the *Shokaku* and the *Zuikaku* of the Fifth Imperial Carrier Division.

Studying the messages at Pearl Harbor, Nimitz concluded that the Japanese were headed for Port Moresby. He and General MacArthur both agreed that the New Guinea base was the key to the entire Allied defense system. Nimitz had often told his staff that when they were able to take the offensive, Port Moresby would be the base from which such a move would have to start.

MacArthur insisted that Port Moresby should be developed into a major base, and he received no argument from Nimitz. But MacArthur had only 200 Army Air Force planes in the whole area, and many of the crews were inexperienced. He sought Nimitz's help with carrier planes. New Guinea was MacArthur's territory to defend, but Nimitz promised carrier support only if he was assigned responsibility for the region. Unfortunately, few carriers were available. The *Saratoga* was at Puget Sound in the state of Washington undergoing repairs, while the *Enterprise* and the *Hornet* were still en route from Japanese waters after launching the Doolittle raiders. Only the carriers *Lexington* and the *Yorktown* were available.

Fletcher's Task Force 17 had no battleships or cruisers, and it was built around the *Yorktown* under Captain Elliot Buckmaster at Nouméa. Rear Admiral Aubrey Fitch, who had left Pearl Harbor on a high-speed dash for the South Pacific on April 16, had the carrier *Lexington* and supporting ships.

A third force of three cruisers, the HMAS *Australia,* the HMAS *Hobart,* and the USS *Chicago,* plus several destroyers, was Task Force 44, under command of British Rear Admiral John C. Crace.

Nimitz ordered Fitch to join Fletcher's group on May 1 at a rendezvous 250 miles southwest of Espíritu Santo in the New Hebrides, and Crace was ordered to enter the Coral Sea on May 4, where he would be reinforced by another American heavy cruiser and a destroyer. These forces were all placed under Admiral Fletcher's control.

Admiral Halsey wired Nimitz that the *Hornet* and the *Enterprise* would arrive at Pearl Harbor on April 25 and leave there within five days. He said he planned to arrive in the Coral Sea by May 15. The ships would be of no assistance, however, unless the Japanese postponed their May 4 invasion of Port Moresby.

Fletcher was ordered to intercept the Japanese to keep the lifeline to Australia from being cut. The first threat was to Tulagi, but the Australian cabinet decided that its small garrison could not possibly ward off a Japanese invasion fleet, and ordered the island evacuated on May 1.

MacArthur's headquarters advised Nimitz that the Japanese were unloading troops on Tulagi May 3, with Japanese warships standing by. After a seaplane base was established, the Japanese withdrew most of their ships.

Fletcher's task force sailed north at high speed, and by the evening of May 4 was a hundred miles southwest of Guadalcanal. Bad weather prevented operations, so he decided to delay his attacks against Tulagi. Unfortunately, he failed to advise Fitch and Crace. Instead, Fletcher ordered all task forces to meet at a new rendezvous at dawn on May 4, 300 miles south of Guadalcanal. And he dispatched the tanker *Neosho* with the destroyer *Russell* to a point near Willis Island, with orders to arrive there by 8:00 A.M. on May 4.

Meanwhile, Fitch's *Lexington* headed at high speed directly to Tulagi despite the bad weather. But now his carrier was 250 miles from Fletcher's *Yorktown,* which could not support his attack on Tulagi on May 4.

During three separate *Yorktown* attacks against Tulagi, only minor damage was caused to a Japanese destroyer and two minesweepers despite inflated pilot reports. When an Army bomber reported that an aircraft carrier and two cruisers under Rear Admiral Arimoto Goto was patrolling the area, Fletcher broke off the Tulagi raids.

Nimitz was disappointed in Fletcher's actions when he learned of the limited damage done to Tulagi, and advised him that more crew training was obviously needed.

Rear Admirals Kuninori Marushige and Goto withdrew their forces to the north while another Japanese strike force left Truk on April 20 under Vice Admiral Takeo Takagi and was ordered to the area. Takagi's two large carriers, with supporting warships, were deemed sufficient to handle any American opposition to the Port Moresby invasion.

Then a number of things went wrong on both sides. First, due to faulty communications, Takagi did not hear of Fletcher's raids on Tulagi until late on May 4. Also, the Americans failed to spot Takagi's Port Moresby invasion forces headed south at high speed, reaching the northern tip of Malaita Island, in the Solomons, by midnight.

The two American task groups under Fitch and Fletcher finally established communications with one another and joined forces at 8:15 A.M. on May 5, while Crace's ships joined Fletcher at the new rendezvous. The word went out to all ships, "Attack and re-attack! Seek out the enemy. Destroy him!"

The *Yorktown's* radar reported unidentified aircraft at 11:00 A.M. in the middle of a weather front, so a Wildcat fighter, unable to see the planes, was dispatched to the spot by fighter direction radar. The pilot reported a four-engine Kawanishi H6K seaplane and shot it down.

Fletcher was concerned that the Japanese pilot had warned his commander of their presence, but evidently the pilot was caught by surprise and never had a chance to radio the alarm.

Vice Admiral Shigeyoshi Inouye, who commanded the 4th Imperial Fleet, was equally concerned when the seaplane failed to return. He suspected it might have been shot down, but he had no idea where it had happened. He ordered his carriers to bomb Port Moresby prior to the landing of Japanese troops.

Until now a huge weather front had enveloped a 300-square-mile area, effectively hiding the American ships. But when the three task groups reached the Coral Sea, the sky over the American ships cleared. Fletcher ordered his ships to take a

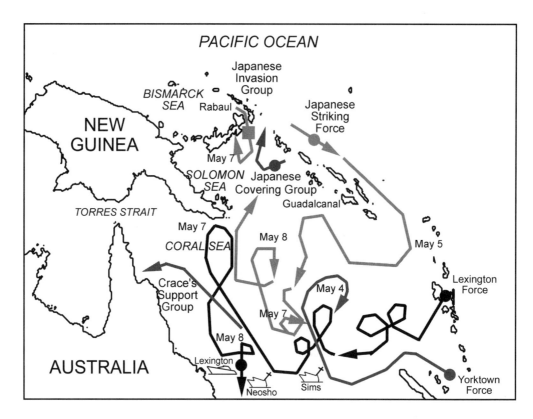

The Battle of the Coral Sea, May 4–8, 1942. (Drawn by Stewart Slack.)

northwest heading to block the Japanese from entering the Coral Sea through the Jomard Passage.

The men in the code room on Fletcher's flagship paused in their work on May 6 as a message came through in the clear. It was in English, transmitted from Corregidor in Manila Bay—where American and Filipino troops had held out until they were forced to surrender—addressed to Pearl Harbor but intercepted by all ships at sea. "Going off the air now. Goodbye and good luck, Callahan and McCoy." The Philippines had fallen. That was the last broadcast from the Navy transmitter on Corregidor.

The Japanese invasion fleet, bound for Port Moresby, with Marushige's supporting ships, now was en route to the Louisiade Archipelago, while Goto's covering force was headed west after passing to the east of the Solomon Islands toward the Coral Sea.

Now that Fletcher had all of the available American ships in the area—plus Crace's ships—under his control, he formed a single task force on May 6 and gave Fitch responsibility for air operations. Unfortunately, again he forgot to tell Fitch.

Although Fletcher's scouts had failed to find the Japanese fleet, he was convinced it was headed for Port Moresby. His assumption was correct, and at one time his scout planes were within a few miles of the Japanese fleet before they had to turn around because of fuel shortages. Actually, Goto's covering force was heading from Bougainville on a southwest course by 8:30 A.M. on May 6, while Takagi's strike force was west of Rennell Island northwest of Fletcher's fleet. An hour later Takagi headed due south to locate the American ships before they could attack the invasion fleet. If they did so, Takagi was confident that Goto's covering force—forming the other half of a pincer movement—could easily destroy them.

Both sides continued to make errors. Incredibly, Takagi did not send out his scout planes to locate the Americans. When Fletcher's ships were finally located at 11:00 A.M. by a Japanese reconnaissance plane from Rabaul, Takagi didn't learn their exact whereabouts until the next day! During this period Fletcher's ships were being refueled and were particularly vulnerable, so Takagi's lack of knowledge was crucial to the success of the coming battle.

Four B-17s of the Army's 19th Group spotted the carrier *Shoho* at 10:30 A.M. They attacked, but were driven off by heavy antiaircraft fire. MacArthur flashed the word immediately to Fletcher.

At 1:00 P.M., a Japanese reconnaissance plane saw an American naval convoy south of Rabaul on a southwest heading, so Inouye realized two of his forces had been discovered by the Americans—but he neglected to inform Takagi. At 3:20 P.M. he ordered the invasion of Port Moresby to proceed.

Now Fletcher learned that three Japanese carriers were south of the Solomons, possibly supporting an invasion of Port Moresby. He learned that a landing force would pass through the Jomard Passage on May 7 and that the attack would begin the next day.

Fletcher ordered his fleet to position itself southeast of Port Moresby at a point where his planes could reach the invasion area. While his ships were being refueled they had been moving away from the area, but now he ordered them to take a northwest heading and ordered his planes to take off at dawn on May 7. With refueling completed, the empty tanker *Neosho* and the destroyer *Sims* turned south to an agreed-upon rendezvous.

Throughout May 6, with neither side aware of the exact position of the other, nature aided the Japanese. The heavy cloud cover that had hidden the Americans now moved over the Japanese ships.

At midnight, the Japanese strike force turned north to refuel, while the invasion fleet approached the island of Misima at the northern end of the Jomard Passage. Goto's covering group moved into position northeast of the island of Deboyne to protect the fleet's left flank. Orders went out to the light carrier *Shoho* to launch its planes at dawn. Fletcher's task force now positioned itself near the southeast tip of New Guinea, 310 miles southeast of Deboyne.

Rear Admiral Chuichi Hara, in charge of Takagi's carriers, became concerned about the vulnerability of his ships while they were refueling. He talked the strike force commander, Takagi, into stopping the refueling, telling him that he suspected an American fleet was in the area. He argued that this fleet had to be destroyed before his carriers could support the landing force. Takagi agreed.

On the morning of May 7, with no definite word about the American fleet, Hara's misgivings increased. He was somewhat relieved when a scout plane reported an American carrier and a cruiser to the east at 7:36 A.M. Unfortunately for Hara, the pilot's report was completely in error. What he had spotted was the tanker *Neosho* and the destroyer *Sims*. Hara ordered the *Shokaku* ("Flying Heron") and the *Zuikaku* ("Joyous Heron") to head for the alleged carrier. Japanese scout planes flew ahead to pinpoint it.

But Rabaul had received a report that an enemy formation, including a carrier and ten other ships, was only 200 miles south of the cruiser *Kinugasa,* and 280 miles northwest of Takagi's last known position. Again, the typically bad Japanese communications failed to get the word to Takagi until later, while his planes were off in search of the nonexistent carrier.

The *Neosho* and the *Sims* were located at 9:35 A.M. and soon overwhelmed by a deluge of bombs, and their sinkings caused a large loss of life. But the pursuit of these hapless ships gave Fletcher precious time to prepare for the coming battle. At 6:30 that morning his fleet was 115 miles south of Rossel Island, the easternmost point of the Louisiade Archipelago.

When Fletcher heard of the sinkings, and the Japanese strike force responsible for them, he debated whether to go after the strike force or remain in position. But his responsibility was to disrupt the Port Moresby invasion, and he rightly remained in position. Despite the risk of reducing the antiaircraft fire from his surface ships, vital for protection of his carriers, he split his force at 6:45 A.M. He ordered Crace's Task Force 44 to continue on its westerly heading to block the southern exit of the Jomard Passage, while he sought the Japanese carriers with his other ships. He justified his action by telling his staff that he expected to fight the bulk of the Japanese forces.

A Japanese scout plane spotted Crace's Task Force 44 on May 7 at 8:10 A.M. as it headed for the southern exit of the Jomard Passage. This time Takagi was informed immediately and he realized that the Americans had split their forces.

Admiral Kenzo Yamada, commander of the 25th Air Fleet at Rabaul, had been expecting an American carrier to appear in the Coral Sea and planned to use his land planes against it. Crace's appearance, however, changed his mind, because he was concerned that Crace's forces could hinder the Port Moresby landings. He ordered bombers at Rabaul and Bunto to attack Crace's ships.

Crace's ships soon had to fight off waves of high-altitude Japanese bombers, who started to brave the heavy antiaircraft fire but then dropped their bombs short and fled the scene. Torpedo bombers swept in, but their torpedoes were easily evaded. Then another flight of high-altitude bombers approached the task force, but not one bomb hit any of Crace's ships.

Crace's jubilation was short-lived when American B-26 bombers from the American Army Air Force base at Townsville, Australia, dropped bombs on his ships. Although no ships were hit, Crace lodged a strong protest to General MacArthur.

The Japanese attacks against Crace's task force served to draw most of the air attacks away from Fletcher's fleet. After hearing untrue reports from their pilots that an American battleship and a cruiser had been sunk, Japanese commanders were jubilant and ceased additional attacks.

At dawn on May 7 scouts from both sides sought to locate one another's carriers. Fletcher became increasingly concerned when the Japanese ships could not be located. He relaxed when a coded message came in that said, "Two aircraft carriers and four heavy cruisers 10 degrees, three minutes north; 152 degrees, 2'7 minutes east." This report indicated that the Japanese were a short distance north of the island of Misima, and 225 miles northwest of Fletcher's force.

Fletcher ordered a course change to reduce the distance, and then ordered the *Lexington*'s bombers and scouts to take off at 9:25 A.M. to attack what he was sure was the Japanese strike force.

The initial report proved to be erroneous. It should have read not four cruisers, but two cruisers and two destroyers and no aircraft carriers. Fletcher believed these ships must be Marushige's supporting warships for the invasion and not Takagi's or Goto's ships. He was disturbed because he had dispatched most of his strike force against the least important ships of the enemy fleet. The true situation was even worse, because there were no carriers in the enemy force, despite the report of his scouts.

Fletcher knew that if he recalled his pilots by radio he would reveal his location. He just hoped that the Japanese carriers could be located while his planes were out, and that his position would remain unknown to Japanese commanders. However, a Japanese seaplane had been shot down at 8:20 A.M. and the pilot had reported the position of Fletcher's task force. Admiral Goto immediately ordered the *Shoho* to launch its planes and attack the American ships.

For the moment the Japanese had the advantage, but then Admiral Inouye at Rabaul—concerned that Crace's ships at the southern end of the Jomard Passage could cause havoc to the Port Moresby–bound troop ships—ordered the landing forces to reverse course temporarily. These ships were advised to take a position north of the passage until the coming sea battle was decided.

Fletcher was unaware of the change in Japanese plans, but his decision to divide his forces now proved to be the right one. Inouye's decision, made with inadequate information (he had not even heard about the inflated reports about the American battleships and cruisers supposedly sunk by Japanese Army bombers) proved to be one of the most decisive factors in what was to follow.

Commander Ault had taken off on the morning of May 7 with the *Lexington*'s planes right behind him. Then the *Yorktown* group followed. They had been briefed that the enemy was approximately 160 miles away, but ninety-two planes rode through a rough weather front in vain. They listened for radio reports from the scouts, but they could not find the Japanese carriers either, only two heavy cruisers and two light cruisers.

Ault, listening to his radio, heard a report from an Australian base that reconnaissance planes had spotted a carrier thirty miles from the cruisers that were providing cover for Japanese troop ships. He ordered his attack planes to change course, and seven minutes later they found the enemy north of Misima Island. Enemy fighters filled the air as the American pilots eased into their dives from 12,000 feet. It became a free-for-all, every man fighting for his life, but they dove straight for the Japanese carrier.

Lieutenant Commander Robert E. Dixon of Scouting Squadron 2 led his bombers in first, and they hit the carrier on the stern and on the middle of its crowded flight deck. Then the bombers and torpedo planes took over in a coordinated attack.

Lieutenant Commander James Flatley shook his head in disbelief. The sight of the dive-bombers pouring bombs into the carrier was so awesome he became almost physically ill. Bombs plowed into the shattered ship every three or four seconds as explosions tore the ship apart. When the bombers from the *Lexington* were finished, the carrier appeared to be in her death throes.

Lieutenant Commander W.O. Burch, commanding officer of the *Yorktown*'s Scouting 5, watching the tortured ship trying to turn into the wind to launch some of its planes, called Lieutenant Commander Taylor and his torpedo squadron, "We're going in, Joe."

"Wait a few minutes. It will be five minutes before we can get there."

"Can't wait. Carrier is launching planes."

He led them down and they added their bombs to the carrier's woes. The damage was so severe the last bomber pulled away without dropping. Ensign H.S. Brown Jr. headed for a cruiser and dropped his bomb squarely on the quarterdeck, while torpedo planes completed the destruction of the Japanese carrier. Heavy smoke almost obscured the stricken ship as the torpedoes struck home. Within three minutes the light carrier *Shoho* slipped between the waves at 11:35. She had been hit by at least seven torpedoes and thirteen bombs. Only 100 survivors were rescued.

On the *Lexington,* Captain Frederick C. Sherman paced the bridge with excitement, paused a moment as the radio blared from the fighting zone, and smiled with satisfaction as a voice shouted, "Scratch one flattop!"

The rugged SBDs had shown they could take it. The planes were riddled with holes, had wheels missing, and their wings were shot up—but they managed to fly home.

The old TBDS, which long since should have been replaced by newer torpedo bombers, also served with distinction. In all, the operation cost six planes and the crews of five. This was a severe blow to the Japanese, because an aircraft carrier had been sunk with most of her planes and crew, and twenty-three planes had been shot out of the sky.

The Battle of the Coral Sea was not over, however. At dawn the next day scout bombers went out to search for the other Japanese warships. Bad weather hid the enemy for some time until Lieutenant Joseph Smith of the *Lexington*'s Scouting 2 found them. He reported, "Two carriers, four heavy cruisers, and destroyers."

"Pilots, man your planes!" came the cry to the ready rooms. The decisive phase of the Coral Sea battle had arrived. Attack groups of both the Japanese and the Americans left about the same time, although neither side saw the other because of limited visibility. The Japanese came in high, while the American pilots hugged the wave tops.

At 11:00 A.M. the *Yorktown*'s scouts and bombers dove at the *Shokaku* while Torpedo Plane Squadron 5 swung in to drop their fish. Lieutenant Commander Burch led Scouting 5 toward the carrier. Ensign J.H. Jorgenson followed quickly behind him, releasing his bomb at 2,000 feet. He saw his skipper's 1,000-pound bomb hit flush on the carrier's deck despite the swirling smoke.

Planes from the *Lexington* struck at the *Zuikaku* but no direct hits were made. When they returned to their own carrier they saw it was doomed. In their absence—despite a protecting screen of cruisers and destroyers and a vigorous fighter patrol—the carrier had received at least seven torpedoes and thirteen bombs from Japanese planes. The *Yorktown* had also taken several direct hits.

After attacks by dive and torpedo planes, fires on the *Lexington* were briefly brought under control, then they flared anew, more fiercely than ever. "Abandon ship!" Captain Frederick C. Sherman's voice broke when he gave the tragic order. The ship was sunk by the task force's own destroyers at 6:54. The fact that ninety-six percent of her personnel was saved was some measure of comfort for her grieving commander.

At Rabaul, Admiral Inouye was in a state of indecision about the true status of the battle. He cancelled the Port Moresby invasion and ordered the troopships to return to Rabaul. He even cancelled a proposed night action against the Americans.

When Fletcher was advised that the Japanese had abandoned the invasion, he refused to believe it. The information was so contrary to his views about the Japanese. He suspected a trap.

Incredibly, in the confusion of planes returning to their carriers that night after dark, Japanese planes tried to land on the *Yorktown*. One of them was shot down before the Japanese pilots realized their mistake.

After five days of intensive fighting, the Japanese had lost one carrier, had two crippled, and smaller ships had been sunk or damaged. They also lost more than

The crew of the *Lexington* is removed during the Battle of the Coral Sea.
(Courtesy of U.S. Navy.)

5,000 men. Besides the *Lexington,* the Japanese had sunk the destroyer *Sims* and the fleet tanker *Neosho.* The Americans lost thirty-three aircraft in air battles, and another thirty-six planes went down with the *Lexington.*

Admiral Nimitz responded immediately with a message to the Fleet. "Admiral Fletcher utilized with consummate skill the information supplied him and won a victory with decisive and far reaching consequences for the Allied cause."

Although the score favored the Japanese, they retreated from the action, and their invasion of Port Moresby by sea was deferred and finally abandoned.

Admiral Nimitz ordered Fletcher to withdraw and head south, a fact unknown to King in Washington, who was unable to learn of the battle's progress because of a radio blackout. When he learned that Fletcher's fleet was headed for Nouméa to refuel, he reacted strongly. He considered Fletcher's action an unwarranted retreat from the battle zone.

King sent Fletcher a blunt message that he must keep the pressure on the Japanese until he was ordered to do otherwise. Fletcher advised King that he would cease refueling if the Japanese transports started to head south, and would position his fleet off the northeastern part of Australia to await orders. King's distrust of Fletcher began during this episode, and he expressed his feelings to Nimitz, who explained that it was he who had ordered Fletcher south. But King remained upset because decoded communications indicated that the Japanese were about to make another attempt to invade Port Moresby, although these reports later proved unfounded.

King now regretted that he had agreed to send Halsey to assist the Doolittle raid, believing strongly that Halsey was his best carrier commander. Both King and

Nimitz were now convinced that the aircraft carrier was the prime offensive weapon in sea war.

The Battle of the Coral Sea was unique because it was the first naval battle fought entirely by air. The greatest significance of the battle came to light later, when it was learned that the Japanese found it impossible to take over the southeastern coast of New Guinea. From this vantage point, they could have severed the Australian–American lifeline.

The crippling of the two Japanese carriers would also prove a severe blow to their plans for future operations. The *Shokaku* had to spend several months in a Japanese shipyard. During the battle the carrier lost thirty of its aircraft, and the air group aboard the *Zuikaku* lost so many pilots that the carrier had to return to Japan to train new crews. Japan was now forced to halt her outward expansion for the time being. As it turned out later, this was to be the limit of her aggression against the Allies in the Pacific.

While Allied commanders worried about Japan's next move, imperial strategists realigned their forces. In Washington Admiral King and Admiral Nimitz at Pearl Harbor fortunately suspected what these plans would be. They calculated the Japanese would strike far from the South Pacific, hoping to use overwhelming force to catch a vital area unprepared. King and Nimitz thought it might be Midway or the Aleutians, or possibly both. They issued orders to increase Midway's defenses.

The *Yorktown* badly needed repairs after the heavy damage she had sustained, and her air groups, and those from the sunken *Lexington,* needed rest. Admiral Fletcher's fleet was recalled to Pearl Harbor during the middle of May, and the *Hornet* and the *Enterprise* came with it.

A task force with the *Hornet* and the *Enterprise* left Pearl Harbor on May 28 under the command of Rear Admiral Raymond A. Spruance. Two days later, Rear Admiral Fletcher's force followed. Although there was uncertainty about what steps the Japanese might take next, everyone expected that a showdown between the two navies was near.

8

The Battle of Midway

Admiral Yamamoto decided before the war that if hostilities erupted between Japan and the United States a great sea battle would have to be fought between the Imperial Fleet and America's Pacific Fleet. He told his staff that to assure victory for Japan a decisive battle must be fought and a great naval victory achieved, otherwise Japan would have no hope of winning such a war. He said his plan involved an attack on Midway, with a diversionary attack simultaneously against the Aleutians, to force the Pacific Fleet to engage in a decisive battle.

After the Battle of the Coral Sea, Lieutenant Commander James Flatly had trained his pilots in a new tactic developed before the war by Lieutenant Commander John S. "Jimmy" Thach to counter the Japanese Zero. No American plane of that period could match the Zero's 5,000 feet-per-minute rate of climb, its tight turning radius, and its higher speed. Thach recommended a change from the standard three-plane formation which forced a pilot to watch his section leader as he prepared to shoot and to try to avoid running into his opposite wingman if the leader made a sharp turn. Thach explained his new concept for attacking a superior fighter like the Zero by saying two two-plane formations could split apart when an enemy fighter focused his guns on one aircraft in a section, permitting the other section to shoot it down while it was preoccupied by the attack against the first section. "If we space two sections of two planes far enough apart they will meet in a half circle when they turn toward each other. And, if an opponent attacks one section, that pilot can also take a head-on shot while his partners, in the other section, will have a free shot at the preoccupied attacker."

The "Thach Weave" succeeded so well during the Battle of the Coral Sea that losses were kept to a minimum. To date in 1942 the situation on the world's battlefronts had been desperate for the Allies. Now the situation changed for the better: Russia succeeded in defending Moscow and Stalingrad, and British General Sir Claude John Eyre Auchinleck won the first battle of El Alamein in the Libyan desert.

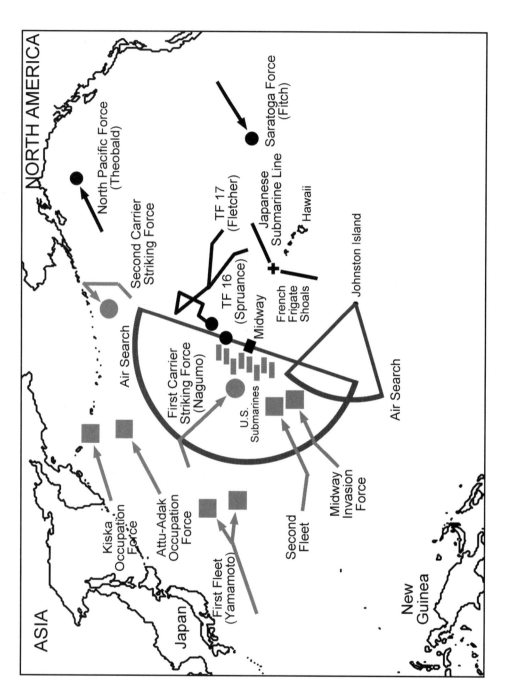

The Battle of Midway: the approach. (Drawn by Stewart Slack.)

King now demanded that aggressive young flag officers with aviation backgrounds be placed in command of carrier groups. He told Nimitz that he was too lenient with his subordinates, and, in particular, criticized Fletcher's actions during the Battle of the Coral Sea. Wilson Brown—whom King had earlier insisted be removed—and Fletcher were not aviators, and King insisted that further changes be made.

Nimitz voiced strong support for Fletcher, requesting that King promote him to vice admiral, but the chief of naval operations refused, saying he did not consider the battle a victory. King was still upset by the loss of the *Lexington,* which had been his first major command, and he told Nimitz that he would never promote or decorate a man who lost his ship.

Nimitz's background was in submarines, but he agreed with King that naval aviation commands had to be reorganized. Nimitz advised King, however, that the present critical situation in the Pacific was not the time to make major command changes. After the Coral Sea battle King and Nimitz agreed privately that the war in the South Pacific would be fought at sea, although technically most of the area was assigned to General MacArthur.

Unknown to the Americans, Rear Admiral Matome Ugaki, chief of staff for the Combined Fleet, proposed an all-out attack on the Hawaiian Islands. This action, he said, would lure the American Fleet into a major confrontation. Occupation of India and Ceylon was considered, as well as an attack on Australia to prevent a counteroffensive. All these suggestions were rejected by his staff because surprise could not be achieved at Hawaii, and the Japanese Army said it had insufficient troops to invade Ceylon and Australia.

The loss of the *Lexington* continued to make King cautious. He told Nimitz that he should be protective of his carriers for the present. He also ordered that the *Lexington*'s sinking be kept secret for the time being, informing the British command only that she had been damaged. General Marshall's staff was also not informed, and even Secretary of the Navy Knox was not told the truth. King privately accused some of Marshall's staff of leaking bad news. Knox, a former newspaper publisher, was often not advised of losses. Three weeks later, however, King revealed the truth.

King was now convinced that the Japanese would strike a major blow in the South Pacific, and demanded that Nimitz leave at least two carriers there. Although Nimitz believed that intelligence data and code breaking reports indicated a major Japanese move in the Central Pacific, he agreed that Halsey's *Enterprise* and *Hornet* should remain in the South Pacific. The *Yorktown* was in a Pearl Harbor repair yard, and the *Saratoga* and the *Wasp* were on the west coast—too far removed from the battlefronts to be helpful in the near future. The Central Pacific, where Nimitz fully expected the Japanese to strike next, was without a single American carrier.

In writing King on May 16, Nimitz told his boss he expected the Japanese to strike Midway, but King could not make up his mind about Nimitz's call for strengthening defenses in the Central Pacific. It was not until the following day, after discussion with his intelligence officers in Washington, that King conceded that Nimitz's forecast was correct. Uncharacteristically, he advised caution, telling Nimitz to employ strong attrition tactics and "not—repeat—not allow our forces to accept such decisive action as would be likely to incur heavy losses in our carriers and cruisers."

Nimitz did not advise King of it, but he decided to take bolder action. He recalled Fletcher from the South Pacific in the middle of May. At Pearl Harbor Fletcher found himself defending his actions during the Battle of the Coral Sea. He responded to Nimitz's queries in a forthright fashion, and convinced him that he had acted in a professional manner.

Fletcher's written report to King had an endorsement by Nimitz: "Dear King, I have had an opportunity to discuss with Fletcher...his operations in the Coral Sea area, and to clear up what appeared to be a lack of aggressive tactics of his force. I hope and believe that after reading the enclosed letter you will agree with me that Fletcher did a fine job and exercised superior judgment.... He is an excellent, sea going, fighting naval officer and I wish to retain him as a task force commander in the future." Nimitz added that he was convinced Fletcher should command the carrier forces in the Central Pacific.

When Halsey returned from the South Pacific, Nimitz was shocked by his haggard look, and he ordered him into a hospital. Halsey had lost twenty pounds and had a severe case of dermatitis. This skin disease almost covered his body, making him frantic by its constant itching.

Halsey recommended his cruiser commander, Rear Admiral Raymond A. Spruance, to replace him. He praised him as a meticulous officer, who consistently displayed outstanding ability, excellent judgment, and quiet courage.

Nimitz agreed. He knew Spruance was a tough, brilliant commander—the type of officer he needed in the coming battle he expected his Pacific Fleet would soon have to fight.

Although the Japanese identified their target only as AF, Nimitz decided to trick the Japanese into revealing its true name, which he suspected was Midway, although King persisted in believing their destination was Hawaii. The commander at Midway secretly ordered a broadcast in the clear that his water distillation apparatus had broken down. A few days later a message was picked up from the Japanese high command that AF was short of water. Nimitz was now convinced his hunch was correct.

The 1st Carrier Force under Admiral Chuichi Nagumo—the man who had commanded the Pearl Harbor strike—left its Japanese port on May 28. He had the carriers Akagi, the Kaga, the Soryu, and the Hiryu.

Vice Admiral Moshiiro Hosogaya commanded a northern force to attack American installations at Dutch Harbor and occupy Attu and Kiska in the Aleutians. He had the carriers Hyujo and the Junyo under Rear Admiral Kakuji Kakuta. This force, under Admiral Yamamoto's plan, would hopefully lure the Pacific Fleet to Alaskan waters, leaving Midway unprotected.

Yamamoto and his staff were on board the huge battleship Yamato, which—with three battleships and the light carrier Hosho and auxiliary ships—moved to position themselves 600 miles northwest of Midway. He selected this position because he could send assistance to either Nagumo or the Aleutian force if it was needed.

Yamamoto ordered submarines to act as scouts, but there was only a limited number available, particularly for Nagumo's force. Bad weather also restricted their operations, and a large number of ships were undergoing overhauls in ports.

As time drew near for the battle, a number of Japanese messages gave Nimitz a clearer idea of the intentions of the Japanese Imperial Fleet. Yamamoto, however, due to strict radio silence, was not kept informed of the activities of his strike forces.

On May 28, the day the last of Yamamoto's ships left their Japanese ports, Nimitz ordered his task forces to leave Pearl Harbor and rendezvous 350 miles northeast of Midway. Spruance commanded Task Force 16 with the Enterprise and the Hornet, and Fletcher was aboard the repaired Yorktown with Task Force 17. But when the task forces joined, Spruance was placed in command, although Fletcher was assigned tactical commander under him.

In Alaskan waters Nimitz placed Rear Admiral Robert A. Theobald in charge of a token force. This action was deliberate, because Nimitz expected the Japanese to make their main thrust at Midway.

A Catalina patrol plane from Midway, flown by Ensign Jewell H. Reid, spotted a Japanese fleet on June 3 and remained in contact for two hours, radioing reports of his sighting. Rear Admiral Raizo Tanaka, commander of the Japanese transports spotted by Reid's plane, advised Yamamoto that their approach had been detected.

Air Force Lieutenant Colonel Walter Sweeney led a formation of nine B-17 Flying Fortresses to attack the invasion ships, which he located 570 miles southwest of Midway. While they swung in for attack, he told his copilot that the sea appeared to be dotted with Japanese ships. But none of their bombs hit the enemy ships.

Meanwhile, Nagumo's carriers and supporting ships were screened by clouds and their exact position was unknown to the Americans. Likewise, Spruance's fleet had not been located by the Japanese.

When Fletcher read Reid's report about the sighting of Japanese ships, he correctly surmised they were part of the invasion fleet and not the carrier striking force. He believed the Japanese carriers would strike at dawn the next day. At that time his own carriers would be 200 miles north and slightly east of Midway. He ordered an intensive search of the entire area. Eleven Navy Catalinas were dispatched at dawn on June 4, while the 7th Air Force's 5th and 11th Bombardment Groups were sent out on search and destroy missions.

Nagumo remained confident he could carry out Yamamoto's mission, and on the morning of June 4 he prepared to launch his planes 240 miles northwest of Midway with orders to attack any American task forces. One hundred and eight Japanese fighters and bombers took off first, followed by an equal number in a second wave. Nagumo did not believe there were any American carriers in the area, but he sent out cruiser-based search planes in the direction of the threat.

Scout planes from both fleets were hampered by heavy clouds, but a PBY with Lieutenants Howard Ady and William Chase found the Japanese carriers. They immediately dispatched the news to Fletcher's flagship, but they gave no information about number or location. Fletcher was livid with anger, but the PBY was under attack by Zeroes as it tried to avoid heavy antiaircraft fire from the ships below. The pilots finally had a chance to report more fully. "Many planes heading Midway, 320 degrees, distance 150."

Later they radioed, "Two carriers and a battleship bearing 320 degrees, distance 180 [from Midway], course 135, speed 25."

Fletcher ordered Spruance to "proceed southwesterly and attack the enemy carriers when definitely located. I will follow as soon as planes recovered."

The Air Force squadrons on Midway were told to attack the carriers and ignore the Japanese transports. Four Air Force B-26s, equipped with torpedoes, and six Navy Avengers on their first combat mission were sent to the last known position of the Japanese carriers. Sixteen Marine SBD dive-bombers under Major Lofton R. Henderson and eleven SB2Us under Major Benjamin W. Morris of VMSB-241 were also sent out. On Eastern Island, in the Midway Islands group, the Army Air Forces 11th, Bombardment Group had twenty-seven B-17s. This was just a sandy island populated by gooney birds and without taxiways or a place to park their airplanes. The crews spent most of their time in the shade under the wings of their Flying Fortresses.

Lieutenant Edwin A. Loberg flew five missions out of Midway. His B-17 bombed a Japanese cruiser, the *Mikuma,* and damaged it. It was later sunk by Navy planes. The members of the 11th Group never did spot the carriers.

When Japanese Zeroes appeared over Midway, they strafed military installations while their bombers attacked major installations. Only Marine fighters of VMF-221 were available to oppose them, and nineteen of their aircraft were outmoded F2A Brewster Buffaloes, while the other six were modern F4F Wildcats.

Major Floyd B. Parks led twelve of his fighters, with an equal number under Captain Kirk Armistead, when the Japanese first struck Midway. The outmoded Buffaloes had no chance to win a victory against the Japanese planes, and only ten survived, with only six Japanese planes shot down. After that initial raid, Japanese planes bombed Midway with impunity, although American antiaircraft fire was heavy.

Nagumo was urged by his fliers to send another heavy raid against Midway to complete the destruction of the island's defenses, but he hesitated. Then a bugle blared. This sound signaled an air raid against his eighty ships. When a destroyer sent up a flag signal, Nagumo knew that an enemy plane was in sight.

Four Air Force B-26 Marauders from the 22nd Group at Midway headed for one of Nagumo's ships. Captain James F. Collins noted during his bombing run that six Navy Avengers were fighting off Japanese Zeroes as they flew just above the waves to drop their torpedoes. These were planes from the *Hornet*'s Torpedo Squadron 8 temporarily assigned to Midway. The enemy fire was heavy and Lieutenant Langdon K. Fieberling in the lead watched with horror as three of the squadron's Avengers exploded and crashed into the sea. He continued on but the fire from the ships was murderous and he crashed into the ocean. The B-26s seemed doomed as they fought their way through heavy fire, and one Marauder was quickly downed. The Japanese fire was so savage that they were all forced to release their torpedoes too soon, and not one scored a hit. One B-26, with flames streaking behind it, released its torpedo and careened over the carrier *Akagi,* crashing into the water on the other side of the ship. When the encounter ended, only two B-26 crews and one Avenger survived.

Nagumo reached the decision he had been pondering. He ordered a second strike at Midway. This meant that planes on the *Akagi* and on the *Kaga* would have to unload their torpedoes—in anticipation of attacks against the American carriers—and reload with general-purpose bombs. The *Hiryu* and the *Soryu* were already equipped with bombs.

Major Henderson arrived on the scene with his SBD Dauntlesses of Marine Scout Bombing Squadron 241 at 7:55 A.M. He decided on a glide bombing attack instead of dive-bombing because ten of his pilots were new, and only three were experienced SBD pilots.

He signaled them to follow as he made a wide circle before dipping down for a run at the *Kaga,* maneuvering on the surface far below him. While Zeroes and Nakajima 97 fighters roared toward them, Henderson led his men in a glide toward the carrier, while flak mushroomed around them, at times viciously shaking their planes. Henderson's plane was hit hard, and he and eight others crashed into the sea. Captain Richard E. Fleming took over the lead and dropped to 400 feet before releasing his bomb. His plane was a sieve, with 179 holes in it as Japanese fighters swarmed around them and antiaircraft fire followed them as they pulled away with only two minor wounds. The other six planes were so badly shot up that they had to be scrapped.

The second group, under Major Benjamin W. Norris, arrived on the scene at 8:20 with eleven SB2U Vindicators. Norris decided he could not break through the defensive barrier around a carrier, so he and his men attacked a battleship. They dove through murderous flak that downed three planes before they released their torpedoes. Norris believed they had scored two direct hits. In reality he had scored near misses.

Lieutenant Colonel Sweeney from the Seventh Air Force's 5th Bombardment Group based in Hawaii concentrated on three carriers from high altitude. Without opposition, the B-17s made long runs but failed to make a single hit.

Nagumo was incensed when a Japanese search plane reported that it had located ten ships that he believed were American, but he failed to identify them or announce their location. He was puzzled by reports of the presence of American carriers, and disturbed that the operation was not going as planned. He and Yamamoto had believed they would have two days to attack Midway and occupy it before the American carriers could put in an appearance. The Japanese expedition to the Aleutians should have drawn the American ships up there to repel an invasion.

But at least one carrier had to be in Nagumo's area, so he advised Yamamoto and the commander of the invasion fleet that he was revising the operation's plan. His planes were now returning from the strike at Midway, and due to the confusion of the battle it took forty minutes to recover them. Nagumo realized his situation was precarious because his carrier decks were filled with planes that needed to be refueled. Fuel lines and bombs littered their decks, and there was an imminent possibility of further attacks by American carrier planes.

Spruance, 155 miles away, was equally concerned. His carriers were without adequate supporting ships, and as yet he had not heard from the *Hornet*'s scout planes, who were searching for the enemy carriers.

When word came about the location of the Japanese carriers, he ordered the *Enterprise* and the *Hornet* to steam toward a position a hundred miles from them before they launched their planes. His chief of staff, Captain Miles Browning, objected. He advised Spruance that in his opinion Nagumo would make a second strike against Midway. In that case, he said, despite the greater distance, an earlier launch might arrive over the Japanese fleet while their planes were refueling on deck.

Fletcher—who had ordered Spruance's carriers the *Enterprise* and the *Hornet* to launch attacks when they came within range of the Japanese fleet—ordered his *Yorktown* to be held in reserve because his reports indicated there were only two Japanese carriers and not four. Spruance had his doubts, but agreed to Fletcher's request. At a point thought to be 200 miles from the Japanese carriers, the American carriers launched their planes. Unknown to the Americans, Japanese planes were now returning from Midway, while their carriers were under attack by Army and Navy planes based on the island.

Fletcher became concerned that an attack on the *Yorktown* might find his carrier vulnerable, with a deckload of airplanes. He ordered its torpedo squadron, half of its bombers, and six fighters to launch at 8:40 A.M. Lieutenant Commander Clarence W. McCluskey led thirty-six SBDs from VB-6 and VS-6 from the *Enterprise*. Lieutenant Commander John S. Thach led six of the *Yorktown*'s F4F Wildcats to protect twelve torpedo bombers and sixteen dive-bombers. Stanhope C. Ring, the *Hornet*'s air group commander, took off with thirty-six bombers and fighters.

Nagumo's indecision now began to paralyze his actions. He had sent most of his fighters to Midway to protect his bombers, and now he did not dare to launch his torpedo planes against the American fleet. Without fighters, he knew they would be easily destroyed.

Rear Admiral Tamon Yamaguchi, in charge of the Second Carrier Division squadrons on the *Hiryu* and the *Soryu,* sent a message to Nagumo. "Consider it advisable to launch attack force immediately."

Nagumo had been impressed by the success of his fighter planes against American torpedo bombers, and he had no illusions that the heavy losses sustained by the Americans would not be duplicated among his own squadrons without fighter protection. He decided to wait until all his planes had returned from Midway and had been rearmed.

He rescinded his earlier order to remove the torpedoes and replace them with bombs for the planes on the *Kaga* and the *Akagi.* He signaled the other carriers, "After completing recovery operations, force will temporarily head northward." Now Nagumo's fleet was steaming away from Midway and posing an additional problem for the Americans.

Lieutenant Commander John C. Waldron, commander of the *Hornet*'s Torpedo 8, had laid it on the line to his pilots prior to takeoff. He told them that if only one man survived, he expected him to get a hit on a Japanese carrier. This tough airman, of Sioux Indian descent, had no illusions about their old Devastator bombers. They were much too slow to survive in combat against Japanese fighters, but he had insisted on the installation of armor-plated bucket seats and 30-caliber machine guns for the rear-seat gunners. Waldron had trained them to the peak of perfection, and now they would go out searching for the Japanese carriers.

On Waldron's left, Lieutenant Commander Eugene E. Lindsey led his fourteen Devastators from Torpedo 6. Six Wildcats, led by Lieutenant James S. Gray, rode above them. They had been launched to provide fighter protection for the *Enterprise*'s bombers, but had become separated in the clouds and joined Waldron and Lindsey's formations.

They searched widely in the area where the Japanese carriers had last been reported. Waldron decided they must have changed course, and following his hunch, swung his formations to the north. They spotted the carriers at 9:20 A.M., and were soon under attack by at least twenty Japanese fighters.

Lieutenant Commander Thach now appeared with six *Yorktown* Wildcats. These American fighters tore into the Japanese coming at them in a stream. Thach's wingman was quickly shot down, but Thach caught the Zero on the outside of the turn and shot it down. When another appeared, he held his fire until he was within range. Just as he was about to turn, he spotted another section leader on his right turning just before he did. His first reaction was that the pilot had made a mistake and was looking on his own tail instead of watching Thach's. Thach swung around and a Zero started to follow, but Thach got a low shot and knocked the Japanese fighter out of the sky. A wild melee of fighters jockeyed for position around Thach, and he began to wonder if they would come out alive. He did not dwell on the matter because he knew their armor plating and self-sealing fuel tanks would provide them vitally needed protection. Even so, American fighters were pitted against seventy-five Japanese fighters and were going down in growing numbers due to the sheer inequality of the contest.

While Thach's planes tore into the Japanese fighters, Waldron led his TBDs through a wall of flak as they swung in for their attacks against the *Akagi.* Soon all but his TBD and Ensign George Gay's were still flying. Then Waldron's Devastator was riddled by gunfire and his plane plunged into the ocean.

A mile east of the *Akagi,* Lieutenant Commander Lance E. Massey arrived with twelve Devastators from the *Yorktown's* VT-3. His pilots were shocked when their leader's plane was shot down and only five other planes were able to launch their torpedoes. Then three more planes were shot down.

Gay, Torpedo 8's sole survivor, continued toward the carrier eighty feet off the water. He was determined to make a hit and release his torpedo just a thousand yards from the carrier. He had taken off that morning for the first time with a torpedo on his airplane, and he had eyed its installation with more than curiosity. He had never even seen a dummy torpedo before, let alone a real one. These were Mark 3 torpedoes, he had been told, whose design dated back to World War I, although they had been modernized with new fins to improve their accuracy.

Gay reduced his TBD's speed to eighty knots, despite the vicious fire from the ship now directed at only his plane.

The Japanese commander's orders to his fighters to concentrate on the torpedo bombers, bringing them down to sea level, was made because he considered them the greatest threat to his carriers. This normally would have been the case, but the American dive-bombers, riding high above the battle, were left unmolested to make their runs.

Gay's Devastator reached the release point despite the fact that it had been hit repeatedly, but primarily by small arms fire. His rear gunner, Robert Huntington, cried out, "My God! I'm hit!" Gay pushed the electrical release to drop his torpedo, but it did not drop. He tried the emergency pull and the torpedo fell away, sliding into the water, and headed for the *Akagi.*

Gay's situation was desperate—with his gunner wounded and his plane so riddled that he had difficulty controlling it—as he bore down on the carrier with its big guns up forward pointing at him. It seemed as if every gun on the ship were firing at him as he came in straight to give the gunners a smaller target. Just a few feet off the water, below the level of the carrier's deck, he fired his forward guns at a gunner firing at him, and the man disappeared quickly from sight. He knew that if he flew to the far side of the carrier the guns would be focused directly on him, so he reached a decision. He climbed above the level of the deck and flew down the flight deck. He could clearly see the captain on the bridge waving his binoculars in rage. Soon he was past the stern, but then he found himself over two cruisers. The fire intensified and he thought grimly that the whole fleet must be firing at his lone Devastator. Now five Japanese fighters attacked him, so riddling his plane that his engine caught fire. Without rudder or aileron control he tried to pancake on the ocean, but his left wing hit first and the TBD cartwheeled into a crash landing. He had left his canopy open during the run, but the impact of the TBD with the water slammed it shut. As the plane started to sink, he tried frantically to open it—otherwise he would be trapped in the plane when it sank. The hood finally released, and he crawled back on top of the fuselage to rescue his gunner, although he was convinced that he was dead. The TBD sank before he could reach Huntington.

Now a Japanese carrier headed straight for him as it recovered its airplanes. He dove into the water, grabbing a seat cushion for added protection as he hid beneath it. The carrier passed him only 1,500 yards away, and he was not detected. He calmed down as he realized that he was safe—at least for the time being—and he marveled at the action unfolding around him.

Lindsey's fourteen Devastators of Torpedo 6 came in slowly against the *Kaga,* but Japanese fighters downed ten of them before they could release their torpedoes. The other four dropped their torpedoes but they all missed.

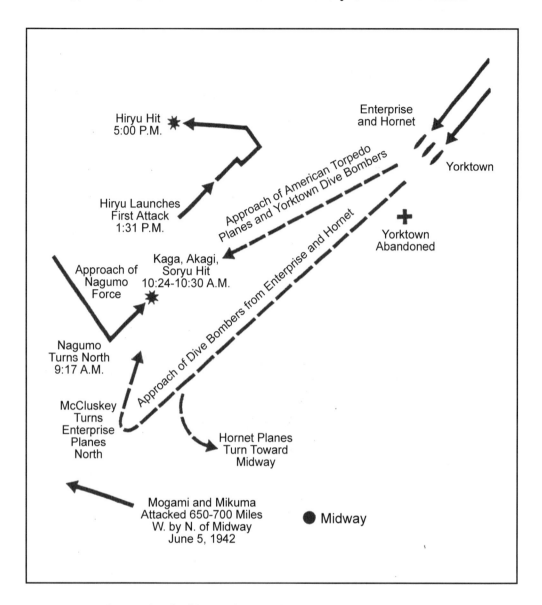

The Battle of Midway: the attack. (Drawn by Stewart Slack.)

Despite their heavy losses, some TBDs fought their way close enough to three Japanese carriers to release their torpedoes. Tragically, all torpedoes missed or malfunctioned. American pilots had been bitter since the start of the war because Japanese torpedoes almost always functioned expertly, but their American counterparts seldom did.

When Spruance learned of the torpedo plane losses, he was shocked. The *Hornet's* Torpedo 8 lost every one of its torpedo bombers. The *Enterprise's* Torpedo 6 lost ten of its original fourteen planes. The *Yorktown's* Torpedo 3 had two survivors out of twelve.

Despite these appalling losses, and the failure to damage a single carrier by torpedoes, the torpedo squadrons had played a critical role. The Japanese carriers had been under such constant attacks that they were unable to launch their own bombers. Nagumo's greatest fear had been the torpedo bombers, so he had instructed his fighter pilots to concentrate on them. They had done so with a vengeance. Now, American dive-bomber attacks, originally scheduled to coincide with the torpedo attacks, had to be made alone. But they had a distinct advantage. The Zeroes largely ignored them.

Nagumo ordered his fighters and bombers to take off after he became aware that the torpedo attacks were over, at least for the time being. The first Zero flew off at 10:24 A.M., just as a lookout called, "Dive-bombers!"

Now SBDs plunged down against the *Akagi* without opposition because the airborne fighters were still searching for torpedo bombers at low altitudes. The *Kaga* was under similar attack.

Lieutenant Commander Maxwell Leslie led his *Yorktown* SBDs in a seventy-degree dive against the *Soryu* with the *Kaga* on its right. Although he had inadvertently lost his own bomb, he fired his guns as they closed on the carriers while his men prepared to drop.

Lieutenant Paul Holmberg concentrated on the *Kaga,* using the red circle on the carrier's flight deck as his aiming point. He hit his release switch at 2,500 feet and then the manual emergency lever to assure positive release of his 1,000-pound bomb. As he pulled away, he watched his bomb hit the deck near the ship's superstructure. Three more bombs hit the carrier and a mass of flames shot high in the air. The *Kaga* had been refueling its planes, and flaming gasoline spread across the hangar deck as the ship became a flaming torch.

Lieutenant Commander Clarence W. "Wade" McCluskey's thirty-six SBDs from the *Enterprise*'s VB-6 and VS-6 failed at first to spot the carriers. When he found a Japanese destroyer harassing an American submarine that was also out to torpedo a carrier, he followed the destroyer, hoping it would lead him to the Japanese fleet. It proved to be a good guide, and he spotted the carriers ahead. He signaled for his pilots to approach from the southwest at right angles to Leslie's *Yorktown* formation, which was attacking the *Akagi,* and concentrate on the *Soryu.*

Lieutenant Wilmer Early Gallaher of VS-6 led the attack against the *Akagi,* and his formation made two hits on its deck. They exploded among the fueled airplanes and detonated the bombs and torpedoes piled nearby. Flaming gasoline added to the horror, and Japanese pilots and work crews died horrible deaths amid the exploding bombs.

The *Soryu* received three hits on its crowded deck, and soon the carrier was a raging inferno as men leaped off her deck into the ocean to escape the tremendous heat and spreading flames. Captain Ryusaku Yanagimoto ordered his men to abandon ship. Despite urging from his men, he refused to leave, and died as it sank.

The *Akagi* turned north in an attempt to launch her planes, but SBD Dauntlesses followed her, coming down in screaming dives from 14,500 feet. The first bombs exploded on deck, and soon the afterdeck was engulfed in flames. Her captain ordered the carrier to take violent evasive action, but the bombs continued to rain down. Five more made direct hits, while three others exploded along her sides, also causing damage.

With her deck in flames, Admiral Nagumo was urged to leave the carrier. Chief of Staff Ryunosuke argued with him, but the admiral refused. Taijiro Aoki, the *Akagi*'s captain, added his voice. "Shift your flag to the *Nagara* and resume command of the force." This time Nagumo agreed, but by this time the fires were so fierce that the admiral had to slide down a rope to escape the doomed ship.

The *Yorktown* under attack during the Battle of Midway. (Courtesy of U.S. Navy.)

Before Nagumo left his flagship he ordered the *Hiryu* to launch eighteen dive-bombers and six fighters against the American carriers. He assigned Lieutenant Michio Koboyashi, a veteran of the Pearl Harbor attack, to lead them.

The *Hiryu* was farther to the north, so it had escaped the first American bombs. Now Admiral Yamaguchi ordered an immediate strike against the American carriers.

The *Hiryu*'s Bakugeki 99 dive-bombers and six Zeroes found the *Yorktown,* and the American ship's combat air patrol immediately attacked while the Japanese planes were still twenty miles away. All but eight bombers were shot down, but the others continued on to the *Yorktown,* fighting their way through antiaircraft fire and pursuing American fighters. Three hits were made on the carrier, but the American fighters managed to shoot down another five bombers and three fighters. Koboyashi was one of them.

Fortunately, the carrier's damage was slight, although one bomb had gone down the stack, knocking the ship out of commission temporarily until engineers rerigged temporary steam lines. It was not long before the *Yorktown* was cruising again at 16 knots.

Commander Thach returned in the heat of the battle and had to hand-crank his wheels down while he orbited the ship. Finally he landed while the battle still raged above him.

"Launch fighters!" This latest call came at 2:30. Sixteen Japanese torpedo bombers, escorted by fighters, had been ordered out by Yamaguchi under Lieutenant

The Battle of Midway. In this overall view by artist R.G. Smith, the first three of
four Japanese carriers have already been set on fire.
(Courtesy of McDonnell Douglas.)

Joichi Tomononaga. Despite massive antiaircraft fire, half of the torpedo bombers
got close enough to the *Yorktown* to launch their torpedoes. The ship's captain,
Buckmaster, marveled at the tenacity of four torpedo pilots who fought their way to
reach their drop points.

Thach attacked one of them just above the waves and opened fire. The plane's
wing was on fire, with its ribs visible, but the pilot held his course until his torpedo
was released, with its wake heading directly for the *Yorktown*. Then it exploded.

Another Wildcat roared off the deck and, with its wheels still down, attacked a
torpedo plane and destroyed it. But the pilot, in turn, was shot down by a Zero. He
had only been airborne one minute. He was later picked up alive and well.

Buckmaster had ordered his ship to maneuver violently against the constant
attacks, and the first two torpedoes were avoided. Two others hit amidships with
such violent impact that he was almost thrown to the deck. He ordered an immediate damage assessment. He was told, "Large holes in our port side, Sir." Then the
firing ceased, and for the time being the attacks were over. A pilot reported to him,
"Got the last of the bombers, Sir." Buckmaster's tired eyes glanced around his ship.
"I'm afraid they've got the *Yorktown*."

The ship was in bad shape, listing heavily to port, and turning out of control in
a tight circle at reduced speed. Admiral Fletcher left the *Yorktown* and went on
board the cruiser *Astoria,* while Buckmaster surveyed the damage. The ship was
about to capsize, so he ordered, "Abandon ship!"

Scout planes went out in search of the *Hiryu,* and Lieutenant Samuel Adams found
her. He gave her position to the *Enterprise* and the *Hornet,* and planes were launched.

The Japanese carrier *Hiryu* burns during the Battle of Midway.
(Courtesy of U.S. Navy.)

The report that Yamaguchi received about the attack against the *Yorktown* confused him. He believed now that two American carriers had been disabled. He ordered his last five dive-bombers, four torpedo planes, and six fighters to attack at twilight. His other crews needed rest after a long day of violent action. In advance, he sent his scout planes out to locate what he believed was the single undamaged carrier.

Admiral Yamamoto, 600 miles from the battle zone, realized that the outcome would be decided long before he could arrive with his fleet to help out. Captain Aoki's request to sink the *Akagi* had shaken him. As yet he had not realized how badly the battle had gone against the Japanese carriers. He delayed permission, agreeing to her sinking later only when the realities of the situation became apparent. He did refuse Aoki permission to go down with his ship.

The *Hiryu* was sighted by the *Enterprise* scout bombers at 4:50 P.M. at a time when the three other Japanese carriers were undergoing their final moments. Lieutenant Wilmer E. Gallaher led a contingent toward it.

Japanese lookouts reported their appearance at 5:00 P.M. The *Hiryu*'s own planes were being refueled for the third attack Yamaguchi had ordered at twilight. Only a few Japanese fighters rose to challenge the Americans, but they shot down three of the attackers.

The dive-bombers headed for the *Hiryu*'s deck, connecting solidly, and their bombs lifted the flight deck and threw it over the navigation bridge. Then more bombs struck near the elevator, and a huge mass of flames erupted and began to spread.

When pilots from the *Hornet* arrived, it was evident that their bombs would be wasted, so they pulled away. They tried to hit the battleship *Kirishima*, but caused little damage.

The *Hiryu* was doomed. Her commander, Captain Kaka, blamed himself for the loss of his ship, and announced that he would go down with it. Rear Admiral Yamaguchi argued otherwise, but his appeal was in vain.

Admiral Yamamoto had believed from initial reports that the American fleet was practically destroyed and was retiring to the east. He ordered Nagumo to destroy what remained of the U.S. Pacific Fleet. But after he realized how utterly incorrect his earlier assumption had been, he sent orders to his Aleutian carriers to return to Midway at high speed and ordered his own fleet commander to proceed to Midway.

But on the evening of June 4, when Nagumo advised Yamamoto that there were three American carriers, and that they were steaming westward—and not in retreat—Yamamoto viewed the news with shocked disbelief. He quickly realized, however, that at last he was getting the truth about the battle. Nagumo advised his chief that he was retiring to the northwest with the *Hiryu* in tow. Yamamoto promptly removed Nagumo as overall commander of the Midway operation. To replace him he selected Vice Admiral Nobutake Kondo, the commander of the Midway invasion fleet.

The *Soryu* sank at 7:20 P.M., even before her crew could be removed, and the *Kaga* followed five minutes later. An American submarine, the *Nautilus*, under Lieutenant Commander William H. Brockman Jr., hastened the *Kaga*'s end by torpedoing her, and 800 officers and men still on board went down with her.

Admiral Spruance ordered his fleet to reverse course and avoid a night battle against possibly superior enemy forces. But his main reason was that he wanted to be near Midway the next morning—June 5—to resist any landing attempt there and also to be in a position to follow the Japanese fleet in its retreat to the west.

An attempt to scuttle the *Hiryu* was made during the night by a work party left behind for that purpose. While they were working topside, Ensign Mandai and a group of men who had been trapped in the engine room cut a hole in the steel bulkhead before dawn on June 5. To their surprise they found the *Hiryu* abandoned except for a work party under Commander Kunizo Aiso and a party of thirty-nine men who had been assigned to scuttle the carrier. They signaled a destroyer to pick them up, but it disappeared without responding.

By morning the carrier was listing badly and down at the bow. After making all preparations for scuttling the ship, the party lowered the ship's cutter and prepared to leave. They found only a few oranges, some sea biscuits and a case of beer to take with them.

The cutter had far too many people in it for safety, but they rigged a stubby mast and used a blanket for a sail, and after hoisting the Japanese flag they set out for Wake, the nearest place occupied by their countrymen. After the cutter was spotted by an American patrol plane, a destroyer was dispatched to pick them up, which they did without a fight.

But the *Hiryu* stubbornly refused to sink, and it had to be sunk by a Japanese destroyer at 9:00 A.M. using torpedoes. Captain Kaka went down with her. All her planes and 500 men, including sixty of the ship's pilots, were killed in action. One of Japan's most able flag officers, Rear Admiral Yamaguchi, was one of those who died.

Captain Kawaguchi, the *Hiryu*'s air officer, was on a nearby ship when the *Hiryu* sank beneath the waves. There were tears in his eyes as he admitted to fellow officers that the Americans would soon win control of the air in the Pacific.

Japanese scouting planes reported the retreat of the American fleet and realized that any possibility of a night action was eliminated. Yamamoto considered his options, including total withdrawal. He knew that by daylight his remaining ships would be vulnerable to American carrier planes. One of his aides suggested that all ships shell Midway before they withdrew. He claimed that the *Yamato*'s guns, and those of the other capital ships, could put the airfields at Midway out of commission and then the island could be occupied by Japanese troops.

Yamamoto's reply was sharp. "Of all naval tactics, firing one's guns at an island is considered the most stupid. You've been playing too much shogi!"

Admiral Ugaki suggested that they wait for the Aleutian carriers and launch another attack. If this proves impossible, "we must accept defeat in this operation but we will not have lost the war."

Yamamoto was now advised that the *Hiryu* was sinking. His great plan to win a decisive victory against the American Pacific Fleet—which, he believed, would force the United States to agree to a negotiated peace—had collapsed. Instead, he was faced with the loss of four carriers and thousands of highly experienced airmen and sailors.

Yamamoto hesitated to make the irrevocable decision he knew was necessary. He sharply rebuffed one officer who asked who would apologize to the Emperor. "I am the only one who must apologize to the Emperor."

At 2:50 A.M. on June 4, Yamamoto cancelled the Midway operation. He wired Admiral Hosogaya, whose 5th Fleet was in Aleutian waters, to give up the operation and return home.

Hosogaya sent Yamamoto a message asking that his northern force be permitted to proceed with its invasion of Adak, Attu, and Kiska. He said they could be taken easily. Yamamoto agreed, but he insisted that Adak be eliminated from the invasion plans.

Attu and Kiska were occupied by the Japanese. The decision was made in Washington not to retake the islands in the near future but to bomb them on a regular basis.

A Zero 21 fighter plane was found in a bog on Akutan Island after it had crash-landed, breaking the pilot's neck. It was removed in late June and brought to the North Island Naval Air Station at San Diego for repairs to make it flyable.

During extensive flight tests starting in October, much was learned about the Zero's performance. Most American planes were supercharged for operations at high altitudes, but the Zero was not. But a combat pilot pointed out that they rarely found Zeroes above 10,000 feet, so he considered that superchargers in American planes were superfluous, and detrimental because they reduced performance. But American pilots were impressed by the Zero's workmanship, although they learned that 50-caliber bullets caused them to disintegrate.

Test pilots learned that the Zero's vertical tail was designed to counter the strong torque on takeoff caused by the plane's extremely large propeller. Thus the plane's pilot didn't have to use excessive right rudder. But this design feature also contributed to a condition that proved to be an advantage for the Americans. At high speeds, the Zero rolled very fast to the right, but only slowly to the left. Thach and others conceived a new tactic for fighting Zeroes to take advantage of this condition. Pilots were advised to attack Zeroes at high speed and roll to the left, which was the Zero's slow-roll direction.

After all test data was evaluated, the F6F Hellcat was redesigned, and later ended the Zero's domination of combat conditions. The finding of this almost intact Zero—the first of its kind to be salvaged—proved of inestimable value to the war effort.

Lieutenant Commander John W. Murphy Jr., in the American submarine *Tambor,* reported that Japanese warships had been located at 3:30 A.M. on June 5. The ships belonged to the fleet under Rear Admiral Takeo Kurita, which had reversed course after Yamamoto canceled the Midway strike an hour earlier. The

Admiral Nimitz inspects Midway Island after the battle. (Courtesy of U.S. Navy.)

American submarine followed the fleet until dawn so that its commander could positively identify the ships. But the Japanese panicked when they spotted the submarine in their rear, and the *Mogami* rammed the *Mikuma,* which was severely damaged. Both ships continued their retreat.

Spruance ordered Navy and Air Force planes at Midway to attack Kurita's fleet. They found the *Mikuma* with no problem because it trailed a huge oil slick.

Captain Richard E. Fleming, with six SBDs, attacked the *Mikuma,* but the anti-aircraft fire was so heavy that Fleming's plane caught fire and crashed on the ship. Fleming later received a posthumous Medal of Honor for his act.

The *Mogami,* which had lost its bow in the ramming, fought off more attacks from the *Hornet* and the *Enterprise,* and managed to escape. It reached Truk safely with its protective destroyers, but the loss of life on all ships had been heavy.

The *Mikuma*—damaged earlier by Lieutenant Loberg's crew in a B-17 from the 11th Bombardment Group—now came under attack by the same SBDs that had attacked the *Mogami.* They were successful, and the ship was sunk.

Meanwhile, valiant efforts had been under way to take the *Yorktown* in tow after it suffered major damage from Japanese bombs and torpedoes. But a Japanese submarine commanded by Lieutenant Commander Yahochi Tanabe torpedoed the carrier. She sank at 6:01 A.M. on June 6. The destroyer *Hammann* was alongside the carrier, and it broke in half and sank at the same time.

While the battle raged, the downed Ensign Gay had a spectacular view as he watched three Japanese carriers burn furiously while he hid beneath a seat cushion that had broken free from his sunken airplane. After he believed he was relatively free of discovery, he inflated his life raft. He had lost a lot of blood because he had a bullet hole in his left arm, a piece of shrapnel in his left hand, and a badly burned left leg. After thirty hours in the water, he became seriously dehydrated.

Lieutenant Shelby O. "Pappy" Cole of VP-43 spotted Gay and radioed for a PT boat to pick him up. When he was advised that it would take three days, he and his crew agreed to land in the water and pick him up. Fortunately, they were not discovered by the Japanese, and the weakened Gay was hauled out of his life raft. He was flown to Midway first and then to Pearl Harbor for an extensive convalescence.

The Battle of Midway had been a decisive defeat for the Japanese—putting an end to their successful offensive—and effectively turned the tide of the Pacific war. Japanese losses totaled four fleet carriers, a heavy cruiser, 258 aircraft, 100 of their experienced carrier pilots, and approximately 2,000 sailors. American losses were forty shore-based and ninety-two carrier aircraft, the carrier *Yorktown* and the destroyer *Hammann*.

Carriers had proved themselves despite overwhelming opposition, and airpower reached maturity in the minds of those who had doubted its effectiveness. It was a great victory, due in large part to the work of the dive-bombers.

During the Battle of the Coral Sea—when the *Shoho* was sunk and the carriers *Zuikaku* and the *Shokaku* were damaged—the Japanese lost one-third of their airpower. Unquestionably these losses contributed to their defeat at Midway.

The Japanese, stunned by this latest defeat, reorganized their strategic plans. They had lost five carriers during May and June, so it became necessary to remove the battleships *Ise* and the *Hyuga* from operations until they could be converted into carriers. They become known as morphadites.

The Japanese at Midway were outguessed, and the Americans won one of history's most decisive battles. The fact that their code had been broken was an asset. It enabled the United States to follow Japanese ship movements with some degree of accuracy. However, American naval commanders did not dare put full reliance on decoded messages because they were afraid of being deliberately misled.

One of the problems of the earlier Coral Sea action was the inadequate number of fighters to both cover the strike groups and provide air defense. At Midway, the terrible inadequacy of the TBD Devastator proved disastrous. The aircraft was totally incompatible in performance with the F4F and the SBD, which prohibited its inclusion into the strike group. The TBD's cruising speed was 110 knots. Complicating the plane's inadequacy was the MK 13 torpedo, which had to be dropped within 1,000 yards of a target at an altitude of eighty feet, and at an airspeed of eighty knots. The aircraft could not carry the 2,000-pound torpedo above 10,000 feet, which increased its vulnerability.

Consequently, the torpedo squadrons flew into the Japanese formations at 1,500 feet. The *Enterprise*'s VT-6 and the *Hornet*'s VT-8 were unescorted by fighters, while the *Yorktown*'s VT-3 had an escort of only six fighters. The limited endurance of the F4F was also a factor.

It was not until the Grumman TBF Avenger came into the fleet that carrier air groups could fly as a strike group in the same airspace with proper fighter escort.

After the battle, the fleet returned to Pearl Harbor flying the Number 2 pennant, marking a victory of which you are proud but not boastful, instead of the Number 1 pennant, which is reserved for great and decisive actions. Spruance was being much too modest.

9

Guadalcanal

There were many people who deserved credit for the Midway victory (Spruance claimed that Lieutenant Commander McCluskey, the *Enterprise's* Air Group Commander, deserved the most credit), but the man most responsible was Admiral Spruance. His experience and intelligent decisions contributed most to the successful outcome of a battle that could easily have gone the other way.

The Japanese people were not told about the Midway defeat, although the landings in the Aleutians were announced. Returning crewmen were not permitted home leave, and the wounded were treated at remote hospitals. Yamamoto isolated himself, and became ill because of the defeat of his fleet.

The *Chicago Tribune* and the *Washington Times-Herald* revealed that the United States had known in advance about the Japanese attack on Midway. Admiral King was furious, believing the Japanese would now realize their diplomatic and military codes were broken, but this did not occur.

The Japanese occupation of part of the Aleutians gave them a battle line that stretched from Alaskan waters to the Solomon Islands in the South Pacific. In considering his next moves, Admiral Nimitz knew that Japanese losses in the Coral Sea and at Midway would seriously restrict their offensive capability. He decided that an all-out assault against Australia was not to be expected in the near future. He discussed his options with King and decided to commit his forces to the South Pacific. Nimitz believed that the Japanese posed the most immediate threat there after their invasion of Guadalcanal on July 4.

Nimitz reminded his staff, "The chain of islands to the east of New Guinea is the logical route to return to the Philippines." But first, he knew Japanese bases in the northern Solomons and on New Britain posed a stumbling block to the Allied command that could become more serious if the Japanese continued their advances.

Nimitz established a new command for the South Pacific under Vice Admiral Robert L. Ghormley and sent precise instructions to him. "Hold the island position. Continue to support operations in the southwest and central Pacific. Amphibious operations should be planned against positions now held by the Japanese. D-day is tentatively scheduled for August 1."

General MacArthur proposed a different plan to General Marshall that the Army chief of staff tentatively approved on June 12. It was contingent upon Admiral King's release of Navy ships for the operation.

These ambitious plans by both commanders ran counter to decisions by the Combined Chiefs of Staff that Europe must have priority on all men and materiel while the Pacific war was waged on a containment basis. Previously, King had supported Marshall's cross-channel invasion and President Roosevelt's insistence that more troops be sent to Europe at a time when Army soldiers were not available in large numbers and still needed extensive training.

Marshall was now tempted to join King in a plea for a limited offensive in the Pacific now that the Japanese fleet had been roundly defeated, but he reminded the Joint Chiefs of Staff that such an offensive must not be undertaken at the expense of the invasion of Europe.

Marshall rejected MacArthur's plan to invade Rabaul as too risky. He recommended that an indirect approach be taken through the eastern Solomons, where Japanese defenses were much weaker. MacArthur's request to command some of Nimitz's ships was flatly rejected. He was advised that any amphibious operations must be commanded by one of Nimitz's officers.

With MacArthur's plan rejected, King directed Nimitz to seize Tulagi in the Solomons with his amphibious forces, despite the president's order not to increase American strength in the area. This was within MacArthur's area of responsibility, and any Navy operation there had to be approved by the president and the Joint Chiefs.

But King used a presidential order that had been issued in March, and never rescinded, that permitted a limited operation in the Solomons. King chose his words to Nimitz carefully to avoid later repercussions. He advised him to "prepare for an assault."

King presented his plan to the Joint Chiefs on June 25 for concurrence only. MacArthur now abandoned his plan to attack Rabaul due to Marshall's opposition. He agreed to support King's plan but he insisted that he command the operation. Marshall supported his position, but King refused.

Admiral King always believed that Marshall was too subservient to MacArthur because he had once served under him, and there was some truth to the claim. But now Secretary of War Stimson insisted that MacArthur head the operation. King and Stimson had long disliked one another. King was convinced that Stimson's long backing of MacArthur was not in the best interests of the war effort, and he flatly refused to agree to MacArthur's request to command the operation. King was on fairly solid ground in his opposition, reminding Marshall that Army officers did not understand airpower or seapower.

When MacArthur was advised of the conflict, he wrote Marshall a letter that the Army had supreme command in Europe, and he expected the same in the Pacific. He threatened to invade the Solomons with or without Navy support—although he did not specify where he would get the ships and troop transports. He told Marshall that if the Navy got its way the Army would be reduced to an "occupation force."

King and Marshall reached a compromise on June 30. They agreed that Nimitz's command would be moved farther west into MacArthur's area of responsibility to include the eastern Solomons and Tulagi. They also agreed that Ghormley would command the assault in the eastern Solomons.

It was further decided that MacArthur would command operations in the western Solomons, eastern New Guinea, and the Bismarck Archipelago. After the Joint Chiefs approved this new Pacific strategy on July 2, the assault on the eastern

Solomons was scheduled for August 1, 1942. At last King had approval for an American counteroffensive in the Pacific.

At a conference in San Francisco on July 4, King told Nimitz, "The eastern Solomons is only the start of our counteroffensive." He told him that after the eastern Solomons were seized, he wanted New Guinea occupied and Truk, Saipan, and Guam invaded. Although he had reached a compromise agreement with Marshall, he advised Nimitz that none of these plans included use of MacArthur's forces. Nimitz looked at him with surprise when King explained that these were his own ideas and had not been approved by the president or the Joint Chiefs.

The brilliant, and often arrogant, Admiral Richmond Kelly Turner had been selected by Nimitz to command the South Pacific's amphibious forces. Turner objected to such grandiose plans, telling King that he knew nothing about such operations. King was sharp in his reply, reminding him that he would soon learn, and that actually no one knew much about amphibious operations.

Turner had come up with a tentative plan, which he now presented to King and Nimitz. It called for the occupation of the Santa Cruz Islands and Tulagi, Florida Island, and Guadalcanal in the eastern Solomons. Next, he proposed the occupation of Funafuti in the Ellice Islands and reinforcement of the Army garrison at Espíritu Santo Island in the New Hebrides. King readily agreed to all of Turner's plans.

Ghormley, assigned to command the South Pacific under Nimitz, fell victim to MacArthur's charm and arguments. He advised King and Nimitz that available forces would be inadequate unless strong reinforcements were sent to the area.

King reacted angrily against Ghormley. He said MacArthur was vacillating and fainthearted. He reminded Ghormley that MacArthur had been ready to attack Rabaul just three weeks earlier.

Ghormley's support for MacArthur quickly dissipated once he realized that he was acting counter not only to Nimitz and King, but also to the Joint Chiefs.

In early July, President Roosevelt recalled former Chief of Operations Admiral William D. Leahy to active duty. Much to King's chagrin, Leahy became a de facto chairman of the Joint Chiefs. Roosevelt tried to soothe King's outraged feelings by saying that Leahy was just a personal aide without command authority.

The more King studied the needs for his ambitious South Pacific operations, the more he realized he had to find more men and equipment. Marshall refused his pleas for both, citing their need for the invasion of North Africa, which had been approved for fall.

King persisted, arguing that the Allies should make at least a limited offensive before the Japanese consolidated their gains and became a new threat to Australia. He believed that Guadalcanal, centrally located in the South Pacific, was ideally suited for air bases, and he expected it to be the key to Japan's plans for an expansion toward Australia. The Japanese had already occupied Tulagi, opposite Guadalcanal, and had set up a seaplane base. When the island's terrain was found unsuitable for land bases, the Japanese quietly began to build such bases on other islands.

Despite his need for more troops, ships, and planes, King went ahead on July 2, 1942, with a limited offensive after the Joint Chiefs issued a directive of approval. Ghormley's South Pacific Command was ordered to take Tulagi and Guadalcanal, and MacArthur's Southwest Pacific forces were given the responsibility for capturing the rest of the Solomons, and later Rabaul. Marshall had strongly resisted these operations, but King argued that the occupation of these islands was the key to stopping Japan's advances and providing bases for future Allied drives toward the Philippines, which would culminate in the ultimate defeat of Japan.

Unfortunately, King and Nimitz failed to comprehend the severity of Japanese opposition, particularly on Guadalcanal, which was not secured until after six months of bitter fighting.

The Japanese had occupied Guadalcanal's north central shore on July 4 and then moved down the east coast of New Guinea. Ghormley and MacArthur organized three major task forces. Task Force NAN, under Rear Admiral Leigh Noyes, was assigned to provide carrier support for the counterattack, with the *Saratoga* commanded by Captain DeWitt Ramsey, the *Enterprise* under Captain Arthur G. Davis, and the *Wasp* under Captain Forrest P. Sherman. In addition there was a second amphibious force for the landings, and a third force to supply aerial scouting, with seaplanes assigned to patrolling. Both top commanders agreed that MacArthur's land-based planes would be used in advance of all operations.

The carriers left Wellington, New Zealand, to prepare for assaults against Tulagi and Guadalcanal scheduled to start on August 7. The amphibious forces split up, with the Tulagi force leaving Savo Island to the north, while the Guadalcanal group headed between Savo and Cape Esperance to make their separate attacks.

Task Force NAN launched its planes before sunrise on D-Day. After some difficulty forming up due to their close proximity, the planes bombed installations at Guadalcanal and Tulagi at dawn. The Japanese were caught by surprise.

Although the operation was code-named "Watchtower," the Marines had a more appropriate designation—"Operation Shoestring." Ghormley had been openly critical because the landings had been made in haste, the maps of the area were totally inadequate, and none of his commanders had any experience in amphibious landings. (This was the first one since 1898, when American forces landed in Cuba during the Spanish-American War.)

Admiral Frank J. Fletcher had expressed his strong misgivings about the operation when he first met with Ghormley. He told his staff he would not expose his three carriers for more than two days while the Marines were landing. It was estimated these landings would require five days. Fletcher had already lost two carriers since the war had started—the *Lexington* and the *Yorktown*—and he had become inordinately protective of the three under his care.

Planes from Lieutenant General George C. Kenney's 11th Bombardment Group and carrier bombers made strikes on the landing areas during the last week of July and the first week of August. There was only an unfinished airstrip on Guadalcanal, and it was repeatedly bombed. Fletcher knew that if it became available for use by the Japanese it could be used to bomb targets in New Zealand and to harass Australia-bound shipping.

sThe initial landings on Guadalcanal by the First Marine Division on August 7 proceeded with little resistance. On Tulagi, however, Allied troops ran into fierce resistance. They were given a profound lesson in the tenacity of Japan's foot soldiers.

At Guadalcanal, the 600 Japanese assigned to fight off the landing force retreated inland, abandoning many of their supplies. Brigadier General Alexander A. Vandegrift noted that the beaches were dangerously cluttered, but he had no alternative. He ordered that the landing party of men and supplies be continued.

In the air, the Japanese offered stout resistance to the planes of Task Force NAN. After scores of attacks, the American fighter strength was down to seventy-eight airplanes.

Fletcher, who had earlier predicted failure of the operation, became so concerned that he sent a message to Ghormley at Nouméa. "Recommend air support groups be withdrawn because of heavy enemy torpedo and bombing attacks." Ghormley approved his request, but General Vandegrift, commander of the 1st Marine Division, bitterly opposed the move. He said it would expose his amphibious forces and troops to massive air attacks. Fletcher agreed to remain in the area, but he pulled away from the invasion front.

After two days of fighting, the Japanese retreated farther inland and the immediate objectives on Guadalcanal were achieved. The Japanese, aware from their scouts of what was happening, quickly sent Admiral G. Mikawa with a task force to a position west of Bougainville. He ordered, "Destroy the American landing and cargo vessels at Tulagi and Guadalcanal."

Meanwhile, in a vicious night action at 1:30 A.M. on August 9 off Savo Island, four Allied heavy cruisers and a destroyer were sunk.

At noon the next day, Admiral Turner ordered his amphibious ships at Guadalcanal to withdraw. This left Vandegrift's 16,000 Marines unprotected and only half supplied. Fortunately the Japanese did not exploit the situation, although the newly renamed Henderson Field was bombed repeatedly after the airstrip was captured from the Japanese. On August 18 a detachment of Japanese landed unopposed, but they were then practically destroyed in two days of vicious fighting.

The limited superiority of American forces in the area was gone. By August 23, the Japanese had several carriers and two battleships in the Rabaul area, and they were determined to drive the invaders out of their beachheads.

Admiral Fletcher had three carriers and one battleship in his task forces, and the *Hornet* was en route from Pearl Harbor with additional warships. His carriers, however, needed to be resupplied. When Fletcher was advised that the Japanese carriers remained north of Truk, he sent the *Wasp* and two heavy cruisers and seven destroyers south to refuel. Unfortunately, his intelligence was faulty: three Japanese carriers and a strong supporting force were headed southward toward Guadalcanal.

Extensive carrier searches found no sign of major Japanese fleet activity on the 24th. Fletcher was astounded, therefore, to receive word from another source that a Japanese carrier and a task force were only 280 miles from his own carriers. Scouts from the *Enterprise* reported even more disturbing news. A larger fleet was only 200 miles away. They identified the smaller carrier *Ryujo* and the newly repaired *Shokaku* and the *Zuikaku*. The news was grim. Fletcher told his staff, "The Japs will do everything in their power to destroy our landings at Guadalcanal and Tulagi."

Yamamoto had assembled an impressive fleet to protect four large transports carrying 1,500 troops for Guadalcanal with Admiral Nagumo in charge. He had the *Shokaku* and the *Zuikaku,* while Rear Admiral Hara had a smaller force with the light carrier *Ryujo.*

Ghormley told Fletcher to use his three carriers to cover the island's sea approaches. Fletcher's ships had been patrolling south of the Solomons. At dawn on August 23, his carriers were 150 miles east of Guadalcanal.

Bad weather prevented either fleet from locating the other. Yamamoto had sent the *Ryujo* to a position 200 miles north of Malaita—and 280 miles from Fletcher's carriers—to divert attention from the rest of his fleet. The *Ryujo* was found by a patrol plane and word was flashed immediately to Fletcher, who fell for the ruse. He dispatched bombers and torpedo planes from the *Saratoga* and the *Enterprise*. Bombing 6 from the *Enterprise* went out to locate the carrier, and was joined by Torpedo 3. Neither squadron hit the carrier.

Commander N.D. Felt of the *Saratoga* led twenty-nine bombers and seven torpedo planes against the carriers, which they found at 4:06. Scouting Squadron 3 attacked first, and the *Ryujo* smoked heavily as they swung away after exploding three 1,000-pound bombs on her deck. Felt looked back as he led his group to their carrier. The *Ryujo* seemed in a bad way, with flames shooting up from its hangar deck.

Then TBF Avengers of Torpedo 8 made passes at the smoking carrier. Two more hits were scored and the fate of the *Ryujo* was sealed. One torpedo that missed the carrier struck a destroyer squarely and it quickly broke in two and sank. When the *Ryujo* sank, fifteen of her bombers and twelve of her fighters tried to reach Guadalcanal. They ran into fighters from Marine Aircraft Group 23 from Henderson Field, who shot down sixteen of them and forced the others to turn back. Marine Captain Marion Carl personally destroyed two Kates (Japanese bombers) and a Zero.

The Japanese lashed back. They sent dive-bombers and torpedo planes against the American carriers. The U.S. combat patrol tangled with the Japanese bombers and fighters, and a brisk battle ensued. Thirty dive-bombers got through, scoring direct hits on the *Enterprise*. Flames shot high above the carrier as the battleship *North Carolina* moved closer to protect her with her guns.

When the attacks ceased, only three Japanese bombers could be seen streaking away, while the surface of the ocean resembled a junkyard of airplane debris.

Planes from the second attack group from the *Saratoga* attacked an enemy group of four heavy bombers, six light cruisers, and destroyers 150 miles north of Malaita Island at 6:05. The TBMs (Martin torpedo bombers), hugging the water and racing swiftly toward the ships despite heavy fire, made one direct hit on a heavy cruiser. The battle encompassed a wide area, as two SBDs landed bombs on the battleship *Mutsu*.

While the *Enterprise* fought off her attackers, her planes sought out the *Ryujo*. They failed to find her, and were instructed to land on Guadalcanal because of damage to their carrier. They remained at Henderson Field for a month flying missions in support of the ground troops.

The *Enterprise*'s damage became more serious as she was repeatedly attacked by Japanese bombers as Fletcher's task force retired southward. She later had to leave for Pearl Harbor because of limited repair facilities in the area. She remained out of action for two months. Japanese surface ships were now free to bombard the Marines at Guadalcanal while the *Ryujo*'s bombers made continuous attacks after their carrier sank.

The Japanese had now lost over ninety airplanes, a destroyer, and a transport in addition to the *Ryujo*. Another aircraft carrier, a battleship, two heavy cruisers, and a light cruiser also suffered damage. Fletcher's carriers lost only twenty planes, and Admiral Nimitz hailed the operation as a major victory. "This action permits consolidation of our position in the Solomons," he said.

One final action completed the carrier action. Five Avengers and two SBDs from the *Saratoga* were led by Lieutenant Harold Larsen in an attack on the seaplane carrier *Chitose*. She was not sunk, but she had to return to Truk with a thirty-degree list.

Admiral Tanaka now ordered his troopships in the north to make a run for Guadalcanal to reinforce the Japanese soldiers, who were fighting desperately to hang on to their inland positions. There were twenty-five ships in this fleet, which was discovered on August 25 by Major Richard C. Mangrum's VMSB-232. His SBDs, and a few planes from another unit, had been sent out to intercept the Japanese ships.

Captain John L. Smith and his MAG-23 (Marine Air Group) Wildcats had been ordered to provide fighter support for Mangrum's bombers, but they had to return to Henderson Field when their fuel ran low. Tanaka's aptly dubbed "Tokyo Express" consisted of the cruiser *Jintsu,* Admiral Tanaka's flagship, eight destroyers, and the troop transports. His ships scattered when the bombers appeared. Lieutenant Lawrence Baldinus placed his bomb on the *Jintsu*'s deck, forward of the bridge between the two turrets, and Tanaka was knocked unconscious. He soon recovered and transferred his flag to the destroyer *Kagaro,* ordering the *Jintsu* to return to Truk.

Ensign Christian Fink of the *Enterprise*-based flight crew at Henderson dropped his 1,000-pound bomb on the transport *Kinryu Maru,* causing havoc on its heavily loaded deck as flames shot high above the ship. Tanaka ordered the destroyers *Mutsuki* and *Yayoi* to remove the survivors.

No other planes scored hits, but they circled the fleet, noting the departure of the *Jintsu* for Truk, while the *Kinryu Maru* burned to the waterline. Only then did the Navy and Marine pilots depart.

Tanaka ordered the remainder of his fleet to continue to Guadalcanal, but they soon came under attack by Colonel LaVerne G. Saunders's 11th Bombardment Group, whose planes sank the destroyer *Mutsuki.*

Tanaka issued orders at noon on August 24 that his fleet should retire to the Shortland Islands, one of the smallest of the Solomon group, located south of Bougainville.

Tanaka's attempt to reinforce the Japanese troops on Guadalcanal had been a costly failure. But the Americans had also paid a price, losing seventeen airplanes, with the carrier *Enterprise* temporarily out of action.

The American strategy was to seize a weakened outpost, build air bases on it to protect it, and then begin a methodical attack on Japanese supply lines. This was a new kind of warfare, fought largely at sea, with the primary goal of eliminating Japanese airpower.

For those fighting on the ground on Guadalcanal, and to a lesser extent at Tulagi, life became intolerable due to the mud when it rained and the rampant malaria and dysentery. The troops suffered food shortages as supplies were frequently interrupted. Each night "Washing Machine Charlie" and "Louie the Louse" made life even more miserable by dropping bombs or dropping flares to illuminate the area so Tanaka's ships could shell Henderson Field and Marine positions. The situation stalemated because neither side could muster the additional forces to reach a decisive victory, as the Marines battled inland in some of the worst jungle fighting of the war. American airpower on Guadalcanal was almost destroyed, at a time when the Japanese were increasing the number of planes at Rabaul.

Conditions stabilized when General George C. Kenney's 5th Air Force received more planes, and a new air force, the 13th, arrived with new groups.

Rear Admiral John S. McCain, in command of all air operations in the South Pacific, sent Nimitz his personal evaluation of the situation. "Guadalcanal can be consolidated, expanded and exploited to the enemy's hurt. The reverse is true if we

Admiral John S. McCain. (Courtesy of U.S. Navy.)

lose Guadalcanal. If the reinforcements required are not available, Guadalcanal cannot be supplied and cannot be held."

The Japanese fought savagely to retake Henderson Field on September 14, but the Marines stubbornly fought to hold the line at "Bloody Ridge." The American

Marines succeeded. A thousand Japanese soldiers died in their attempt to defeat the Marines, with a loss of only forty Americans.

The fighting on Guadalcanal was often hand-to-hand, with the battle lines changing only a few feet at a time. Meanwhile, the *Hornet* and the *Wasp* fought to protect their supply line at sea, with the outcome constantly in doubt.

The two carriers were the only ones operational in the South Pacific on September 15, when they were ordered to escort six transports with the 7th Marine Division to Guadalcanal. They had to pass through the infamous "Torpedo Junction" between Espíritu Santo and Guadalcanal. Japanese submarines attacked, hitting the battleship *North Carolina* with a torpedo, and the destroyer *O'Brien*. The latter was sunk but the battleship limped away. Then three torpedoes hit the *Wasp,* and she quickly sank. Except for the Marine Pacific Air Wings on Guadalcanal, only the planes of the American carrier *Hornet* were left to face the Japanese with their large carriers the *Shokaku* and the *Zuikaku,* and the light carriers *Zuiho* and the *Junyo*. Brigadier General Ross E. Rowell had only fifty-eight operational aircraft on Guadalcanal by October 1, whereas the Japanese at Rabaul had three times that number.

Admiral King had followed events in the South Pacific with growing anxiety. His faith in Ghormley had gone down ever since the Savo Island disaster and the loss of four cruisers and a destroyer. Once he became convinced that Ghormley lacked aggressiveness, he ordered Admiral Halsey into the South Pacific. Halsey was named as Ghormley's subordinate until Ghormley's actions could be more carefully reviewed. Ghormley was another surface admiral who did not understand airpower.

Japan's 6th Imperial Cruiser Division moved toward Guadalcanal on the night of October 11. They had been ordered to shell Marine positions and then to quickly withdraw.

Meanwhile, Rear Admiral Norman Scott shepherded 6,000 American Army troops to Guadalcanal. They arrived just in time to help the embattled Marines fight off a major Japanese offensive.

The Japanese had the same idea of landing fresh troops. But first they bombarded the American positions. An early attempt by the *Shinyu Maru* to land 600 Japanese marines had failed when the ship was torpedoed. Now a stream of Japanese combat ships and merchant ships entered the upper Solomons and New Britain areas.

Lieutenant General Harukichi Hyakutake, commander of Rabaul's 17th Army, was ordered to command the proposed 20,000-man army on Guadalcanal. He planned to lead a new offensive himself and then attain his major goal of capturing Port Moresby.

Originally the Japanese had 16,000 men on Guadalcanal, and with 4,000 additional men Hyakutake expected to reach a parity with the 21,000 U.S. Army and Marine Corps troops now stationed there.

American airpower was built up on Henderson Field on October 9 when forty-six aircraft of MAG-14 and twenty more Wildcats were assigned to Major Leonard K. Davis's VMF-12. This buildup was crucial because at sea the *Hornet* was the only carrier available to maintain the seaplanes while repairs to the *Enterprise* and the *Saratoga* were rushed at Pearl Harbor.

Nimitz ordered Allied submarines to concentrate their operations around the Bismarck Archipelago.

The *Saratoga's* twenty-four Wildcats of VF-5 had been left at Guadalcanal's Henderson Field before she left the area. But Lieutenant Commander Leroy C. Simpler's squadron had suffered crippling losses flying missions out of Guadalcanal. All units on the field had suffered equally, and not just in combat. Operational losses were horrendous, with eight in one day.

Marine Major John L. Smith's VMF-223 had been reduced to himself and seven other pilots. He had been the first Marine pilot to shoot down a Japanese plane near the island, and now his total number of kills was nineteen. He was ordered home to receive a Medal of Honor. This was well deserved because his squadron had destroyed eighty-three Japanese planes, and like most pilots in the Pacific theater, they had been rushed into combat with inadequate training.

Major Richard Mangrum's VMSB-232 had lost eleven SBD pilots and gunners, and seven other pilots and four gunners had been wounded. When he was finally sent home, Mangrum was the only one who could still walk.

The first planes of VMSB-141 under Major Gordon A. Bell came to Guadalcanal on September 23, and by early October its complement of twenty-one pilots was complete. VMSB-231 completed the defenders of Guadalcanal.

General Hyakutake opened his campaign to retake Guadalcanal on October 13, telling his men their actions "would decide the fate of the entire Pacific." Captain Joseph J. Foss, executive officer of VMF-121, was one of those who denied them that chance by shooting down his first Zero, but certainly not his last. But Henderson Field came under such heavy bombardment that it became temporarily unusable due to its cratered runway and the loss of irreplaceable fuel in the field's tank farm.

The battleships *Haruna* and the *Kongo* created havoc with their shelling of the field, killing forty-one people, including five pilots. One of these was Major Gordon Bell, who had just brought in a flight of thirty-nine SBDs. Only four of the planes were in flyable condition after the shelling, and sixteen out of forty Wildcats were destroyed. B-17s from Espíritu Santo suffered from attacks, and two of their eight planes were destroyed. The rest were quickly flown out of the field despite its poor condition.

On the ground the fighting turned vicious, with constant air attacks by both sides as Japanese warships moved closer to shore to bombard American lines night and day. Both sides again sought to rush supplies and men to Guadalcanal. Hyakutake was convinced at dawn on October 15 that Henderson could be taken. He had five transports of troops and supplies waiting ten miles offshore to land, and he ordered their unloading in daylight.

The Americans were aware of the troop transports, but there were only three SBD dive-bombers available on Henderson's bomb-pocked field, and only one managed to take off. Ground crews had worked through the night to get more planes repaired. Once they were ready, fuel was drained from wrecked airplanes into the newly repaired planes. Incredibly, twelve more SBDs were flyable by dawn. A number of Wildcat fighters and a few of the discredited F-400s also became available. (This prewar model of the P-39 fighter had proven deadly to its pilots because it was underpowered, and had insufficient firepower against Zeroes.) Then a PBY-5 Catalina joined the attack force, which now numbered 21.

The Catalina was Major General Roy S. Geiger's personal airplane. The 1st Marine Aircraft Wing's commander called it "The Blue Goose." Major "Mad Jack" Cram commandeered it, taking off with two 2,000-pound torpedoes under its wings. Catalinas were not equipped to drop them, and his superiors objected when he asked permission at Espíritu Santo to load them. He did so anyway, arguing that there were no Avengers to carry torpedoes. Somehow he jury-rigged them to the Catalina and devised a mechanical system for their release.

The SBDs attacked the transports first, with Cram far behind in his lumbering PBY-5. He soon came under attack as thirty Zeroes fired at the American planes. The navigator's hatch cover disintegrated in a hail of bullets, but Cram aimed his Catalina at a transport and released his two torpedoes. They smashed into the transport's side, ripping it open.

The Zeroes concentrated on "The Blue Goose" as Cram swung away from the scene, attempting to dodge them with a number of violent maneuvers that the plane was never designed to withstand. He arrived over Fighter Field 1 with a Zero still on his tail, and his Catalina in desperate shape.

Lieutenant Roger Haberman of VMF-121 had developed a bad engine during the fighting at the beachhead and was returning to base. Arriving at the same time as Cram's plane, and seeing his predicament as his wheels were about to touch down, he jammed the throttle forward and took off to let Cram land first. Before the Zero's pilot realized what was happening, Haberman was on his tail and shot him down.

When General Geiger saw the condition of his plane, he threatened to court-martial Cram "for the deliberate destruction of Navy property." Later, after he had had a chance to appreciate the incredible courage of the pilot, he awarded him a Navy Cross.

Meanwhile, the transports came under increasing attacks by B-17s from the New Hebrides that sank one of them, while Navy and Marine pilots destroyed the other two. With the beaches jammed with Japanese troops trying to get ashore, fighter pilots swept back and forth on strafing missions. The other transports fled after 5,000 troops were landed.

Of the twenty-one planes that took part, three SBDs and four Wildcats were lost. Geiger's air strength now was down to fourteen aircraft, and only nine were fighters.

Admiral Nimitz became so disturbed by the news from Guadalcanal that he wrote King and the Joint Chiefs of Staff. "It now appears that we are unable to control the seas in the Guadalcanal area. Thus supply of our positions will only be done at great expense to us. The situation is not hopeless, but it is certainly critical."

Marine General Vandegrift wrote Admiral Ghormley that his command must receive maximum support of air and surface units. Ghormley dispatched some of VMF-212 under Lieutenant Colonel Harold W. Bauer, along with seven SBDs. Geiger's force now grew to twenty-eight—still pitifully small considering the threat he faced. The new planes arrived just as nine Japanese Vals were attacking the seaplane tender *McFarland*. It had just delivered precious aviation fuel and was on its way out of the area with medical evacuees from Lunga Point.

Bauer was low on fuel, but he did not hesitate. He attacked the Vals and soon four of them were shot down. Now he had to leave the scene and the shocked Japanese let him depart. They thought he was a whole squadron! (Bauer was later killed in the middle of November after shooting down eleven Japanese planes. He was honored posthumously with a Medal of Honor.)

Ghormley became concerned by the increasing Japanese activity south and east of the Solomon Islands and in the vicinity of the Santa Cruz Islands. He wired Nimitz that his forces were inadequate to meet this threat.

When King was advised, his doubts about Ghormley's aggressiveness seemed to be justified. He urged Nimitz to replace him, citing the fact that he had permitted inadequate night dispositions around Guadalcanal. He claimed that they permitted the Japanese to control the area after dark, although Allied forces seemed to have some control by day. King charged that Ghormley's inefficiency and lack of aggressiveness had contributed to a series of disasters.

Nimitz had agreed to Halsey's subordinate position to Ghormley as task force commander. Now he recommended to King that Halsey replace Ghormley. King quickly agreed.

Nimitz wired Halsey, "You will take command of the South Pacific area and South Pacific forces immediately." After reading the dispatch, Halsey exclaimed, "Jesus Christ and General Jackson. This is the hottest potato they ever handed me."

King was correct about Ghormley, and Nimitz had waited too long to remove a commander who rightly or wrongly had lost the confidence of his men. His command had suffered very heavy losses because he had lost control of the situation.

Nimitz advised his Pearl Harbor staff of his action, saying, "for his effect on morale, Bill Halsey is worth a division of fast carriers."

Halsey called Vandegrift and Admiral Turner to a conference, making it clear that he intended to move fast to control the situation. He asked Vandegrift, "Can you hold?" Vandegrift replied, "Yes, I can hold, but I have to have more active support than I've been getting."

When Admiral Turner was asked to furnish greater support, he replied that the Navy had been doing all it could. He said he was losing transports and cargo ships at an alarming rate. "I need more warships to protect them." Halsey told Vandegrift, "You go on back there. I promise to get you everything I have."

Halsey's command of the fleet was greeted with enthusiasm, and morale received a sharp lift. Even more important, the *Enterprise* left Pearl Harbor for the South Pacific on October 16 with a task force under Rear Admiral Thomas C. Kinkaid. He was ordered to rendezvous with the *Hornet* group under Rear Admiral George D. Murray and report to Halsey with all speed.

Lieutenant Loberg, who had left Hawaii with the 11th Bombardment Group the 1st of July for duty in the South Pacific, now operated out of Espíritu Santo. During these hectic days, both Air Force officers and enlisted men worked together to maintain their B-17s despite limited spare parts and the lack of replacement crews and planes.

They were sent out during the early morning hours of October 23 to locate the ships that had been shelling their field, but failed to find them. After refueling, they headed out again for a five-hour flight. They found no warships, but instead ran into a Catalina pursued by a four-engine Japanese flying boat—a Kawanishi 97.

Loberg ordered his crew to battle stations and dove at the Kawanishi with forward guns firing. The Catalina gratefully took off for its home base. The huge Japanese craft momentarily disappeared in the clouds, but when it broke into the clear Loberg suddenly found himself right alongside it, not fifty feet away. As the two planes jockeyed for position, gunners on both planes fired. Loberg's B-17 shuddered with the impact. The two planes were so close that the crew of the Flying Fortress could see the Japanese gunners huddled over their machine guns, including one cannon. The B-17 was repeatedly hit, and bullets ricocheted off the cockpit armor behind Loberg and Bernays K. Thurston's seat on the right.

The Kawanishi now made a tight turn and Loberg did likewise, trying to keep away from the cannon in the Kawanishi's tail. A cloudburst momentarily halted the battle, but then it was resumed as the two planes fought one another in and out of the clouds. It became a guessing game as to where the other plane would appear. The Japanese pilot tried to keep close to the Flying Fortress to prevent it from flying underneath and raking it with its top turret. Loberg knew that the flying boat had no defensive guns underneath. The two planes rolled and made tight turns to avoid giving the other an advantage in an unprotected spot.

Loberg's navigator, Lieutenant Robert D. Spitzer, was hit by cannon fire that sprayed the cockpit, and Lieutenant Robert A. Mitchell, the bombardier, was struck by fragments from an armor-piercing shell that buried itself in his machine gun, putting it out of commission. Mitchell was dazed, his head bowed over his machine gun with a look of shock and amazement. He tried to use his gun but it refused to function. He lifted off the top cover to see what was jamming it, but the cover itself had jammed and would not release.

Mitchell crawled out of the bombardier's compartment and stood between Loberg and Thurston. Blood streamed from one eye and dripped down his chest. "My foot hurts. I can't stand on it."

Loberg was too busy to respond, moving his B-17 closer to the Japanese flying boat. He was astonished to find that its fuselage was riddled with holes. Spitzer called out, "He's smoking! One of the motors is gone!" Loberg noticed that one of the big plane's propellers was windmilling. A few moments later Spitzer yelled, "He's down!"

Loberg glanced down. He spotted the Japanese plane on the ocean's surface as smoke and flames billowed above it. The fuselage began to buckle and black objects—which had once been members of its crew—began to appear in the water. The plane soon sank, although fuel kept burning on the surface as black smoke drifted downwind.

The next day the *Hornet* and the *Enterprise* groups met northeast of New Hebrides at noon. Halsey placed Kinkaid in charge of a consolidated group he named Task Group King.

Kinkaid met with his staff, reminding them that they were in for trouble because their group was only half the size of the Japanese group. "At least," he said, "we're twice as strong as the Japanese think we are." Kinkaid ordered his new group to travel along the northern shores of the Santa Cruz Islands, and then swing east of San Cristobal, the southernmost island in the Solomons group.

At Truk, Admiral Yamamoto began to assemble ships to make an all-out effort to retake Guadalcanal. He ordered three carriers, five battleships, fourteen cruisers, and forty-four destroyers to form a fleet to fight a decisive battle with the American fleet. He told his staff that his intelligence indicated the Americans were northeast of the lower Solomons in the vicinity of two island groups—the Stewart and Santa Cruz Islands.

Meanwhile, on Guadalcanal, General Hyakutake prepared to make three simultaneous attacks at widely separated points. The main attack from the south was assigned to Lieutenant General Masao Maruyama and was established to coincide with Hyakutake's attack against Henderson Field.

The battle began on October 23. It was supported by Japanese fighters and bombers from Rabaul and Buin, a Japanese base on Bougainville. From Henderson, twenty-four American Marine and Navy Wildcats rose to contest the invaders.

Captain Joseph J. Foss of Marine Fighter Squadron 121 led his men against sixteen Japanese bombers and twenty-five fighters. One of his Wildcats soon got into trouble, with a Japanese Zero on his tail as he was preoccupied by another Zero. Foss dove for the tail-end Zero and shot it down. The Japanese plane disintegrated, and Foss had to try desperately to avoid the debris. He personally shot down four more Japanese planes that day, and his squadron totaled twenty Zeroes and four bombers. This was the first of a series of kills; Foss got another five the next day, bringing his total to sixteen. He later earned a Medal of Honor for his exploits.

In the air on October 25 the Japanese suffered more heavy losses, losing twenty-two planes to American fighters, and another to antiaircraft fire.

Three hundred miles east of Guadalcanal, Yamamoto's supporting force, including Nagumo's Third Fleet and Kondo's Second Fleet, awaited word to join the action. This site had been chosen to prevent American ships from coming to Guadalcanal's assistance, and to prevent the escape of survivors of the ground fighting.

Halsey had divided his fleet into three task forces with orders to rendezvous north of the New Hebrides. Task Force 16 had the *Enterprise,* and Task Force 17 had the *Hornet.* Task Force 64 had the battleship *South Dakota.* Rear Admiral Kinkaid, on board the *Enterprise,* was assigned as tactical commander, with 171 aircraft at his disposal.

But Nagumo had four carriers and 212 aircraft. He also had four battleships, compared to one for the Americans; twelve Japanese cruisers, versus six for the Americans; and twenty-four Japanese destroyers compared to twelve American destroyers.

At first Nagumo was overconfident, believing the *Enterprise* was still out of action and that his superior fleet would have no trouble against the much smaller American fleet. But when he did not hear from General Hyakutake that Guadalcanal had been captured, some of his confidence evaporated. He sent Hyakutake a radio message, urging him to speed up his ground victory because he was running low on fuel. Some of his confidence returned when he received a report—which later proved to be false—that Henderson Field had been overrun. During this period he refueled and ordered his fleet to proceed to Guadalcanal. Unknown to Nagumo, the Americans had turned back waves of Japanese troops in savage attacks, and were now forcing the Japanese to halt their drive. Unfortunately for the Americans, Japanese reinforcements then arrived on the island.

While fighting continued on the ground, scout planes from each fleet searched for the other. Some of Nagumo's ships were located after midnight on October 25 by a Catalina search plane. Nagumo, however, still had no idea where the American ships were cruising. But once he knew he had been discovered, he changed course, and finally one of his search planes sighted the American ships on a northwest heading. Now scout planes of each side almost simultaneously reported the positions of the opposing fleets.

Kinkaid ordered his carriers to launch their planes. "Pilots, man your planes!" came the call, and pilots rushed for their flight decks.

Lieutenant Stockton B. "Birney" Strong and Ensign Charles Irvine left their search sector when they heard a report identifying the *Shokaku* and the smaller *Zuiho.* Strong did not hesitate. He ordered his wingman to join him in an attack on the

Zuiho. There was an eerie atmosphere as they prepared to dive, with no combat patrols above the ship and no antiaircraft fire. During his dive, Strong noted with surprise that there were no aircraft on board, then realized they must all be in the air.

Strong released his bomb, and so did Irvine. They struck aft, ripping open the *Zuiho*'s deck, leaving antiaircraft guns pointed in unnatural directions after they were knocked off their mountings. Captain Sueo Chayashi radioed Nagumo that his damage was so severe that he had to leave the scene and would not be able to recover his aircraft.

Strong and Irvine left in a hurry, while dispatching by radio the details of the sighting and their bombing. Now the *Zuiho*'s fighters swarmed around them and the two pilots headed for the deck, careening away just above the wave tops. In the process they shot down two Zeroes as they fought them off for forty-five minutes before the Japanese broke off and returned to their ship. The two American planes landed safely on the *Enterprise* after their incredible feat.

Once the Japanese fleet was located, Kinkaid ordered the *Enterprise*'s Avengers, SBDs, and Wildcats to attack it. En route, they were attacked by Japanese fighters, and two Avengers were shot down. They had been flying at low altitude, which proved to be a severe disadvantage to the Wildcats, and eight of the nineteen American planes were shot down.

After the Americans fought their way out of this entanglement, they failed to find the Japanese carriers. They bombed what they reported was a battleship, but it proved to be the cruiser *Chikuma.* It suffered heavy damage and limped away to safety.

Lieutenant Commander Mamoru Seki had taken off at 5:15 A.M. with a large number of Japanese fighters and bombers. His fighters had attacked the planes from the *Enterprise,* but they were no longer able to protect the Vals and Kates due to a shortage of fuel. He ordered his bombers to continue their flights to locate the American carriers. They found the *Hornet* and raced to drop their bombs and torpedoes, but the carrier's Wildcats went after them.

Commander Lawrence L. Bean, the ship's medical officer, had gone on deck to stretch his legs after the last planes left the *Hornet.* Striding back and forth, he stopped abruptly when the loudspeaker blared, "General quarters!" He hurried to the sick bay. The phone rang, and he picked it up quickly. "Twenty-four bogies, sir. They've disappeared now in the clouds."

The *Hornet* shook under Bean's feet. He heard a loud explosion forward of the sick bay. Before his unbelieving eyes the deck lifted up. Two more blasts came, then the guns stopped firing and a deathly stillness pervaded the ship. The deck slanted beneath Bean's feet and the ship began to list. Thick smoke billowed around the deck.

Seki led his bombers in the attack on the *Hornet,* but his aircraft had careened off the carrier's stack, crashed into the flight deck, and exploded with its bombs. Other Vals continued their runs, and bombs were dropped accurately as Kates swept in low with their torpedoes. Another plane now deliberately smashed into a gun gallery and exploded in a ball of flames that enveloped the forward elevator shaft.

"Now hear this!" Captain Mason ordered: "Prepare to abandon ship."

Bean hurried to the deck just as a flaming *kamikaze* plane crashed into the communications tower. He was sickened by the sight of the Japanese pilot's body burned to a crisp sitting on deck alongside his plane in the same position he bad occupied in its cockpit.

Captain Mason ordered evasive maneuvers to avoid dive-bombers so that further attacks did no additional damage. Then two more torpedoes struck simulta-

A Japanese bomb splashes astern of a carrier during the
Battle of the Santa Cruz Islands. The intensity of the air action
can be seen by the amount of antiaircraft fire.
(Courtesy of U.S. Navy.)

neously and the *Hornet* listed fifteen degrees. Mason, realizing his ship was doomed, issued the final call: "Abandon ship!"

Men scrambled over the side as violent explosions tore the ship apart.

Nagumo's warship, the *Shokaku*, was 150 miles away when Lieutenant James F. Vose, flying with a *Hornet* formation, located it. Zeroes above the carrier fought off waves of American planes in a whirling dogfight, but four 1,000-pound bombs hit the carrier's deck and set it on fire.

More than 200 Japanese planes were involved in attacks on the American fleet. One of the first American ships to go down was the destroyer *Porter,* evacuating men from the *Hornet* after it was hit by a torpedo.

The *Hornet*'s planes were advised by radio to land on the *Enterprise* while their ship was towed away from the scene. Six Japanese torpedoes hit the crippled ship, although she still refused to sink. American destroyers were ordered to sink her but they had to fire hundreds of rounds before she finally sank.

Meanwhile, Nagumo's flagship the *Zuiho* was fleeing to Truk, and the admiral turned command over to Admiral Kakuji Kakuta on the *Zuikaku*. The air battles continued as American pilots searched for the *Zuikaku* and the *Junyo* of Kondo's advance force. But Vals and Kates located the *Enterprise* and managed to land three bombs despite the valiant efforts of the carrier's fighters. Forty-four men died in raging fires aboard the carrier, but Captain Osborne B. Hardison prevented more hits by his skillful maneuvering of the ship.

The battleship *South Dakota* shot down several planes and possibly saved the *Enterprise,* which came under repeated heavy attacks. A Kate hit the deck of the destroyer *Smith* and almost destroyed her.

The Battle of the Santa Cruz Islands ended with the Japanese claiming a great victory. They were wrong. Although the *Hornet* and the *Porter* were sunk, the Japanese had suffered serious damage to two carriers, forcing them out of action for months. They had lost sixty-nine bombers and fighters along with their experienced crews. It was becoming evident to American fliers that Japan's early veterans were being lost, and that the new pilots replacing them were not as efficient or aggressive as the early professionals.

Kinkaid's ships retired independently on a southwesterly course. His carriers had lost seventy-four planes, with twenty-three pilots and ten crewmembers. Despite his losses he had routed a much stronger fleet. Most importantly, the Japanese assault on Guadalcanal was brought to a halt. Later it became evident that the Japanese had lost their last chance to defeat the Americans on Guadalcanal.

10

Tokyo Express

The loss of the *Hornet* upset Admiral King. His carrier strength had been too thin prior to the battle, but now it was down to a dangerous level. No reinforcements were available because the Allies were preparing to make landings in North Africa. Only the crippled *Enterprise* and the *Saratoga* were available to face what was still formidable opposition in the South Pacific.

General Hyakutake's position on Guadalcanal grew progressively worse, as more of his men became sick, with many dying under the persistent American attacks. Admiral Yamamoto assembled a large force of battleships, cruisers, and destroyers from Kondo's Second Fleet. It was ordered to shell Henderson Field around the clock. He also ordered Tanaka's "Tokyo Express" to make a run to Guadalcanal, with the Japanese Army's 38th Division as much needed reinforcement for Hyakutake's battered army.

While eleven troop transports moved down the "Slot," northwest of Guadalcanal, Kondo's warships positioned themselves north of Savo Island to provide distant cover while a large force of battleships and smaller ships bombarded Henderson Field and nearby American positions. Vice Admiral Gunichi Mikawa provided protection for the troop transports. His force was equipped with only cruisers and destroyers.

Rear Admiral Daniel J. Callaghan had only five cruisers and eight destroyers to defend the Guadalcanal area. Although he was aware of these Japanese movements, and suspected their intentions, his force was totally inadequate to challenge this latest threat.

Admiral Kinkaid's carrier task force, with the *Enterprise,* returned on November 13. It had been a fast turnaround for the carrier, whose battle damage had been quickly repaired.

On the night of November 12–13, Callaghan's ships searched the Savo Island area as they entered Lengo Channel. Three groups of Japanese ships had been reported in the vicinity. Actually, Vice Admiral Hiroaki Abe was on his way through the area with the battleships *Hiei* and the *Kirishima,* the cruiser *Nagaro,* and fourteen destroyers. The unequal contest began at night as the Japanese tried to force their way through the Solomons to provide a safe passage for their troopships.

Callaghan's ships blasted the Japanese ships at close quarters and Callaghan and almost his entire staff were killed when the *San Francisco* was repeatedly shelled. The battle became a nightmarish encounter for both sides. The *Hiei* was crippled and two Japanese destroyers were sunk. Admiral Abe now was forced to cancel the Guadalcanal bombardment. Although the Japanese force outnumbered the Americans, they were turned back and almost routed. But Callaghan's command lost five ships and 700 men, and the new cruiser *Juneau* was torpedoed the following day with the loss of 664 men, including the five Sullivan brothers.

When Yamamoto learned that Abe was retreating from the battle, the news so infuriated him that he relieved him of command. Tanaka was ordered to turn back from his express run to Guadalcanal and head for the Shortland Islands.

Callaghan's remaining ships were also in full retreat, as Kinkaid's task force rushed to cover them.

Japanese troopships, which had been ordered to turn back, received orders from Tokyo to turn around and proceed with their original plans to land their troops on Guadalcanal.

The *Hiei* was still ten miles north of Savo with a five-destroyer escort when Marine pilots in SBDs and Avengers found her. Two Avengers released torpedoes into her side while SBDs roared down in dives to release their bombs on the stricken ship.

The word went out to sink the *Hiei,* and throughout the day Navy and Marine aircraft from Henderson Field went back and forth. Eight Zero fighters that tried to protect the *Hiei* were shot down. Finally the huge *Hiei* could take no more punishment, and it sank beneath the waves. It was the first Japanese battleship to be sunk by the Americans.

Yamamoto reinstituted his original plan to send in warships on the night of November 13–14 to shell Henderson. Cruisers from Rabaul fired a thousand eight-inch shells at the airfield, but their fire was so erratic that little damage was done. Japanese ships were constantly harassed by PT boats from Tulagi.

The Guadalcanal command sought air assistance from MacArthur, but he sent only eight P-38s to Henderson. He could have furnished many more planes, but he did not choose to do so. His action embittered Marines, who charged that the general was deliberately withholding equipment and planes despite their desperate situation. MacArthur earned the moniker "dugout Doug," which would stick throughout the war and afterward.

Tanaka believed that his Tokyo Express could now bring his troopships to Guadalcanal, telling aides the Americans were powerless to stop him. He sent 11,000 troops of Lieutenant General Tadayoshi Sano's 38th Division through the Slot on the night of November 14. The light carrier *Hiyo,* a few Zeroes, and a dozen destroyers provided a screening force for added protection.

Lieutenant Doan "Big Red" Carmody, in an *Enterprise* scout plane, reported the ships' presence in the Slot 120 miles distant and cruising toward Guadalcanal. (Elements of Air Group 10 from "Big E" were temporarily based on Henderson Field.) After Carmody made his report, he joined Lieutenant W.E. Johnson in an attack on the transports. They were opposed by Zeroes, and both pilots missed their targets.

Now every plane in commission from Henderson Field and the *Enterprise* went out to attack the transports. Admiral Kondo kept his two carriers to the north, ordering only a few Zeroes to take part, and then only if the situation became desperate.

Navy and Marine pilots dropped their 1,000-pound bombs on the vulnerable transports without opposition, and torpedoes slammed into the ships' hulls. Wildcats quickly eliminated the few Zeroes and fired into the crowded decks. The slaughter sickened the pilots as troopships sank beneath them. And Tanaka still refused to reverse his course. Even when his transports lay dead in the water, their decks aflame, with dead and injured men tumbling out of their sides ripped open by torpedoes, and destroyers frantically trying to save some of the men, he still refused to concede defeat.

American pilots choked down their emotions as they machine-gunned the survivors without mercy. They were determined that these Japanese soldiers would not get to Guadalcanal and provide reinforcement for their own beleaguered troops.

The *Enterprise*'s VS-10 was the last squadron to leave Henderson, as they were led to the scene by Lieutenant George G. Estes. To their surprise they found that four transports had survived the carnage and were now beached, disgorging their troops. Three thousand Japanese soldiers tried desperately to reach shore through the churning waters. At least that many had already died, and another 2,000 were rescued but failed to reach land. Some troops got ashore, but only 260 cases of ammunition and 1,500 bags of rice were landed. All other supplies were destroyed, including the transports, which were bombed the next day.

Kondo's action in remaining insensitive to the plight of his countrymen was a disgrace, and he was relieved by Yamamoto.

The previous five days had been among the bloodiest of the war, and the Japanese had lost two battleships, nine heavy cruisers, four destroyers, and twelve transports and cargo ships. American losses included two light cruisers, seven destroyers, and damage to a battleship. Incredibly, only six SBDs and two Wildcats were lost.

Halsey was ecstatic. He sent a message to his command. "Your names have been written in gold letters on the pages of history, and you have won the everlasting gratitude of your countrymen. My pride in you is beyond expression."

Admiral King was equally pleased—and relieved. He had feared the worst. He asked Congress to promote Halsey to four-star rank, and his request was approved.

While Japanese leaders debated what new steps to take, their soldiers on Guadalcanal felt betrayed and abandoned. They had lost heavily since the Americans invaded the island in August, and—unlike the American high command—the Japanese leadership had not found ways to reinforce its troops. The Japanese commanders realized that any further attempts to reinforce their troops would undoubtedly result in even worse losses.

Although it was not soon apparent, the Japanese had threatened the eastern Solomons for the last time. The great goal of occupying Australia and cutting the lifeline between it and the United States had ended in abysmal failure. The Japanese needed a respite, but so did the Pacific Fleet. Many of its warships needed extensive repairs, and its crews badly needed rest.

The Japanese made one final attempt to reinforce Guadalcanal on November 30, but it was thwarted by Task Force William, under Rear Admiral Carleton H. Wright. In the process, Wright lost a heavy cruiser, and three others were seriously damaged.

On December 31, 1942, Admiral Osami Nagano, chief of the Naval Staff, and General Hajime Sugiyama asked for an audience with Emperor Hirohito. It was granted. They made their apologies for their failure to retake Guadalcanal and asked permission to abandon the island. The Emperor granted it.

During the first week of February 1943, the Japanese secretly withdrew their forces, and Allied airpower was too meager to detect their departure. During a sweep of the island to the west by Major General Alexander E. Patch's Army troops on February 9, they were surprised to learn that the Japanese had left. Everywhere there were signs of a hasty departure.

The Papuan campaign had been concluded three weeks earlier on January 2 by Lieutenant General Robert I. Eichelberger's command.

By late January almost all the fighting was over in eastern New Guinea, but it had taken 30,000 troops—half of them Australian—to dislodge 12,000 Japanese. Guadalcanal had been won with far fewer troops under much worse combat conditions because airpower had proved so decisive. Although Allied airmen were consistently outnumbered in the air, the skill and tenacity of Air Force, Navy, and Marine airmen made the difference. The United States Marines on the ground were superbly led, and they and their Army comrades who joined them later achieved victory in October during some of the fiercest battles ever fought. Teamwork between all of America's services brought about the decisive defeat of the Japanese, who until now had thought themselves invincible. It was a source of pride that the Americans had proved to themselves they could outfight the Japanese at any time and any place. Midway had been the turning point of the war, and now Guadalcanal marked another major milestone.

The geographic position of the United States put it squarely between two wars that had little in common. Air operations on the Atlantic side—except for participation in three amphibious operations—were essentially used in blockade operations and to protect ships delivering raw materials to America's factories and war munitions and reinforcements to its allies in Europe and North Africa. In the Pacific, it had been a matter of halting an enemy advance that, in a few short months, had spread over all of the Western and parts of the Southern and Central Pacific, and then carrying out the bitterly contested task of driving the Japanese homeward across the broad expanse of an island-dotted sea.

In the Atlantic, the desperate struggle to reduce losses due to U-boat operations slowly made headway because of a combination of factors. The British Navy had pioneered techniques against the German submarine menace and the U.S. Navy adopted them, along with the naval innovations of other Allied nations.

The British had learned that control of the Atlantic shipping lanes could be regained through the use of sonar, the escort carrier, and night-flying airplanes. These techniques underwent further refinement and the U-boat's defeat was credited finally to a combination of escort carriers, destroyers, and destroyer escorts. After these ships were equipped with sonar, early detection of a submarine—plus aggressive action—proved the final answer to the problem.

During the second month of the United States' entry into the war, German submarines sank more than a ship a day. One month later, in February 1942, the losses doubled, and again increased at an alarming rate in March.

At the start of the war, the USS *Long Island* was the only escort carrier in the Navy. When escort carriers were ordered into mass production in May of 1942, there was no time to build them from the keel up. Merchant ships and tanker hulls proved satisfactory to the task, and when these converted ships became available in sufficient numbers, the impact on ship losses was considerable. Allied losses to submarines in 1943 were forty percent less than in 1942.

While the fight to keep the shipping lanes open continued during the spring of 1942, Task Force 99 made a dangerous trip through the Mediterranean. The carrier *Wasp,* under Captain W. Reeves Jr., delivered more than forty Spitfires and RAF pilots to beleaguered Malta through waters that were almost a fascist lake. Once the mission was performed with success, it was repeated.

Prime Minister Winston Churchill promptly conveyed the appreciation of the British Government when he wired, "To the captain, and ship's company of the USS *Wasp.* Many thanks to you all for the timely help. Who said a Wasp couldn't sting twice?"

While Generals Bernard Montgomery and Erwin Rommel fought one another across North Africa, each side gaining a temporary advantage, Malta became a dangerous "thorn in the side" of Axis supply lines. When Tobruk fell on June 21, 1942, and the British army dug in at El Alamein for a last stand only seventy-five miles from Alexandria, Malta was the last Allied base in the Eastern Mediterranean. Therefore it took on renewed significance. The new squadrons of British fighters proved vital to the defense of the area.

Before major operations could proceed to recapture the Axis-dominated lands, the long sea-lanes to England and to Russia's Arctic had to be made safe. The story of the men, ships, and planes who succeeded finally in reducing the German undersea fleet is a lengthy one of courage and perseverance. Volumes could be written about their achievements, but a few examples will illustrate the depth of their activities.

Fleet Air Wing 7 was the first to fly the neutrality patrol before America's entry into the war. Despite vicious Atlantic storms and heavy fog, they flew daily in their Catalinas and Hudsons, guarding convoys and hunting U-boat packs.

The U.S. Navy insisted on positive evidence before it would credit a victory. Captain Dan Gallery, Commander of the Fleet Air Base in Iceland, told his fliers, "Bring home the skipper's pants. That's the only way we can credit a claim." Although he spoke in jest, there was almost enough truth in his comment to make the crews somewhat resentful after they had been denied several claims.

A few days later, Lieutenant Robert Hopgood briefed his PBY-5A crew before takeoff. He said finally, "We will pick up a convoy heading seaward for 'torpedo junction.'" The ocean between Iceland and the United States had earned this dubious distinction. The patrol plane rode 500 feet off the water while rain squalls lashed at her windshield and a forty-knot wind bumped them around. Then Hopgood saw a ship ahead of them.

Copilot Lieutenant Bradford "Tex" Dyer took a quick look. "Must be the lead destroyer of the convoy." Hopgood peered through the rain. "It's a submarine!" he shouted. Hopgood headed for it, noticing as they drew closer that there were men on the conning tower. He dropped their depth charges and watched them straddle the sub, with many hitting squarely.

"She's badly damaged," Dyer said. Hopgood watched the U-boat blowing out compressed air and leaking oil while men rushed frantically on deck. Hopgood pulled away quickly as he saw holes appearing in the wings of his flying boat. He kept the submarine in sight for two hours while remaining out of range of its guns.

When the German submarine pulled up to an Icelandic ship, Hopgood watched in astonishment as the German crew boarded it and then tried to ram the submarine's bow. "Must be in danger of sinking," Hopgood told Dyer.

A Lockheed Hudson bomber and patrol aircraft. (Courtesy of Lockheed.)

Flak burst around them, so Hopgood swung to a safer distance until a British destroyer charged up to the scene as the Germans tried to scuttle the submarine. They were too late, and the destroyer picked up the German crew and captured the submarine.

Dyer blinker-signaled to the destroyer's bridge. "How many prisoners?"

"Fifty-two" they blinkered back. "Nice work. Now find one for us."

"We haven't got the skipper's pants," Hopgood said with a grin, "but we've got the skipper. That should be even better."

"It better be!" Dyer said, remembering the times when they had been denied a claim.

The USS *Bogue* was one of the first escort carriers to see service in the Atlantic as it operated with the British to escort convoys from the vicinity of Newfoundland to the western approaches to Iceland. It was monotonous and dangerous work week after week, but these escort carriers performed a vital mission with their Grumman Wildcats and Avengers. When the final total was in, 500 submarines had been sunk in the Atlantic, and the U.S. Navy verified 126 of them. Without this mass destruction of U-boats, there would have been no invasion of Europe, because even in 1943 Germany had 450 submarines available to destroy troop and supply ships.

The *Bogue* was not only one of the first escort carriers to take part in the deadly battle, but it also won a Presidential Unit Citation for sinking sixteen German U-boats.

Although the battle against the U-boats had not been won by the fall of 1942, the invasion of Africa to clear Axis forces out of the rest of North Africa was set for November 8. In a three-pronged amphibious operation, landings were made on the coast of French Morocco while American and British troops, escorted by the Royal Navy, landed near Oran and Algiers. They were supported by the American carrier *Ranger* and the escort carriers *Sangamon,* the *Suwannee,* and the *Santee,* in limited but effective operations.

The escort carrier (CVE) *Card* came in for extensive action in the Atlantic during the summer of 1943 when it operated offensively against a U-boat concentration eighty miles northwest of the Azores. Its work was typical of other CVEs in the fight against German U-boats.

Lieutenant A.H. Sallenger in his TBM picked up two large submarines, running 500 yards apart for mutual protection. His fighter escort radioed he was having engine trouble, so Sallenger found himself all alone in the attack. He picked out the closest sub and dropped two depth bombs for a perfect straddle. He watched the submarine wallow, twisting awkwardly in the water. He reported to his ship, "Both subs on surface. Second one trying to assist the first."

Captain A.J. Isbell, commander, Task Group 21.14, ordered, "Vector three planes to the area." The *Card* was only seventy-eight miles away, so the additional torpedo bombers arrived shortly as the undamaged submarine submerged.

Sallenger immediately attacked while the other planes sought the first submarine. Sallenger called on the radio, "Subs have all the guns aboard we've been hearing about. There's no question about it, they don't know how to use them. Their firing is rotten. It was all around me but no hits."

During the night, the *Card* remained north of the concentration, so the next day it was 150 miles from the attack zone. Sallenger had the rear starboard sector and Ensign J.P. Sprague was along in his fighter plane.

When Sallenger hadn't checked in after an hour and a half, Captain Isbell became concerned. After three hours with no word, everyone was upset. Isbell delayed launching additional planes until they could try to reach the two planes by radio. He feared they had either crashed into one another or had been shot down.

Isbell was relieved when a search plane spotted Sallenger and his gunner in a rubber lifeboat in the midst of a huge oil slick.

"Send the *Goff* over to pick them up," he ordered. Sallenger immediately called the captain following his rescue. "The submarine is here," he said. "You'd better get out."

"Thank you," the captain said. "We're already rendezvousing."

Sallenger later related what had happened. "I'd been cruising along at 800 feet with less than a mile visibility on the return leg to the ship, when I picked up the two submarines on the port bow. Sprague went in first to strafe while I swung over them. Both subs were firing and just as I approached, two shells hit in the bomb bay. One completely knocked out all the electrical gear and the radio equipment. I figured I was in a perfect position to release, but the bombs failed to drop. I did a fast turn and came back to use the emergency, but then my right wing was hit and it caught fire. I managed this time to drop two depth charges and got a perfect straddle."

Sallenger described how he drew away with his plane in flames and landed three miles away in the ocean. He and his gunner got out and released their life raft. He had a chance to watch Sprague strafe the submarine, but then he lost sight of the fighter.

When Sallenger learned that his radioman was still below in his compartment, he dove through the bottom of the plane and tried to drag him out. He worked quickly, but he could not reach the man's body. When the plane started to sink, he had to get out hurriedly. He almost drowned because the airplane was twenty feet below the surface when he freed himself.

The captain asked him, "Still think the sub's gunnery is lousy?"

"No, sir! I'd say they're quite effective with those guns."

The next day other escort carriers sank the troublesome submarine a hundred miles to the south. When it was spotted still on the surface, a 500-pound bomb made a near miss and scattered the gun crews. Then depth charges cracked the pressure hull in the engine room and the submarine surrendered. Forty-four survivors were picked up, and the U-boat captain said the other submarine undoubtedly had gone to the bottom during the first day's attack.

In 1943 the strength of the Pacific Fleet grew daily, as Halsey acquired the largest number of ships he had ever had. The Fleet consisted of three new battleships, *the Washington,* the *North Carolina,* and the *Indiana;* plus four old ones; the carriers *Enterprise* and *Saratoga;* and the auxiliary carriers the *Chenango,* the *Suwannee,* and the *Sangamon,* which added their potency to the available striking forces.

While the battle for Guadalcanal was waged, the Japanese continued to build air bases in the Solomons, including one at New Georgia's Munda Point and others on the smaller islands of Kolombangara and Vella Lavella. To the north, Bougainville had five airfields, with the largest, at Kahiki, on the island's southern tip.

John H. Towers was promoted to vice admiral in October 1942, and was named Commander of Air Operations in the Pacific under Nimitz. Due to a conflict of personalities with King that went back years, Towers had been deliberately prevented by King from getting a top command in the Pacific. He had been kept in a subordinate position as chief of the Bureau of Aeronautics for years.

Fortunately, Nimitz, a former submariner with no aviation experience, was well aware of Towers's brilliance as an air strategist, and he persuaded King that he needed Towers. King reluctantly decided that Towers's undoubted talent was desperately needed. As Nimitz sought to emulate the early successes of Japanese carrier aviation, and hopefully surpass them by using carriers to spearhead Pacific operations, he turned to Towers to develop the necessary strategy.

It was an almost impossible situation for Towers—the Navy's most distinguished airman—but he dutifully served without rancor under Nimitz. His memorandum to his boss called for the development of a carrier force to dominate Navy actions in the Pacific. But he could only suggest, and he was never given any authority to command an operation until the war was almost over. King absolutely refused to permit Nimitz to give him that responsibility until 1945. Despite King, Towers was in a position—as Nimitz's deputy commander for air operations—to recommend aviators whom he had known and often trained through the years for top command positions. Therefore he exerted a special influence on all of Nimitz's operations because the Commander in Chief, Pacific, held Towers in such high esteem that he followed his advice in almost every instance.

Towers was the third naval aviator to learn to fly prior to World War I, and he had long resisted the theory that battleships should form the line of battle, with carriers

John H. Towers when he was a rear admiral.
(Courtesy of U.S. Navy.)

off to one side to support the main battle line. He had waged a continuing fight for aviation since the days when Navy commanders believed airplanes were useful only for reconnaissance and spotting gunfire. He had long preached the doctrine that airplanes were the Fleet's main striking force, but only the early Japanese successes of its carriers had forced a change in top-level Navy strategy. King and Nimitz were now believers, but without any knowledge of how to develop a vital new strategy. Fortunately Nimitz backed Towers, giving him the authority to develop his theories for use of fast carriers in task groups, which Towers had long advocated.

What Towers now conceived was a task group with four carriers—with approximately 224 airplanes in each group—as the core for all major operations. Battleships, cruisers, and destroyers were relegated to the status of support ships. This concept would have been considered heresy just a few years earlier, but it now proved to be an unbeatable concept, although it would take time before new carriers could be built to prove its maximum effectiveness.

King still had doubts about an airman in command of a carrier task force at sea; he believed such officers were specialists, when what was needed was a man with broader command experience. King trusted few men in such positions. These included Halsey, Marc A. Mitscher, and John McCain. With the exception of Towers, other flag officers were too young and inexperienced.

In retrospect, if Towers had been given a major command at sea, there is no doubt he would have ranked with Halsey, Spruance, or Mitscher—but throughout most of the war King refused to give him that chance. It was shameful on King's part to let his personal animosity overrule his better judgment. But this quiet, outstanding airman served the Navy and the nation to the best of his ability, giving Nimitz the benefit of his brilliant mind, as carriers were developed as primary attack forces.

When the Allied leaders met at Casablanca on January 14, 1943, they agreed to General Marshall's proposal that the available men, supplies, and equipment for Europe and the Pacific theater be divided between the two major war fronts at a ratio of seventy to thirty. It was also agreed that the war in the Atlantic should have top priority to reduce the enormous shipping losses, and that a Combined Bomber Offensive would be initiated against Germany. The Americans reluctantly agreed to the British proposal to invade Sicily and then Italy. Admiral King assured Churchill that he would find the necessary warships to support the invasion. The proposed 1943 cross-channel invasion, which Churchill still opposed, was postponed. The Americans agreed with the postponement if Churchill would approve an intensification of the war in the Pacific, which he had opposed. The British Prime Minister had insisted that only a holding action should be maintained in the Pacific.

King fought for permission to increase the pace of the Pacific war and to go on the offensive. Although Churchill never gave in, the British chiefs of staff voiced their approval.

At the end of the conference, agreement was reached to continue Pacific operations with forces already committed to the region, to keep up the pressure against the Japanese, and to achieve a position of readiness for an eventual full-scale offensive against Japan once Germany was defeated.

King used one phrase in the agreement to suit his own ends—that adequate forces should be allocated to the Pacific and Far Eastern theater. He used it as an excuse to give Nimitz anything he needed to support his activities in the Pacific. As soon as he returned to Washington from Casablanca, King sought support from General Marshall and General Arnold.

Now that the pace of operations in the South Pacific had been reduced, men who had experienced heavy combat over an extended period of time were rotated home as replacements became available.

The F4U-1 Corsair made its first appearance in the theater on February 12, 1943, when VMF-124 arrived for duty. The inverted-gull-winged fighter quickly de-

An F4U-1D with VMF 212 in 1944.
(Courtesy of Ling Temco Vought.)

veloped a bad reputation. On carriers, during a three-point landing to engage the arresting wire, the nose remained so high that the pilot's visibility was drastically reduced. The plane's oleo gear also proved too stiff, causing the already nose-high airplane to rebound into the air. There were accidents involving planes and deck crews until the tail hook was lengthened and the oleo gear's stiffness was reduced. Until these features were corrected, the Corsair served largely with the Marines on shore. It was faster than any Japanese plane, and had a longer range than other American fighters.

★ ★ ★

At Yamamoto's Rabaul headquarters, he personally approved a plan called "I-go Sakusen" and scheduled it for early April.

Now that Nagumo had been removed from command, Vice Admiral Jisaburo Ozawa was placed in charge of the First Carrier Division. He was ordered to strip the *Zuikaku* and the *Zuiho* of planes and fly them to land bases, and come to Rabaul. This order was given because Yamamoto feared American carriers and land-based planes might destroy his dwindling carrier strength. Planes from the Second Carrier Division's *Junyo* and the *Hiyo* were ordered flown to a base at Ballale, south of

Bougainville. Yamamoto, with 350 planes available, now called for an attack April 7 against Guadalcanal. Several American ships were sunk off Guadalcanal and Tulagi, although the Japanese attack was repulsed. Japanese plane losses were heavy. Thirty Zekes and Vals were shot down, with the loss of only seven American planes.

A secret message was sent in code that Admiral Yamamoto would personally inspect bases at Ballale, Shortland, and Buin on April 18. The details were precise, and when the American command decoded it, it was decided to intercept Yamamoto's plane and shoot it down. Captain Thomas G. Lanphier Jr., assigned to Major John W. Mitchell's 339th Fighter Squadron, intercepted Yamamoto's transport and its fighter escort. The transport was shot down, along with another transport carrying high Japanese officials.

The Japanese high command was so shocked by Yamamoto's death that they did not immediately announce it to their people. When his ashes were returned to Japan, the media reported that "while directing general strategy on the front lines in April of this year, engaged in combat with the enemy and met gallant death in a warplane."

The Americans purposefully kept silent, not wanting to tip off the Japanese that their military code had been broken.

Yamamoto was given a state funeral, and his death shocked the Japanese people, who had always revered him. A veteran admiral with little imagination and ability, Mineichi Koga, was appointed to succeed him.

In the Aleutians, the Japanese who occupied Attu and Kiska were constantly bombed by the Navy's Patrol Wing 4 and the 11th Air Force, making their lives truly unbearable. The situation was resolved in May of 1943 when the 7th Infantry Division invaded Attu and overwhelmed the Japanese defenders in hand-to-hand combat. After the outnumbered Japanese made their final charge, the survivors held hand grenades to their heads and killed themselves.

In the middle of August another invasion force headed for Kiska, but the Japanese had departed on July 28. Their withdrawal was credited to their need to defend the Kuriles.

Allied leaders met in Washington between May 12 and 25 at the Trident Conference. Admiral King made a proposal that they consider plans to retake the Philippines. He cited three methods: a straight approach from Hawaii through the Central Pacific, or a deviation north or south of that line. His personal preference, he said, was a drive through the Central Pacific. He admitted that the Japanese Fleet would first have to be destroyed, and the Marianas taken, which he called the key to success of the entire strategy. He claimed these islands were strategically located to attack the Philippines, the China coast, or northwest directly at Japan. Destroy the Japanese Fleet, he said, and the Pacific would belong to the Allies.

He called Japan still dangerous and unpredictable, but he said he had been surprised that she had not been as aggressive in the past months as he had expected. He insisted again that the time was right to accelerate the Pacific war.

The British chiefs were impressed by King's arguments, and for the first time agreed to seriously consider his proposals. Roosevelt and Churchill approved his suggestions, and he was given written authority to initiate his master plan for Japan's defeat with a drive through the Central Pacific.

General Marshall proposed that Nimitz be named Supreme Commander in the Pacific, but King refused to consider such a change. Although King appreciated Nimitz's talents, he strongly believed that he was too inclined to compromise with the Army. Perhaps his objection was more personal. As Supreme Commander Nimitz would report directly to the Joint Chiefs of Staff and not to King.

General MacArthur wrote Marshall to protest King's proposal for a Central Pacific drive, objecting to the transfer of his forces to Nimitz, including two Marine divisions under his command. Instead of a drive through the Central Pacific, MacArthur said, all resources should be sent to him so that he could capture Rabaul, which he called "the great strategic prize."

MacArthur had raised a fundamental principle, and the Joint Chiefs' support for King's plan suffered as a result. A compromise was reached that ordered MacArthur to bypass Rabaul and continue his advance along the northern coast of New Guinea. He was advised that he would only have to release one Marine division—the 2nd—for Nimitz's Central Pacific drive. Planners had recommended that Marshall release the 27th Infantry Division to Nimitz so MacArthur could retain the 1st Marine Division. The plan was approved when it became evident that additional forces were now available, with a growing abundance of supplies to fight the Pacific war using two separate routes to the Philippines.

General Arnold agreed to provide additional Air Force groups in order to permit Nimitz to seize the Gilbert Islands on November 1, 1943. The Marshalls had been considered first, but bases on Tarawa and Makin in the Gilberts were needed to support an invasion of the Marshall Islands.

The Joint Chiefs of Staff met in Washington on June 3 to study a detailed basic operational plan for Admiral Nimitz. It called for landings on Rendova Island by the end of the month, and on New Guinea at Viru Harbor, Segi Point, and Wickham Anchorage. Although the landing at Viru was delayed a day, the other landings proceeded on schedule and the whole operation was completed on August 5.

The Japanese were on the defensive and they knew it. They could only reinforce their garrisons ahead of the Americans and fight a war of attrition. They failed to realize the tenacity of the Allies and the tremendous pace of American industry, which was now producing war materiel at an unprecedented rate, as they tried to convince themselves that the Americans could not afford to risk their naval power at the rate it had been used up in the Solomons.

America's mass production techniques now showed their greatest strength. Inasmuch as it usually took five years from the start of design of a new aircraft until it was delivered to the Fleet, Navy officials for the most part "froze" all aircraft designs. They concentrated production on aircraft that were in the design and development stage during the 1930s in order to get the maximum number of airplanes into service at the earliest possible time. These airplanes were improved throughout the war with protective armor, self-sealing fuel tanks, and more horsepower for their engines to permit greater combat loads.

There was only one adverse effect from this decision to freeze design work. Jet engine development had to be turned over to the British so the United States could remain the "Arsenal of Democracy." This was necessary because production of reciprocating engines remained critical until near the end of the war. As a result, American postwar development of jet engines was delayed.

The first naval aircraft to be equipped with a turbojet engine was authorized on January 7, 1943, with the issuance of a letter of intent to McDonnell Aircraft Corporation for engineering development of two VF airplanes. Two Westinghouse turbojet engines were later specified, and the aircraft was designed as the XFD-1. It was not particularly successful, but years later it served as the prototype for the Navy's FH-1 Phantom fighter.

The Joint Chiefs were now more convinced than ever that an island-by-island campaign in the Pacific would needlessly prolong the war. In effect, MacArthur's theater of operations was further downgraded, and he was ordered to specify his future operations so forces could be allocated between himself and Nimitz.

After King received a formal directive on July 20, 1943, to open his proposed Central Pacific drive, he recommended that the attack on the Gilberts be moved to November 14 to coincide with MacArthur's plans for advancement along the New Guinea coast.

After the capture of Munda and the consolidation of New Georgia, Kolombangara with its 10,000 Japanese troops was bypassed. Vella Lavella was needed to establish another major airfield, so it was successfully invaded on August 14.

The Fifth Fleet was organized on August 5, 1943, under Vice Admiral Raymond A. Spruance. It was scheduled to increased to eleven carriers by late fall, including the new heavy carriers *Lexington* and the *Yorktown* to replace the older ones, and the newly commissioned *Essex* and the *Bunker Hill*. The old carriers *Saratoga* and the *Enterprise* joined the Fifth, along with five new light carriers—the *Independence,* the *Princeton,* the *Belleau Wood,* the *Monterey,* and the *Cowpens.* Spruance was also given eight escort carriers to provide close support of ground troops. With all carriers fully manned, Spruance would have 900 combat aircraft when he began operations against the Gilberts, Marshalls, and Marianas.

After Spruance's abilities were demonstrated in the Battle of Midway, he had been assigned as Nimitz's chief of staff. Now he was placed in command of the Central Pacific Forces under Nimitz. King wholeheartedly approved of Spruance's promotion, convinced that he was the best flag officer in the Navy. With King's backing, Spruance easily won his third star.

The brilliant tactician Rear Admiral Richmond Kelly Turner was named Amphibious Commander. Nimitz and King had early recognized his ability.

Lieutenant General Holland M. Smith was considered by King as the top expert in the field of amphibious landings, although Nimitz was not as enthusiastic about him. Smith was put in charge of the landing, but his belligerent nature and tendency to quarrel with others were marks against him.

Bougainville, gateway to Rabaul, was the last Japanese stronghold in the Solomons. It was invaded on November 1. The operation was hastened by an all-out air attack by planes from Admiral Sherman's carriers the *Saratoga* and the *Princeton.*

After the Marines waded ashore at Empress Augusta Bay, Admiral Halsey, Commander of the South Pacific area, became concerned that the Japanese might strike from their base at Rabaul, imperiling the Marines on the beachhead. He had received inconclusive evidence that Vice Admiral Kurita's Second Fleet had left Truk for Rabaul. As he studied the situation, it seemed conceivable that Kurita might head for Bougainville.

While the majority of the American forces were in the Central Pacific under Spruance preparing for attacks against the Gilberts, Halsey had only two carriers available. The old *Saratoga* and the smaller *Princeton,* plus their escorting warships, under Rear Admiral Frederick C. Sherman, would have to be used.

Halsey flashed word to Sherman aboard the *Saratoga,* "Attack Rabaul!" After the task force proceeded to launch positions on November 5, 1943, the scouts reported, "Simpson Harbor is jammed with ships refueling."

"It's a perfect setup," Sherman told his staff.

The *"Sara"* launched first, then the *Princeton.* As they neared Rabaul a hundred Japanese fighters bore down upon them, and the carriers' Hellcats fought savagely to protect the bombers.

The fact that Rabaul Harbor was surrounded by mountainous terrain forced the torpedo bombers to come down to masthead height. The TBF Avengers roared across the water barely thirty feet off the surface and headed straight for a group of ships. After their "fish" were released, they climbed steeply to clear the hills. They kept their throttles to the fire wall and their engines labored mightily under maximum power.

The horseshoe-shaped harbor was covered with antiaircraft sites on both sides, and the torpedo planes took a beating. The scene was one of confusion as planes twisted high in the sky evading Japanese Zeroes, while others dove steeply toward the crowded harbor. When it was over, nearly half of Kurita's ships were hit, although none was sunk. Ten American planes failed to return to their carriers.

Halsey wired Nimitz that he wanted to strike Truk next, but his request was denied.

A week later Admiral Alfred E. Montgomery's carrier group turned from the Central Pacific and stood by at Bougainville to join Admiral Sherman's group for another attack on Rabaul. Commander Emrick of Air Group 9 aboard the *Essex* warned his pilots, "Teamwork is the keynote. I don't want any 'burning of the blue' for personal glory."

This time bad weather so hindered operations that only limited damage was done to the ships and facilities at Simpson Harbor. But Admiral Kurita's fleet had been so badly damaged by the first attack that he had to reject plans to attack the Marines at Bougainville.

The Japanese on New Guinea found themselves in the uncomfortable position of being outflanked. While Japanese commanders wondered when their turn would come, Allied commanders turned toward the Central Pacific as a more lucrative battleground for future operations. It had remained quiet for almost a year, possibly due to Yamamoto's death. It was more likely, however, that heavy Japanese carrier losses had spread their navy too thin throughout the broad reaches of the Pacific Ocean.

One technical problem was finally solved when American torpedoes were equipped with new firing pin mechanisms to assure detonation. Since the war began many American torpedoes had either exploded prematurely or not at all. Pilots and submarine commanders had protested loudly about the situation. The early Mark 6 exploder in fleet-type submarines was activated by contact or by magnetic detonation. Extensive tests had been run and it was at last discovered that these early torpedoes were running eleven feet deeper than they were set.

While Halsey was moving ahead in the Solomons and MacArthur was conducting a drive along the New Guinea coast, the Joint Chiefs approached the Pacific war from a wholly new viewpoint. Submarines were proving remarkably successful in sinking Japanese ships, and the earlier strategy of an island-by-island campaign seemed to be needlessly prolonging the war and increasing casualties.

11

A Basic Strategy

The Joint Chiefs prepared to attend the Quadrant Conference at Quebec from late August into September of 1943 convinced that the execution of the Pacific war should be speeded up. Before they left Washington, King told his fellow chiefs that the Pacific theater was receiving only a fifteen percent allocation of America's men and materiel. He claimed this had put American forces six months behind schedule because it was only half the thirty percent allotment promised earlier by the Combined Chiefs.

The Joint Chiefs met with President Roosevelt on August 14 prepared to agree on a basic strategy. They agreed that a cross-channel invasion of Europe should be held in 1944, which Churchill had indicated he would oppose. They agreed also that the British should be urged to provide continuing support for the China–Burma–India theater to prevent the Japanese from transferring their forces from there to the Pacific theater.

In Europe, Sicily had been successfully secured by invasion on July 10, and the government of Italy was obviously in a state of disarray. Churchill pressed for an invasion of the Italian mainland despite vigorous opposition from the Americans. The British prime minister sought such an invasion, he said, so the Allies could occupy Central Europe before the Russians did.

Roosevelt and the Joint Chiefs sharply disagreed with Churchill that an invasion of Italy was advisable for any reason. King charged that it would deprive American Pacific forces of the means to carry the war to Japan. He called for a master plan to defeat Japan, and he received the support of the American delegation.

A unified statement was approved on August 16, generally along the lines suggested by the Americans, that an invasion of France should take place on May 1, 1944, and that it would have priority over the Mediterranean theater, although an invasion of Italy was approved. The final agreement approved by the Combined Chiefs called for a concurrent invasion of southern France to coincide with the invasion of Normandy. It was agreed that operations against Japan would be intensified within twelve months of Germany's defeat.

It was a compromise, but to King's disgust it perpetuated the secondary nature of the war in the Pacific. He did get permission to invade the Gilberts, Marshalls,

169

Carolines, Palaus, and Marianas. MacArthur was ordered to bypass Rabaul and move into northwest New Guinea.

King was secretly pleased that he had won more than he had hoped to achieve at the conference. Regardless of the directives approved there, he went ahead with his own master plan to defeat Japan.

Around the world the Allies were achieving successes. Although the Mediterranean theater had been downgraded, Salerno was scheduled for invasion on September 9. In the Pacific, MacArthur's men were advancing in New Guinea, and Nimitz was preparing to invade Tarawa. The war in the Atlantic was slowly being won, permitting an increasing quantity of supplies to reach Europe. Allied bombers were becoming increasingly successful as their numbers increased, and Germany was being bombed around-the-clock. On the Eastern Front, Russia had stopped the German armies and was preparing a counterattack.

Admiral Towers kept insisting to Nimitz that the defensive use of American carriers was self-defeating. He recommended that all task force commanders in the Pacific be headed by an aviator. King and Nimitz refused. Towers's campaign to change the status quo gained increasing acceptance by almost all airmen of stature, and Nimitz bore the brunt of their criticism. He blamed Towers for instigating opposition to his plans. Captain Forrest P. "Ted" Sherman, Towers's chief of staff, intervened. He recommended that a letter be drafted for Nimitz's signature to quiet the uproar. The letter acknowledged the aviators' views, and stated that aviation units should be run by commanders with extensive aviation experience or by commanders whose staff members had such experience. The emphasis was placed on integration of air and nonair commands in preference to the wholesale removal of present commanders and their replacement by airmen. Nimitz agreed with these views, but he reminded Towers that there was an insufficient number of qualified officers of flag rank to suddenly replace all task force commanders in the Pacific Fleet.

Towers replied in a memorandum on October 4, saying that aviation in the Pacific Fleet was sound but that he recommended that a deputy commander in chief, with a rank directly below that of commander, should be established in each major command in the Pacific Theater. He said either the commander or his deputy should be an aviator. "I contend that the Navy as a whole has not been progressive in its attitude toward application of aviation in naval warfare." He reminded Nimitz that a few Navy airmen had forced the Navy to adopt naval aviation. He admitted that these officers had aroused resentment, and that he regretted that it still was in evidence. "This is not conducive to good teamwork," he said. He reminded Nimitz that these aviation-minded officers would have been outstanding in any capacity in the Navy. He urged Nimitz to use his influence to give these officers positions of responsibility suited to their abilities to permit them to establish policy and actions worthy of their backgrounds.

The controversy had started with a poll taken by Admiral Harry Yarnell, former commander of the Asiatic Fleet, who had returned to active duty in the summer of 1943 after four years of retirement. The poll solicited ways to use Naval Aviation more effectively, and it had received strong backing from many in the Navy. He had recommended that the deputy chief of naval operations for air be given the rank of full admiral, with a seat on the Joint Chiefs such as Air Force General Arnold occupied. He even called for Nimitz's removal, and his replacement by an airman, or at least that Nimitz's operations officer for air be raised to the rank of rear admiral.

King remained opposed to any changes in top command positions. He never varied in his view that such positions should be manned by flag officers with general Navy backgrounds and not specialists.

Nimitz compromised—because a sustained controversy would be inimical to successful operations—by appointing two key staff members. A deputy commander in chief was assigned to advise on air matters and a position of operations officer was established.

The grievances of the airmen were valid, but so were some of the arguments made by Nimitz and King. In particular, there were still far too few air officers of flag rank qualified for top combat assignments. Many officers had the ability, but had been held back through the years from reaching these positions. But Towers was grateful because airmen had had a chance to express their strong views about mistakes of the past, and their ideas for the future. Now they were given the chance to display their abilities as new drives began in the Central Pacific. Many of these men made the difference between success and failure.

Towers wrote Nimitz on August 21 that fast carriers should be the principal offensive elements in the Fleet, providing direct air support for amphibious operations. This use of carriers to gain air superiority and provide close support to ground troops on the scale he proposed was unique. "Carrier operations are highly specialized," he reminded Nimitz, "and should be conducted by officers thoroughly trained therein. To be 'air-minded' is no substitute for long aviation experience." He reminded Nimitz that his position as Commander, Air, Pacific, gave him no authority over air operations, and that he could not possibly handle logistics and maintenance without some authority over air operations. In conclusion, he said, "Operations and logistics cannot be divorced."

Nimitz was disturbed by Towers's letter. He understood his desire to get a combat command or at least put an aviator like Admiral Finch in command of the Central Pacific Force instead of Spruance. Barring that, Towers said, Spruance's chief of staff should be an aviator. Nimitz flatly refused all such requests.

Rear Admiral Kelly Turner, Nimitz's amphibious commander, had earlier supported Towers. He wrote another letter on October 4 pressing the recommendation in his previous letter.

During the summer and early fall of 1943 fast carrier operations performed so well that Nimitz asked Towers on October 12 to recommend an officer whom he could appoint as CINCPAC's planning officer. Towers suggested several officers, but emphasized that Captain Forrest P. Sherman was most qualified. Nimitz agreed.

Bougainville, the last and largest Japanese stronghold in the Solomons, was invaded on November 1, and landings were supported by Rear Admiral Frederick C.

Sherman's carriers, the *Saratoga* and the *Princeton*. Halsey restricted operations to the center of the island after landings at Empress Augusta Bay to establish perimeters for airfields only. He believed that capture of the whole island, defended by 40,000 Japanese troops on the island's northern tip, would be too costly. The Japanese resisted savagely, and their air losses were heavy. But not one American pilot was lost.

The Japanese had won vast territories during the early part of the war, including French Indochina, Malaya, the Philippines, the Netherlands East Indies, and most of Burma. The Imperial High Command was now divided about future courses of action. The Army recommended holding the line, but the Navy vigorously opposed defensive operations. Their arguments for a defense in depth was sound, except that their resources were no match for their ideas. Finally, the view of the Japanese Navy prevailed, and they planned a showdown with the American Pacific Fleet.

Admiral Mineichi Koga, Commander in Chief of the Combined Fleet since Yamamoto's death, told the Imperial Command, "Our only hope lies in a decisive naval engagement."

Prime Minister Churchill had often talked about Europe's "soft underbelly," favoring a move up the "boot" of Italy as one of the best pathways to Hitler's "Fortress Europa."

The bloody fight at Salerno on September 9, 1943, proved that there was no such place as a "soft underbelly." The port was taken on the following day with heavy casualties. The Allied drive slowed down as the 5th and 8th Armies found their way up the mainland. It was not until January 14, 1944 that they gained a strong front beyond Naples.

In the Ukraine, the Red Army swept the Germans toward the Dnieper River by the fall of 1943.

The skies over Europe were filled with British and American bombers assaulting Germany with around-the-clock mass air attacks. Since midsummer of 1943 Liberators of Fleet Air Wing 7, under Captain William H. Hamilton, had been given the responsibility to neutralize German submarine and E-boat activity, particularly in the Bay of Biscay. This vital preinvasion work was necessary for the Normandy invasion now scheduled for early June 1944.

Army fliers had previously done much of the work, but the Navy assumed the thankless task of flying hundreds of hours searching for submarines or small surface boats in the Atlantic. It was tedious work, but the U-boat effort was reduced at least twenty-five percent.

In the Pacific Admiral Koga reached a decision. He decided to keep his fleet at Truk, hoping the Americans would seek them out and be destroyed.

While MacArthur's Southwest Pacific Forces continued their advance toward the Philippines, Nimitz's Central Pacific forces cut across the Pacific by way of island groups. Plans to capture one island after another were abolished. It was agreed that key positions would be invaded, while others would be left to stagnate in the backwaters of the war.

The Gilbert Islands came first. Backed by a powerful 5th Fleet, carrier groups launched attacks two days prior to the invasion on November 20, 1943. The fighting was bitter, due to the lack of continuous close support by the carriers, but the key atolls of Makin, Tarawa, and Apemama finally fell. The 27th Infantry Division went ashore on Makin, and occupied it in four days. The Marine 2nd Division invaded Tarawa the next day. Fast carrier forces under Rear Admiral Charles A. Pownall swept the seas, neutralizing airpower as far south as the Marshalls.

Rear Admiral Keiji Shibasaka had 5,000 troops on Tarawa in almost impregnable underground defenses that he thought were impervious to bombs or shells. Shibasaka had said that if the Americans ever tried to invade Tarawa a "million Americans could never take the island in a thousand years."

At first, the Americans encountered massive resistance because the preinvasion bombing had been insufficient to destroy the island's defenses. The Marines had to take underground bunkers one at a time, tossing grenades through gun ports and pouring gasoline into them and igniting their interiors. All but 146 Japanese died, including their commander, but Marines suffered 3,301 casualties, including a thousand deaths.

For the amphibious forces, this was their baptism of fire, and they learned most lessons the hard way. Towers had stressed that the fast carriers should be used for close support. This was done, but not to the extent necessary to vastly reduce the casualties. Instead, the fast carriers under Rear Admiral Pownall swept the ocean area as far south as the Marshall Islands to prevent reinforcements from getting to the Gilberts. The Japanese had only limited airpower, so opposition in the air was insignificant. But the escort carrier *Liscome Bay* was sunk by a Japanese submarine, with the loss of 644 of her men.

After Admiral Halsey's South Pacific forces occupied the center of Bougainville, and MacArthur's Southwest Pacific Forces took Finschhafen on New Guinea, Rabaul was in jeopardy. Until now it had been an important bastion in Japan's Greater East Asia Co-Prosperity Sphere.

The land war became bloodier as troops advanced along the New Guinea coast and in the Solomons. Japan found it difficult to reinforce her advance positions and the losses became increasingly critical.

Generalissimo Chiang Kai-shek of China joined President Roosevelt and Prime Minister Churchill in Cairo on December 3, 1943, for the Sextant Conference. This was the second time the Allied leaders had met in Cairo, but this time the grand strategy to end the war was on the agenda. General Dwight D. Eisenhower was named to head the 1944 invasion of the European continent in June.

At this meeting, it was reaffirmed that a two-pronged advance would be made in the Pacific toward the main islands of Japan. Admiral Nimitz's Central Pacific Forces and General MacArthur's Southwest Pacific command were ordered to continue their advances and head for the Philippines. These decisions had been approved earlier, but once again the suggestion was made that MacArthur be named Supreme Commander in the Pacific. King again raised strong objections, and his view prevailed.

MacArthur's forces were specifically ordered to advance along the New Guinea–Netherlands East Indies–Philippines axis, while Admiral Nimitz's Central Pacific forces engaged in an island-hopping campaign. All agreed that once the Marianas were in Allied hands, they would be used to base B-29s for the strategic bombing of Japan.

In advance of Eisenhower's invasion of Europe in June 1944, the principals at the conference agreed to continue the Combined Bomber Offensive approved at the Casablanca Conference. Massive raids against German cities and industrial sites were ordered to pave the way for the ground forces. They also agreed that amphibious forces would attack Anzio on the coast of Italy to outflank German forces and trap them before beginning a drive toward Rome.

King went along with these decisions, but he warned that Pacific strategy should remain flexible to take advantage of an early defeat of the Japanese fleet, or unexpected withdrawals from the South Pacific. Either event, he claimed, could dramatically change the situation.

King called for a conference of his Pacific commanders in San Francisco to discuss future operations, and Nimitz and Halsey met him on January 3, 1944. Although King insisted that the Marianas were the key to victory in the Pacific, Nimitz and his staff disagreed—a position supported by General MacArthur. But King persisted, saying their occupation would block Japanese lines of communication to the Caroline Islands and that their central position would permit advances to the China coast. Nimitz's war plans officer, Forrest P. Sherman, made the suggestion that Truk be bypassed. In regard to the Marianas, Nimitz and his staff told King that they disagreed with the importance that King attached to the Marianas, but King was unmoved by their arguments.

Nimitz and Sherman met with MacArthur's staff later in the month, and Lieutenant General Richard K. Sutherland, MacArthur's chief of staff, called again for all resources to be pooled and a return to the Philippines to be planned, then moving on to the coast of China.

When Nimitz passed this proposal to King, King was outraged, believing that Nimitz had betrayed him. King wrote him a memorandum saying, "I have read your conference notes with much interest, and I must add with indignant dismay." He reminded Nimitz that advocates of pooling failed to consider when Japanese occupation and use of the Marianas and Carolines was to be terminated. "Southwest Pacific advocates," King said, "fail to admit that sometime this thorn in the side of our communications to the western Pacific must be removed."

Admiral Towers had said that the reason for taking the Marianas was primarily to provide B-29 bases for bombing the home islands, but King insisted that this was not true.

Such a circumstance had long been promoted by General Arnold, and with King's backing. But King believed, and rightly so, that the capture or neutralization of the Carolines should be speeded up to clear the lines of communication with the northern Philippines.

King stated his position in blunt terms. "The idea of rolling up the Japanese along the New Guinea coast, throughout Halmahera and Mindanao, and up through the Philippines to Luzon, as our major strategic concept, to the exclusion of clearing out Central Pacific lines of communication to the Philippines, is to me absurd. Further, it is not in accordance with the decisions of the Joint Chiefs of Staff."

He reemphasized his own strategies for Japan's defeat, saying somewhat petulantly that he thought Nimitz agreed with him. He said Japan should be forced into her inner ring of defenses, which included her home islands, Korea, Manchuria, and the Shantung Peninsula. He said it was imperative to capture Chinese bases and utilize their vast manpower to win the war when it was necessary to invade the main islands of Japan. But first, he said, ports on the China coast must be taken. Nimitz expressed his strong disagreement with King's views, particularly about the need to occupy Chinese ports.

King ordered Nimitz to make Luzon the objective of his forces, but first to clear the Carolines and Marianas of Japanese troops and occupy Peleliu in the Palaus. He said such operations would destroy Japan's lines of communication to the Netherlands East Indies and those to the east of the Philippine Islands. Then, he said, MacArthur's forces could advance without any threat to their flanks to make an attack on Mindanao. King said Mindanao was primarily valuable as a base to permit occupation of the rest of the Philippines. He said he foresaw many difficulties in their total occupation.

MacArthur, in a letter to General Marshall, still fought for a single command in the Pacific, with himself as its Supreme Commander, so he resisted King's independent drives in the Central Pacific and the South Pacific. He referred to them as a weak strategy that would delay victory over Japan by at least six months. He advised the Army chief of staff that the occupation of the Carolines and the Marianas would not obtain a major strategic objective or help in an assault on the Philippines. He repeated again his firm conviction that all assault forces should be given to him if Japan was to be defeated in the shortest possible time.

Marshall presented MacArthur's views to the Joint Chiefs. King defended his position, saying MacArthur's recommendations would prove indecisive. He reminded the Joint Chiefs that the Combined Chiefs had already approved the dual drives. He told Marshall, "Tell MacArthur to obey orders."

Marshall uncharacteristically procrastinated, unsure of the proper course to pursue with so many conflicting recommendations. When in doubt, he followed his normal course: he tossed the problem to his planners for further study.

Admiral Nimitz met on January 27, 1944, with senior officers at Pearl Harbor, including MacArthur's air boss, General George C. Kenney. The latter presented a strong case for not using the Marianas to base B-29 Superfortresses. He called their operations just a "stunt." Most Pacific commanders agreed, and there was little support for occupation of the Marianas. Even Admiral Towers agreed with MacArthur that the Central Pacific forces should proceed by way of the Bismarcks, Admiralties, Palaus, and Philippines. He claimed that air and submarine bases, once they were established in the Philippines, could intercept Japanese shipping more easily. In his view, he said, any invasion of the Chinese mainland could best be mounted from there. Although Nimitz agreed with Towers, King was adamantly opposed, insisting that the Central Pacific drive should continue through the Marianas chain of islands.

King was frustrated by his Pacific commanders and he turned to General Arnold, knowing how much the Air Forces chief wanted the Marianas for his B-29 Superfortresses. Arnold strongly supported King among the Joint Chiefs because the Marianas were only 1,500 miles from Japan's home islands.

The Marshalls were next. Dive-bomber pilots sprawled in their reclining leather chairs in their ready rooms trying nonchalantly to disguise their unmistakable tension as they prepared on January 29, 1944, to strike targets in the Marshall Islands.

Lieutenant Commander I.M. "Ike" Hampton told his Bomber Squadron 6 aboard the *Enterprise,* "We have our targets assigned. Make every bomb count."

The sun was only an intimation of the dawning of a new tropical day as the loudspeakers rattled with a firm voice. "Pilots, man your planes!" Sober-faced men hurried to the aft flight decks of their carriers. A fighter plane on the *Enterprise* moved up to the line and unfolded its wings. Lights flicked on. The flight deck officer jerked his flashlight in a sweeping motion and the pilot pushed the throttle forward.

At intervals, other fighters followed, then the torpedo planes, and finally the dive-bombers, which would be used as long-range artillery on precision targets facing the ground troops.

While the carriers rode the long swells, their planes winged toward heavily defended Kwajalein Atoll. The first wave struck enemy ships in the lagoon, a huge anchorage able to accommodate a large fleet.

The SBDs had been assigned the task of knocking out Japanese shipping, which was so important to keep supplies flowing to the island bases. Loaded with 1,000-pound bombs, they dove through heavy flak from the ships and shore batteries, releasing close to the ground to ensure direct hits.

After the last plane had left the carriers, one officer said, "It's almost uncanny. Either we're damn good or the Japs are damn stupid. Here we are on top of one of their biggest bases in the Marshalls and they apparently aren't putting up their full opposition."

For those in the task force it was a period of tense waiting and wondering. An hour later, word flashed back, "The enemy has been taken by surprise and the harbor is full of ships."

Some of the planes attacked enemy planes and installations at Roi airfield on the heel of the boot-shaped Kwajalein, but those who arrived first found the lagoon such a choice target they passed up shore installations. Off Roi, a large tanker and a light cruiser rode peacefully at anchor. In the toe of the boot, just off Kwajalein, and 40 miles away, a dozen large merchantmen rode at anchor beside interisland vessels and small ships.

"Ike" Hampton eyed the huge assortment of ships and decided the Japanese must have rushed a large group of men and supplies to the Marshalls after the Gilbert Islands were captured. "It's a dive bomber's paradise," he said over the radio to pilots behind him. For the next forty-five minutes the Japanese found it an exploding hell.

While flak burst heavily all around him, Hampton saw the light cruiser anchored off Roi shudder and then explode with a mighty roar after an SBD made a direct hit on her stern. Seconds later, another violent explosion came from the ship as a torpedo connected. Hampton watched with awe as a sheet of flame rolled over the cruiser's deck and the ship began to list. Bombs from other SBDs slammed into a tanker nearby and it exploded with a roar.

Hampton saw another light cruiser heading north as he arrived midway between Roi and Kwajalein. He watched its desperate maneuvers, and just as it turned to the left, a thousand pounder caught her on the stern. The cruiser completed the turn, then two more explosions ripped her insides. She lay in the water without movement, and not one gun replied to the steady attacks.

Arriving off Kwajalein, Hampton figured there must be 70,000 tons of shipping in the harbor. The flak was intense, but none of the ships was under way. It was a perfect setup. He gave the signal to attack. Lieutenant Commander Donald B. Ingerslew followed closely, as did the others. When they pulled out, there were fires on five different ships.

Lieutenant John F. Philips planted his bomb on the stern of a large ship. When he circled away, it was going down fast, with only the bow sticking out of the water. Philips followed an oil slick that seemed a mile and a half wide in the center of the lagoon until he came upon the much-bombed light cruiser. His attention was diverted momentarily to the north by a tremendous explosion. Flames from a burning tanker off Roi shot a thousand feet into the air while black smoke formed an anvil-shaped thunderhead at 10,000 feet.

The bombers winged back to their carriers, protected by the fighters, but only a few Japanese Zeroes had appeared by the end of the strike.

Six heavy and six light carriers in four groups of Task Force 58 under Mitscher opened the campaign on January 29, 1944, to capture the Marshall Islands. Heavy air attacks were made on Maloelap, Kwajalein, and Wotje. The defending enemy air forces were eliminated the first day, and complete control of the air was maintained by carrier aircraft during the entire operation.

Eight escort carriers, attached to the Attack Forces of the Joint Expeditionary Force, arrived two days later on the morning of D-Day on January 31, 1944. The carriers provided cover for the landings and engaged in antisubmarine patrols while they assisted the two fast carrier groups after the troops landed on Kwajalein and Majuro Atoll with close air support. They provided the same services when Marines landed on Roi and Namur on February 1. An air/ground control ship was used for the first time with great effect to coordinate aircraft to areas most in need of support.

Days of continuous air and sea bombardment had killed at least half the Japanese defenders on the ground. "Red Beach" at Kwajalein, which American troops used in their landings, was reduced to rubble before Army soldiers of the 7th Infantry Division set foot on it. Kwajalein was secured after 7,000 Japanese died in its defense. American casualties totaled 372. It was not as bloody a battle as Tarawa because Kwajalein had received a deadly "softening up" before amphibious forces landed.

To effect the neutralization of Wake Island during the Marshalls operation, two squadrons of Coronados from Midway made the first of four night bombing attacks on January 30. These 2,000-mile round trips were completed on February 9.

While the Marshalls were being invaded two fast carrier groups in the west kept Eniwetok Atoll neutralized until the initial objectives were achieved. Their early achievements permitted the second phase of the campaign—seizure of Eniwetok—earlier than the planned date of May 10. Landings were scheduled to be made on February 17, with the ground action supported by aircraft from one fast carrier group and one escort carrier group.

Many men expected to return to Pearl Harbor after the Marshall campaign, but they were disappointed. They anchored at Majuro, the new base in the Marshalls, where they prepared to attack Truk. Spruance built his Truk striking force around three fast carrier groups with six heavy and six light carriers.

For years Truk had been a mysterious place where the Japanese based their southwest Pacific fleet. Japan had acquired it in 1914 from Germany, and even as far back as World War I had developed it secretly into a great naval base. No American had seen it in all those years, but naval commanders knew there was no better base in the world.

Truk, a series of 200 islands enclosed in a forty-mile-wide lagoon, has low-lying reefs. Only five entrances permit ships to get inside the reef, but once in, the largest fleet in the world can be accommodated with ease. The main islands also had sufficient level land area, which the Japanese used for four air bases. The forest-covered

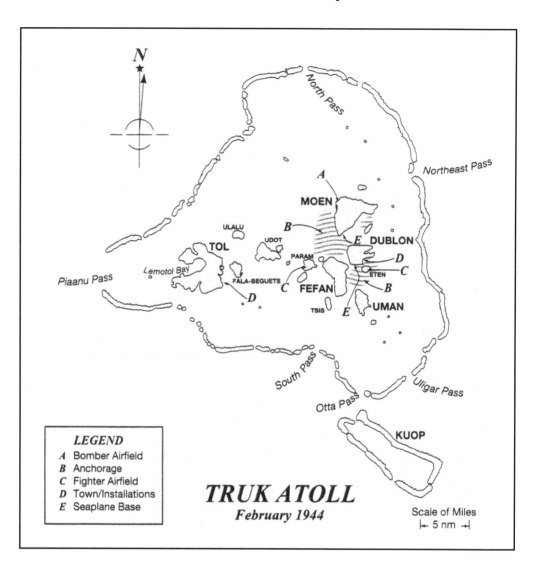

Truk atoll, February 1944.

mountains concealed coastal artillery and antiaircraft guns. The Japanese had made Truk their major fleet base in the Pacific, and the principal anchorage for its Combined Fleet.

Truk lies 1,800 miles southeast of Tokyo, and 750 miles north of New Guinea. It was in an ideal position to bar the Allied way back to the Philippines, the China Coast, to Japan.

Air Group Commander Lieutenant Commander William R. "Killer" Kane walked over to Lieutenant James D. "Jig Dog" Ramage, executive officer of Bombing 10 on board the *Enterprise*. Ramage and several officers were on deck tossing around a medicine ball.

"We're going to hit Truk," Kane said. "Wow!" Ramage replied. Truk had the reputation of being an impregnable Japanese fortress. Kane explained that the attack on Truk would be made to cover the seizure of Eniwetok on February 17 in the western Marshalls. He said the task force would attack at dawn the day before from a position ninety miles to the northeast of Truk.

At a later briefing, Ramage learned that Task Group 58.1 under Rear Admiral John W. "Black Jack" Reeves Jr. would send the *Enterprise,* the *Yorktown,* and the *Belleau Wood,* while Task Group 58.2 under Rear Admiral Alfred E. Montgomery would provide planes from the *Essex,* the *Intrepid,* and the *Cabot,* with Rear Admiral Frederick C. Sherman's 59.3 sending the *Bunker Hill,* the *Monterey,* and the *Cowpens.* The strike would have 276 fighters, 167 bombers, and 126 torpedo planes, for a total of 569 aircraft. Meanwhile, the *Langley* would cover landings on Eniwetok.

It was believed that there were 161 aircraft in operational readiness to defend Truk, and another 180 were undergoing various stages of repair or had no pilots available for them.

The strike force approached Truk from the east on February 16. It was a clear, cool day as they flew toward Truk at 12,000 feet. First, Kane led seventy-two fighters in a sweep to destroy Japanese fighters, while bombers prepared to follow at various times during the day. The Japanese lost seventy-five percent of their aircraft that first day—most of them below 1,000 feet, and many before their wheels lifted off the runway.

Unknown to the American command, Admiral Mineichi Koga had prudently withdrawn his fleet to the Palaus when the first air assaults struck Eniwetok. Only two light cruisers and four destroyers were still there, plus forty-six cargo ships.

VB-10 skipper Lieutenant Commander Dick Poor led the first *Enterprise* strike. Ramage was in charge of the second, taking off at 0900 at the head of twelve fighters, twelve dive-bombers, and eight torpedo planes.

Ramage's rear seat gunner, ARM1/C (Aviation Radio Man First Class) David J. Cawley, spotted their targets after they arrived at the center of the lagoon. They were in an anchorage adjacent to Dublon Island. Unlike most islands in the Pacific, which are generally flat, Truk Atoll is volcanic in origin, with rather high peaks.

Ramage prepared to roll in on his dive when heavy antiaircraft fire surrounded them. He now learned over the radio that most of the Japanese fleet had left Truk, but he was advised there were still plenty of ships to attack. He glanced anxiously at their F6F fighters as they started to tangle with the Japanese Zeroes. They seemed to have the situation under control.

A green-brown Zero sped by on Ramage's right on a parallel course as his SBD-5 passed the outer cay. Then he nervously watched as a huge mushrooming cloud floated upward 300 feet as an ammunition ship caught a direct hit.

Ramage picked out a tanker as his target because it was the largest in a group of a dozen ships. He signaled to his wingmen, Lieutenant (Junior Grade) William W. "Bill" Schaefer and Oliver W. Hubbard, to fall back. This was done to maintain the defensive "V" as long as possible. Ramage split his dive flaps and pushed the nose over and headed down. He placed his "pipper" just forward of the ship's bridge and manually released his 1,000-pound bomb at 2,000 feet. He turned left during pullout and looked back to see where his bomb had hit. He noted the tanker was covered with smoke and water splashes. He and the other dive-bombers left the tanker in a sinking condition and one bomb had detached the stern from the ship. He learned after the war that he had helped to destroy the cargo ship *Seiko Maru.*

Ramage brought his SBD low over the lagoon to strafe other ships, but ran into an old Japanese cargo ship that had been converted to a flak ship. All its guns seemed to be directed at them, but they brought their own guns to bear and escaped without injury. Ramage looked back to check on the rest of his squadron, who had followed him in. One SBD failed to make its pullout and crashed into the lagoon. He learned later that it was flown by Lieutenant (Junior Grade) Donald Dean and ARM2/C J.J. McGorry. The plane evidently sustained heavy antiaircraft fire prior to its pullout.

The second division pulled out safely, but found itself under attack by four Zeroes and a Rufe seaplane fighter. ARM2/C Howard F. Honea, gunner for Ensign Bob Wilson, shot down one of the Zeroes and damaged another.

That afternoon Ramage led his squadron on a second strike and set fire to the 13,000-ton *Hoyo Maru*. Lieutenant Lewis L. "Lou" Bangs's division scored two hits on the aviation stores ship *Kiyo Sumi* and left it in a sinking condition.

Ramage spotted a Japanese cruiser on the north side of Truk Lagoon as TBF Avengers dropped bombs on it. Later he noted that the cruiser was low in the water and barely making headway. They had already dropped their bombs, so he got on the guard channel to report the incident. "Any strike leader from 51 Bobcat, there's a damaged Japanese cruiser just to the north of the lagoon. Come sink it."

He was shocked to hear a transmission from "Bald Eagle," Mitscher's call sign. "Bobcat leader. Cancel your last. Do not, repeat do not sink this ship. Acknowledge."

This unusual order was later explained. Spruance wanted his surface ships to use the Japanese cruiser for target practice. They seldom had a chance to engage an enemy ship in a surface-to-surface exchange. They welcomed such an opportunity, which came rarely in the Pacific war.

Torpedo Squadron 10 was assigned to make a night attack on February 16–17 against the remaining shipping in Truk Lagoon. The squadron's TBFs met heavy but inaccurate fire, and two oilers and six cargo ships were sunk and another six damaged. This was the first night bombing attack in the history of carrier aviation. The *Enterprise*'s squadron used twelve radar-equipped TBF-1Cs. The attack, made at low level, scored several direct hits on ships in the harbor. During the campaign, night fighter detachments of VF(N)-101 flying F6F-3s and F4U-2s equipped with radar were assigned to five carriers and, while not widely used, were on occasion vectored against enemy night raiders.

After the Hollandia–New Guinea operation in late April, Nimitz ordered a second strike against Truk. There were no floating targets, but the Japanese still had a major navy yard, aircraft service facilities, and other military installations based there. Although the decision had been made to bypass Truk, it was a potential operating base. Mitscher's airmen finished the job they had started in February by smashing everything of value to the Japanese during strikes against Truk on April 29 and 30. This time the weather was bad and getting worse as the strike progressed, and antiaircraft fire was intense. Ramage's squadron on the third strike suffered damage, but VB-10 sustained no losses. Others were not so fortunate, and Ramage reported after his strike that a number of American planes had been spotted in the water. He was ordered to report to Admiral Reeves, who said, "I think we have run into the law of diminishing returns." He called Mitscher on the bridge-to-bridge circuit to recommend further operations be canceled. He said the benefits of continuing the raids were not worth the risk. Mitscher canceled the strike fifteen minutes later and intensified rescue operations.

These final airplane strikes shot down fifty-nine aircraft and destroyed another thirty-four on the ground. Only twelve Japanese aircraft remained in service on Truk. The Americans lost twenty-six aircraft, but half of their forty-six downed airmen were rescued.

Rear Admiral Reeves recommended to Mitscher a third Truk strike on May 18. There were no important targets left and losses were bound to occur. His dispatch said, "Would I be stretching my glide too far to recommend you detach Task Group 58.3 to strike Truk?" Mitscher replied, "I will not be badgered into an unwise decision!" And that ended the matter.

The once impregnable Truk was reduced to impotence by this series of strikes. Two light cruisers, four destroyers, three auxiliary cruisers, two submarine tenders, two subchasers, an armed trawler, an aircraft ferry, and twenty-four merchant ships (including four tankers) were destroyed. The Japanese also lost about 275 planes. American losses were light by comparison, with twenty-six planes lost. The *Intrepid* suffered torpedo damage and eleven of her crew were killed.

Truk was now finished as a first line of defense for the Japanese. They were forced to rely upon the Marianas and the western Carolines. This last raid ended the South Pacific's tenure as the primary area of operations.

Simultaneously with their operations, Admiral Spruance split his forces, sending a task group under Admiral Reeves and the *Enterprise* to strike at Jaluit in the Marshalls. This island had been bypassed repeatedly, but planes of "Big E" gave her a thorough lashing on February 20. The other two task groups attacked Japanese air bases in the Marianas.

The element of surprise was lacking on February 23 as waves of Japanese planes rose to fight it out.

Pilots dropped bombs all across the revetment areas on Saipan, Tinian, Guam, and Rota. They had the satisfaction of seeing most of the Japanese planes explode as they rushed past. Meanwhile, planes high above this action made photographic strips of each of the main Mariana Islands under protection of the Navy's fighters and bombers, which continued their operations to neutralize Japan's island air defenses.

Three waves of Japanese bombers rose to attack the American ships, but most were either shot down or driven off by Navy fighters. The heavy losses sustained by the Japanese on this first day—with sixty-seven aircraft shot down and 101 destroyed on the ground—resulted in only limited opposition when the Americans returned on the following day.

Admiral King warmly praised the operation against the Marshall Islands, citing it as the most successful of all Pacific operations. In particular he praised the almost perfect timing in the execution of joint plans. Nimitz expressed his own satisfaction to members of his command, saying he was particularly pleased by the low losses. He now agreed with King about an invasion of the Marianas, but with some reluctance. He conceded that their capture would provide bases for a straight and much shorter route to the Japanese mainland compared to MacArthur's longer and more hazardous roundabout approach through the Philippines.

The stage was set to think in broader terms about ending the war now that Japan's outer defenses had been cracked. Unappreciated at the time was the fact that Japan now found it impossible to fortify her inner defense line, which ran through the Marianas.

The effective use of land, sea, and air forces had resulted in victory against the Japanese in the Marshall Islands. It was a lesson that all American commanders needed to learn—that amphibious forces, protected by carrier aviation, could move into vast areas of the Pacific Ocean and destroy an enemy stronghold. Airmen like General Mitchell, General Arnold, and Admiral Towers had long preached such a doctrine, but few in the surface Navy had paid much attention. Now most leaders were convinced, although some failed to appreciate the need for a new and bolder strategy and to stop fighting the war on an island-by-island basis. MacArthur was one of the worst offenders. His mind-set against airpower was anchored in the past, when it was believed that only the infantry could decide wars.

Despite Nimitz's misgivings, it was now the turn of the Marianas. He had one valid objection—that there were no harbors in the Marianas suitable to serve as fleet anchorages. This was true, but this island chain was ideal to build bases for very heavy bombers like the B-29 Superfortress. There was some opposition in the Navy due to this fact. Most Navy officers did not believe the Air Force doctrine that strategic bombing by itself could force an end to the war. They pointed out, correctly, that it had not proved decisive in Europe, where an invasion was considered a prerequisite to defeat the Nazis. They overlooked the fact that there were still insufficient heavy bombers in England and the Mediterranean to be decisive. Some of these officers were opposed to B-29 bases for more parochial reasons. They did not want to give the Army Air Force an advantage that its strategic war advocates had long sought to prove—that airpower on its own could end the war by bombing Japan into submission.

Since August of 1943 the combination of seapower and airpower had brought vast destruction to the perimeter defenses of the Japanese empire. Amphibious forces had freed 800,000 square miles of Japanese-held territory. The islands now in Allied hands had three good fleet anchorages and countless airfields in the extreme western portion of the Pacific Ocean.

Success had come because the Japanese were outnumbered in ships and planes, and because of the tenacity and fighting abilities of men in all services. The great losses sustained by the Japanese in the last two years had drained her capacity to overcome pilot and plane losses in combat.

In early March the Joint Chiefs reached a decision about future operations. Nimitz was ordered to seize more Japanese-held territories, including the Marianas. The Joint Chiefs agreed with King that Rabaul should be isolated and that MacArthur should proceed westward along the northern coast of New Guinea to seize Mindanao in the southern part of the Philippines by November 15, 1945. Nimitz was advised to bypass Truk, invade the Marianas on June 15, isolate the Carolines, and invade the Palaus on September 15 in support of MacArthur's Mindanao invasion. Nimitz was also ordered to invade Formosa on February 15, 1945, and if necessary to attack Luzon by air on that date. The Joint Chiefs further agreed that landings should be made on the China coast.

Fortunately, wiser heads prevailed later, and the invasions of Formosa and the China coast were cancelled. Losses at both places would have been prohibitive, with no strategic advantage to be gained by the capture of either one.

Mitscher was anxious to wipe out Kurita's remaining ships, so Task Force 58 was built up to eleven carriers. His concern was that Kurita might try to oppose the landings at Hollandia scheduled for April 21.

Task Force 58 struck Palau on March 30 and 31, and also hit Yap and Woleai. Koga, apparently tipped off, had evacuated his nineteen ships and fled north. A thousand American planes worked over his ships remaining in the harbor for two days, sinking two destroyers, four escort ships, and twenty-two auxiliary vessels and destroying 157 aircraft. All carriers reported the loss of only twenty-five planes. Equally important, the harbor was denied to the Japanese for six weeks.

MacArthur now prepared to move his Southwest Pacific Forces 400 miles north up the New Guinea coast to capture Hollandia. The isolation of Rabaul, and the recent occupation of the Admiralties, made this bold stroke not only sensible but also appealing.

"I need maximum naval and air support," MacArthur told Nimitz. Nimitz agreed, and Kinkaid's 7th Fleet was reinforced by ships from the Central Pacific. He also ordered Mitscher to assign his fast carriers to direct air support before and after landings were made.

Task Force 58 arrived off Hollandia on April 21, a day before the landings on the north coast of New Guinea. The force of five heavy and seven light carriers was organized into three groups. Planes were launched a day before Lieutenant General Walter Krueger's 6th Army planned to land. Targets were scarce because Army bombers had given the area a saturation bombing. But preliminary strikes attacked airfields around Hollandia, at Wakde and Sawar, and covered the unopposed landings the next day at Aitape, Tanahmerah Bay, and Humboldt Bay. After the troops were on shore, eight escort carriers of Task Force 78 under Rear Admiral R.E. Davison flew cover and antisubmarine patrols to protect the ships of the attack group during the approach to Aitape and provided support for its amphibious assault. Carrier aircraft destroyed thirty enemy aircraft in the air and another 103 on the ground.

With the limited air opposition, Admiral Mitscher decided to withdraw his carriers and put them to better use against Truk. Shore installations were primary targets, but he also hoped to reduce the enemy's airpower.

American fighters swept Truk the morning of April 29. The Japanese 22nd Air Flotilla, with 156 planes available, were appalled when only eleven came home after the vigorous two-day attack. Truk received its worst pummeling of the war when bombers dropped 748 tons of high explosives, devastating the base. Later, reconnaissance photos showed 423 buildings destroyed and three small ships sunk. Truk was finished as a major Japanese base.

On the second day, Task Group 58.1 under Rear Admiral J.J. Clark was detached from the main force to fly protective cover for a cruiser bombardment of Satawan, and on May 1 supported the bombardment of Ponape with bombing and strafing attacks.

Japanese supply lines to the Southwest Pacific remained critical, so three weeks later carrier planes were ordered to strike Marcus Island and Wake Island, because both islands had large concentrations of enemy patrol and bombing planes. Mitscher hoped the Japanese might have become complacent, particularly at Marcus, which was less than a thousand miles from Japan. He felt sure that they expected the American Pacific Fleet to remain in the southwest Pacific for some time after the invasion of Hollandia.

Before dawn on May 19, the men of the three-carrier task force under Rear Admiral A.E. Montgomery took their battle stations. The planes, jammed tightly on the aft flight decks, trailed purple flames from their exhausts. At short intervals they took off for the Navy's third raid on Marcus Island.

While men on the carriers anxiously watched for the return of the first planes, every eye was alert for an attack by Japanese bombers. The first American formations landed, refueled, and headed back with new bombs and ammunition. For hours planes ran a shuttle service between the carriers and the island until the Japanese defenses appeared to be shattered. Huge fires and explosions rose above the tiny island until they could be seen by crews twenty miles away.

On the 23rd, Montgomery shifted his planes from Marcus to Wake and five composite bombing, strafing, and rocket strikes were made. These were limited operations—unlike the dramatic strikes at great bastions of the Pacific—using only a small element of the main carrier striking forces. But they were the kind of attacks that provide slow attrition, and eventually win a war.

Admiral Soemu Toyoda, Commander in Chief, Imperial Combined Fleet, had issued stern orders on May 7 to his naval forces, "Concentrate majority of our forces for a decisive battle. The enemy's fleet must be smashed with one blow south of Truk."

During the middle of May, the Japanese First Fleet assembled in open anchorage between the Philippines and Borneo. It was an impressive fleet, because Admiral Ozawa had nine carriers and five battleships.

General MacArthur was anxious to capture the Admiralties so they could be converted into an Allied naval and air base to cut off the Japanese supply center that was hampering his operations against enemy troops to the south. The mountainous islands at the head of the Bismarck Sea would provide him, he said, a necessary operational base to return to the Philippines. Capture of the largest islands, Manus and Los Negros, would—in his descriptive words—"cork the bottle."

Seizure of the islands came at a high cost in casualties for the amphibious forces, but by May 12 the entire Bismarck–Melanesia area was effectively bottled up, with 100,000 battle-hardened Japanese veterans inside.

Admiral Mineichi Koga, Commander of the Combined Imperial Fleet at Palau, realized Truk was no longer a safe anchorage. He ordered Admiral Kurita to withdraw his 2nd Fleet to Palau and dispatched Vice Admiral Jisaburo Ozawa's carrier forces to Singapore.

Toyoda studied his intelligence reports during the early part of June, hoping to outguess the Americans on their next move. He was astonished when he learned on June 11 that Task Force 58 was attacking the Marianas. He knew what the loss of these strategic islands would mean for Japan. The inner defenses of the Japanese homeland had to be held at all costs. The Marianas, directly across the shortest routes to Japan proper and the Philippines, could become huge airfields for B-29 Superfortresses. He was aware that they had flown their first mission out of India on June 5 against Bangkok. If the Marianas were lost, the homeland would be in dire peril because most of its main islands were only 1,500 miles away.

12

Marianas Turkey Shoot

Five days after the invasion of Normandy on June 6, 1944, by Allied armies in Europe, Vice Admiral Mitscher led his Task Force 58 to the Marianas. He had seven heavy and eight light carriers to support the invasions of Tinian, Saipan, and Guam. On June 11 a late-afternoon fighter sweep destroyed one-third of the Japanese planes in the Marianas. On succeeding days, bombing and strafing attacks were made against shore installations and against ships in the immediate area. Carriers roamed widely to make attacks on airfields and shipping in the Bonin and Volcano Islands to the north. This was done to keep the Marianas isolated before American troops landed on Saipan on June 15. Then the carriers returned to support the ground troops.

After the Marines landed, Toyoda decided to fight back, although he would have preferred an encounter closer to home.

Task Force 58 had left Majuro on June 6 and joined Admiral Spruance's Fifth Fleet of 644 ships. Spruance told him, "Your job is to clear the air near the islands." Mitscher led his armada with its fast carrier task force of experienced veterans into action with a sureness born of past accomplishments.

Unknown to the Japanese, the Third and Fifth Fleets were the same. When Halsey was in command, it became the Third Fleet, but when Spruance was in charge it was identified as the Fifth Fleet. In each instance, the task forces and groups remained the same, but what changed were their designations. Only the leaders at the top changed positions. This unique propaganda move baffled the Japanese, who believed they faced two huge fleets.

Spruance's plan called for bombardment by surface ships and bombers for four days prior to the invasion. Saipan was scheduled for invasion on June 15. Guam was next on July 21, and Tinian was last on July 24.

On June 15, when the first troops hurried out of their amphibious vehicles on Saipan, American planes concentrated on close support, knocking out machine gun nests, big guns, and the small tanks used by the Japanese. Spruance was pleased with the start of the operation, but he grew more concerned by the lack of reports

This aerial view of the USS *Lexington*—the "Lady Lex"—was taken in March, 1944. (Courtesy of U.S. Navy.)

about the Japanese fleet. Ships and submarines were constantly on the prowl in the Central Pacific to report major ship movements.

While troops swarmed ashore on Saipan, Spruance received a dispatch from a submarine. "Large enemy task force sighted. Heading from San Bernardino Strait at 20 knots."

"Looks like they're coming," Spruance said, with considerable relief. He turned to Mitscher. "I'll give you all the cruisers and destroyers we can spare. Your fast carriers will have to go out and meet them."

Mitscher returned to his flagship, a new carrier named the *Lexington,* replacing the one lost during the Battle of the Coral Sea. He sent a communication to all ships, saying, "Believe Japanese will approach from a southerly direction under shore-based air cover near Yap and Ulithi. Probably will attempt to operate near Guam. May come from the west. Scouts should consider all contingencies."

Fifth Fleet Commander Spruance now released his major battle plan. "Our air will first knock out enemy carriers, then attack battleships and cruisers."

On the night of June 18, Spruance received new word about the enemy fleet and ordered Mitscher to head west and attack with three of his four carrier groups.

Admiral Ozawa, who headed the enemy fleet, planned to launch his planes when he was 350 miles from Saipan, out of reach of American carrier aircraft. He had instructed his attack group commanders to shuttle planes between the carriers and Guam, where they could refuel, load more bombs, and attack the Americans on their return to their carriers.

Fighter pilots from the *Monterey* intercepted two Japanese Judy dive-bombers at sunrise on June 19, and quickly splashed one into the ocean.

All carriers were alerted. "Pilots, man your planes!" came the call over the squawk boxes.

Commander Ernest M. Snowden stood on the *Lexington* watching the vapor trails of the Japanese planes, which, in the distance, appeared only as dots in the sky. The air was so clear he could see the fighter combat air patrol planes from the American carriers rising to meet a large force of Japanese bombers headed for Admiral Lee's surface ships at Saipan. Then, in a whirling dogfight, with guns chattering in rapid succession, the Japanese bombers started to flame and plummet into the ocean.

The Hellcats dominated the battle, and in a day-long battle against Japanese planes from carriers and shore bases, they destroyed 402 enemy planes during a series of attacks, many as they landed to refuel at Guam. American losses were thirty planes. One pilot aptly described it as a "turkey shoot."

It was a bitter blow for Ozawa—not only the loss of his planes and experienced crewmen, but the fact that the *Shokaku* was sunk by the submarine *Cavalla,* and his flagship *Taiho* was also sunk. He gathered what was left of his forces and withdrew. He sailed northeastward for 250 miles and rendezvoused with his supply ships. He wired Toyoda the results of the battle, giving an accurate picture of his shattered carrier groups.

Additional losses during the retreat brought his total complement of aircraft down to 102. Of this number, seventeen were fighter-bombers, eleven were torpedo bombers, forty-four were fighters, and thirty were nonbombardment types. Ozawa was impressed by the Americans' strength. He expected pursuit, and was hoping to forestall an engagement before June 22 so he could build up his attack capability.

Spruance issued orders on June 20 for Mitscher to set out in pursuit before Ozawa could reinforce his fleet. When the Japanese admiral sent his supply ships to Okinawa, he led his fighting ships toward the Philippines. Mitscher disengaged the carriers *Essex,* the *Cowpens,* and the new *Langley* of Task Group 58.4 to cover the Marianas operation while he headed westward with the rest of Task Force 58. After great difficulty, search planes finally reported Ozawa's ships 250 miles to the west.

Mitscher discussed the situation with his staff. "It's a long flight for the planes," he said. "Let me have your opinions." After much discussion Mitscher said, "We can make it." He paused thoughtfully. "It will be a tight squeeze." They all agreed. At the appointed hour, Mitscher called, "Launch 'em!" The bullhorns echoed throughout the decks of the carriers. "Start engines!"

Commander Bernard M. "Smoke" Strean climbed into his fighter at 4:30 P.M. to lead the force of 216 planes of the first wave. After they had cleared, another large group would follow. When the planes were airborne, Mitscher received sobering news from his scout planes. Ozawa's ships now were sixty miles farther west than first reported. He immediately cancelled the second strike.

One by one they headed into the late afternoon sun. During formation the usual gay banter was there, but it seemed more subdued to Commander Strean. He knew that many felt, as he did, that it was a one-way trip.

For two hours Strean's radio was quiet, not a voice cracking a joke or asking for instructions. Then, at 6:40, he heard, "Ships sighted. Look at that oil slick!"

A few Japanese fighters headed in for an attack on the bombers. Strean called, "Leave them to us."

While the fighters tangled with the Zeroes, torpedo bombers headed straight for the ships, pilots wincing as the vicious flak streamed toward them, some even from the big guns of the fleet.

Lieutenant Charles W. Nelson of the new *Yorktown*'s torpedo group picked out an enemy carrier and, with Lieutenants John D. Slightom, James R. Crenshaw Jr., and Carl F. Luedemann, swung in for the kill.

Lieutenant George P. Brown, temporary commanding officer of a *Belleau Wood* squadron, chose another carrier. It looked like one of the big *Hayatake* class, the kind that had been converted from Pacific luxury line ships. He signaled to Lieutenant Benjamin C. Tate and Warren R. Omark to follow.

Brown's small section had been left alone while the other planes of the squadron were engaged in a glide-bombing attack. He decided to seek safety in the clouds until they could get closer to the carrier, but when the clouds drifted away he found they were headed for a large group of battleships, cruisers, and destroyers only 5,000 yards away.

Their TBFs shook violently as a hail of steel ripped into wings and fuselages. "Break it up!" he called to his comrades. "Come in from different angles."

A carrier, fully alerted on the far side of the warships, turned in a tight circle while the massed fire of her guns was brought to bear on the three Avengers. The sturdy airplanes took terrible punishment as explosions rocked them repeatedly. Tracers tore through Tate's cockpit, slashing his right hand, and then a large shell ripped open his fuselage. He hung on grimly, never leaving his prescribed course.

Brown's plane was in bad shape, as shells exploded on all sides. "Fire!" a crewman yelled. He blanched in horror as the flames licked the fuselage, and while others bailed out, Brown dove the plane, hoping to put out the fire. With almost a sob of relief, he saw the flames die down and the plane released its torpedo straight for the carrier.

As Brown's Avenger roared across the fleet it seemed as if every gun were trained on his fire-blackened TBF. When he saw Tate and Omark had not yet made their runs—instead of hurrying to the safety of the fringe of the fleets—he turned back, flying straight down the length of the *Hiyo* to draw their fire away from the other pilots.

Tate and Omark, thrilled by Brown's heroic action to help them, swung in and dropped their torpedoes. They ran the gauntlet almost without opposition and then searched for Brown. To their astonishment his riddled, fire-blackened plane had cleared the scene and was headed due west.

Omark overtook him. After almost touching wingtips, Omark signaled for Brown to turn back. Brown waved a bloodied and shattered arm. Omark felt a sick sense of futility as he noted Brown's khaki shirt was covered with blood and he seemed about to pass out. While he watched with growing anguish, Brown's plane fell off, and dropped steeply out of control.

The carrier *Hiyo* was a mass of flames as Omark turned toward his own task force. He watched the big ship roll slowly over and sink.

The battle had taken forty minutes. At 7:20 P.M. the last American plane headed back while Ozawa's ships fled westward.

Ozawa's losses were severe. During the action one of his three largest carriers was lost, along with two oilers. His carrier planes were now reduced to thirty-five.

The American fliers, who had already lost twenty planes in combat in this first Battle of the Philippine Sea, flew their planes eastward with growing concern. Mixture levels were lowered until the engines coughed in protest. The planes still had

Vice Admiral Mitscher aboard the *Lexington* during
the Saipan battle in the Marianas. (Courtesy of U.S. Navy.)

300 miles left to fly after using up their reserve fuel during the relentless attacks on Ozawa's force.

Mitscher's hands gripped the arms of his swivel chair until the knuckles showed white. Task Force 58 was blacked out—not a light showing—because Japanese submarines had been detected.

Mitscher—his eyes red-rimmed, his face haggard with worry about his pilots and crewmen—often gazed toward the western horizon as the radio crackled with reports from the returning planes. He swallowed tightly as one pilot reported, "Almost out of gas. Going in."

Then came agonizing cries from several pilots who were lost in that awesome black void to the west. They needed position reports to home in. Mitscher's leathery face showed the anguish going on within him. The thousands of men in the huge task force, and the vital carriers needed to pursue the war against Japan, should not be sacrificed even for their pilots. Turning on the ship's lights would save scores of men struggling to get back safely, but lurking Japanese submarines would find the carriers easy pickings.

Mitscher's slight figure shifted uneasily in the skipper's chair, and his jaw was set stubbornly to control the quivering of his lips as he listened to the frantic cries for help coming in over the radio. These were his men. He had sent them out on this dangerous dusk attack, and they had fought well.

Mitscher's normally soft voice vibrated with emotion as he spoke, "Turn on the lights!" The whole fleet lit up like Times Square in New York. Starshells were fired, exploding high above the ships so pilots could see them from afar.

While destroyers prowled through the fleet searching for enemy submarines, hauling wet pilots out of the sea after their planes crash landed short of their carrier, the planes streamed in. Dozens dropped into the ocean, so low on fuel they could not make even the last mile. The sea seemed covered with flickering lights as pilots waved their waterproof flashlights so destroyers could find them more quickly.

Approximately one hundred planes failed to land safely after the attack, but Mitscher's prompt action—along with the outstanding rescue efforts—saved all but sixteen pilots and thirty-three crewmen.

By the time Guam was secured on August 10—the last of the three islands to surrender to the American command—carrier aircraft had sunk 110,000 tons of enemy ships and destroyed 1,223 aircraft. The Japanese had lost most of their best pilots and crewmembers, and their carrier aviation declined steadily after this action. The loss of planes was severe enough, but they could be replaced far more readily than the experienced airmen. The ground fighting had been intense and losses were heavy on both sides, but the Japanese lost far more than the Americans.

The four groups of Mitscher's Task Force 58—plus two divisions of escort carriers—had been instrumental in the early conclusion of the Marianas campaign. Guam, in particular, had received the heaviest aerial bombardment ever meted out to a Pacific stronghold.

The close support on all three islands—the battle for control of Saipan lasted 24 days—was exceptional, as planes bombed enemy concentrations and artillery posts, and heavily strafed infantry positions. The Marines took all this for granted, but close air support for Army troops was greeted warmly. They had not been used to it.

The program for controlling the Central Pacific went into high gear. Three groups of Mitscher's fast carriers struck targets in the western Carolines during the latter part of July. During the first week in August, Mitscher's forces again raided the Bonin Islands, which had been under periodic attack since June 15. This chain was important because it was on a direct air route to Japan.

After Guam was secured on August 10, groups of Task Force 58 retired in turn to advance fleet bases for brief periods of rest and replenishment. This initiated a practice that became a standard operating procedure during future extended periods of action.

In Tokyo, Japan's Imperial High Command viewed the future with growing concern. Many officers in the Japanese Navy realized that Japan had lost the war. The American Navy, now more powerful than the Japanese, would soon feel free to steam up to the main islands, almost without opposition.

What caused the high command the most anguish was the loss of the Marianas. These island bases, only 1,500 miles from Japan, were already earmarked for B-29 operations. Japanese militarists knew that every city was vulnerable to mass raids and eventual destruction.

During the Marianas operation, the chief of naval operations in Washington on June 24 called for a drastic reduction in the Navy's pilot training program. King ordered the transfer of some students already in preflight training to other duties, while retaining enough trainees to maintain a course in these schools that had been extended to twenty-five weeks. The reductions were the result of the discontinuance of the War Training Service Program in August, the closing of flight preparatory schools in September, and the release of training stations.

On July 6 the Bureau of Aeronautics authorized the Douglas Aircraft Company to proceed with the design and manufacture of fifteen XBT2D airplanes. This single-seat dive-bomber and torpedo plane had been designed jointly by Bauer and Douglas. Through subsequent development and model redesignation, this aircraft became the prototype for the AD Skyraider.

After two years of undisputed reign as the only effective dive-bomber in the U.S. Navy, the SBD Dauntless now had a more modern competitor in the Curtiss Helldiver. The Dauntless went out of production in July 1944.

Admiral King met with Nimitz at Pearl Harbor in the middle of July, and then flew on to Saipan. There he told Spruance, "You did a damn good job here." King used the occasion to deflect criticism aimed at Spruance's decision not to release Mitscher to attack Ozawa on the 18th or 19th of June. "No matter what other people tell you, your decision was correct."

King sought advice from Spruance and Turner about an invasion of Formosa and—to his surprise—learned that both opposed it. Such an operation, they claimed, would interfere with the occupation of the Philippines. Spruance stressed that if Formosa must be taken for strategic reasons, Luzon should be invaded first to provide a fleet anchorage at Manila Bay. He reminded King that the Marianas had no suitable harbor.

"What do you recommend now that the Marianas have been taken?" King asked Spruance. "Okinawa," was the reply. King looked at Spruance with surprise, "Can you take it?" "I think so if we can find a way to transfer heavy ammunition at sea." He explained the problem to King. "Okinawa, in the Ryukyu Islands just south of Japan, is 1,400 miles from here, and there is no suitable anchorage adjacent to the Ryukyus."

Spruance also recommended the seizure of Iwo Jima, prior to the invasion of Okinawa. He emphasized that the occupation of Okinawa would complete the blockade of Japan, and he asserted that the war could be won without an invasion of the main islands.

These provocative thoughts had a strong impact on King, because he had become convinced that Formosa must be taken—along with bases on the coast of China—in order to successfully end the war.

During the Marianas operation President Roosevelt had been nominated for a fourth term on July 20. Afterward he announced that he would meet six days later with the Pacific commanders at Pearl Harbor.

King placed Nimitz in charge of the conference and returned to Washington. Before his departure he advised Nimitz to make up his own mind about an invasion of Formosa. Nimitz had long been opposed to such an operation, and he continued to resist King's pleas to seriously consider its merits.

Admiral Turner had long recommended carrier strikes against Japan's main islands, and before King departed he recommended they be started in late January of 1945. King agreed to the strikes, but insisted they begin in late 1944.

Roosevelt arrived on time for the Pearl Harbor conference, but MacArthur delayed his appearance in a deliberate attempt to upstage the president. While the president cooled his heels, MacArthur arrived with much fanfare. He resented his orders to appear at the conference, telling his staff it was just playacting for political reasons.

MacArthur circulated a letter at the conference from King that after MacArthur's command invaded the Philippines an independent British command would take over defense of Australia and the East Indies.

The president hosted a dinner in his quarters that night with MacArthur, Admiral Leahy, and Nimitz. Afterward they gathered in the living room before a large map of the Pacific. Roosevelt maneuvered his wheelchair closer to the map. "Douglas, where do we go from here?"

"Leyte, Mr. President, and then Luzon."

Nimitz then briefed the president, dutifully presenting King's reasons for an invasion of Formosa. He stressed the importance to Japan of maintaining access to its oil fields in the East Indies. He said he hoped the Chinese army would secure bases along its coast to bomb Japan and provide launching points for a possible invasion of the main islands.

MacArthur was opposed to King's strategic plans and the general sharply disagreed with such a viewpoint. "You cannot abandon seventeen million loyal Filipino Christians to the Japanese in favor of first liberating Formosa and returning to China," he said. "American public opinion would condemn you, Mr. President, and it would be justified."

MacArthur again stressed the importance of delivering the Filipino population from Japanese occupation, and the necessity of freeing American prisoners. He reminded the president that Japanese bases on Luzon were too "far away to be attacked by fields on Mindanao. "American airfields are needed on Leyte or Mindoro, or both." Then, he said, his forces could land at Lingayen Gulf and be in Manila in five weeks.

MacArthur refused to accept the possibility of high losses. "Good commanders do not turn in heavy losses."

This last barb was aimed at Nimitz. MacArthur's losses had been light compared to those suffered by Nimitz's command in its Central Pacific drive. But MacArthur's forces had met only light resistance, due primarily to the fact that the Central Pacific campaigns had kept the Japanese from massing on MacArthur's flank.

The president was disturbed by the obvious differences between his top commanders in the Pacific, and he was not convinced by MacArthur's views about the necessity of returning to the Philippines. Therefore, he refused to commit himself.

MacArthur, concerned that his comments might indicate a serious disagreement with Nimitz, assured the president that "we see eye to eye, Mr. President."

The president was not fooled by his comment. MacArthur believed he had won all the arguments about future operations, telling an aide, "We've sold it!" After Roosevelt returned to Washington, he told King that he now favored an attack on the Philippines. King's wrath turned on Nimitz, unfairly blaming him for not pressing his case for Formosa and the China coast. He was even more irate when General Marshall told him he doubted the merits of occupying Formosa.

The fast carrier task forces were reorganized on August 5 into the First and Second Fast Carrier Task Forces, Pacific, commanded by Vice Admiral John S. McCain. Of course, they were the same carriers. During the first week of August Mitscher again raided the Bonin Islands. They had been under periodic attack since June 15 because they were on the direct air route to Japan.

In the middle of August Nimitz moved his Pacific Command to Guam without seeking King's approval. King was upset by the transfer, but he had to accept Nimitz's reasoning that he would be closer to the war front.

Amphibious commanders in Europe finally adopted a strategy that their counterparts in the Pacific had found most effective in assaulting a hostile shore. Landings were made in southern France on August 15 as part of a four-pronged drive. Meanwhile, the cross-channel invasion of Normandy that began on June 6 culminated in the liberation of Paris on August 25. Two American and seven British escort carriers of the Naval Attack Force under Rear Admiral T.H. Troubridge of the Royal Navy supplied defensive fighter cover over the area. They also spotted for naval gunfire, flew close support missions, and made destructive attacks on enemy concentrations and lines of communications. Allied troops landed between Toulon and Cannes, and advanced rapidly up the Rhone Valley. Marseilles surrendered on August 28, and the naval war in the Mediterranean virtually came to a close.

The Navy's first night carrier air group, CVLG(N)-43, was commissioned at Charleston, Rhode Island, on August 24. Its component squadrons VF(N)-43 and VT(N)-43 were commissioned the same day. The latter was the first of the night torpedo squadrons.

Prior to an invasion of the Philippines, MacArthur asked for the invasion of Peleliu in the Palaus and Morotai in the Moluccas.

Simultaneous landings by Central and South Pacific Forces were preceded by wide-flung operations by four carrier groups of Task Force 38 under Vice Admiral Mitscher. The task force committed only part of its strength to direct support and operated principally as a covering action. Task Group 38.4 under Rear Admiral Davison opened the campaign with attacks on the Bonin and Volcano Islands on August 31. Then the entire Fast Carrier Task Force struck the Palau area between September 6 and September 8—leaving Task Group 38.4 to maintain the neutralization of Palau—while it moved against the Philippines with fighter sweeps over Mindanao's airfields on September 9 and 10 and air strikes in the Visayas between the 12th and 14th. Here Task Group 38.1 under McCain separated to hit Mindanao on September 14 and to support landings on Morotai by Southwest Pacific Forces the next day. The landings were preceded by bombing and strafing attacks and were supported by Task Group 38.1 aircraft and by six escort carriers of Task Group 77 under Rear Admiral C.A.F. Sprague. For the landings on Peleliu by Central Pacific

Forces on September 15, preliminary air attacks were made by Task Group 38.4 and four escort carriers under Rear Admiral W.D. Sample's Carrier Unit One. Continued support was given by the same fast carrier group for the invasion itself and continued until the 18th, and by ten escort carriers until the end of the month operating as Task Group 32.7 under Rear Admiral R.A. Ofstie. Carrier support was also provided for landings on Anguar on September 17, Ulithi on September 23, and the shore-to-shore movement from Peleliu to Ngesebus five days later. This support included strikes by Marine Corps land-based units from Peleliu, the first which, VMF(N)-541, had arrived on September 24. Following the action at Morotai, Task Group 38.1 rejoined the main body of the fast carriers, which then launched strikes on airfields and shipping around Manila for two days starting September 21, with strikes also against airfields, military installations, and shipping in the central Philippines. In a month of intense action, carrier planes destroyed 893 enemy aircraft and sank sixty-seven war and merchant ships totaling 224,000 tons.

Landings were made on Morotai by Southwest Pacific Forces with only token opposition. It was a different story at Peleliu. After twelve days of bitter fighting by Central Pacific Forces the island was in American hands.

MacArthur had considered Yap, but decided to abandon that invasion. It could be bypassed because it was one of the enemy's lesser strongholds. The small group that landed on Ulithi—the largest atoll in the western Carolines—found the island unoccupied by the Japanese, who, the natives said, had fled a month before.

Ulithi's thirty islands on a reef formed a magnificent lagoon for anchoring a large fleet, and the U.S. Navy quickly established an anchorage on the northern part.

A Navy pilot, who was shot down over Leyte, was rescued by guerrillas, who told him there were no Japanese troops there. After his rescue, Halsey interviewed the pilot and advised Nimitz in Quebec, who was attending the Octagon Conference. Halsey reported the pilot's information and recommended that the landing for the Philippine invasion be moved up and that the Palaus be bypassed. He strongly recommended that such an invasion be made in the middle of the Philippines instead of the southernmost island of Mindanao.

Nimitz did not totally agree with Halsey, advising him that the invasion of the Palaus was too far along to be cancelled. He brought these matters before the Combined Chiefs of Staff, who agreed with his analysis of the situation, as did Roosevelt and Churchill.

MacArthur, involved at the time in the Morotai invasion, could not be informed of the change in strategy. His chief of staff, General Sutherland, agreed on his behalf, but he disagreed with Halsey's comment that the Japanese air force in the Philippines was a "hollow shell operating on a shoestring." He reminded Nimitz that Allied ground-based fighter cover could not protect the invasion fleet off Leyte, so carriers would have to take over the assignment.

The Combined Chiefs authorized that Leyte, the eighth largest island in the Philippines, should be invaded October 20, two months earlier than originally planned.

In advance of the upcoming invasion of Leyte, Mitscher's task force groups made a high-speed run to a launch position seventy miles off Central Luzon. The ships moved ahead warily, and at midnight on September 30 their aircraft arrived undetected over Manila Bay.

Commander Jackson D. Arnold, the new *Hornet*'s Air Group commander, led the attack. While Commanders Jack Blitch and Grafton B. Campbell of the new

Wasp headed the dive-bombers, torpedo bombers circled Manila Bay, then swept against the shipping.

Bombers from the *Wasp*'s Air Group 14 concentrated on the old Dewey dry-dock, which the Navy once had brought from the East Coast of the United States through the Panama Canal for use in the Philippines. It was hit seven times.

In coordinated attacks, Admiral Bogan's fliers of the *Intrepid*'s Air Group 18 and the *Bunker Hill*'s Air Group 8 struck Clark Field. Admiral Sherman's crews, meanwhile, headed for Nichols Field, including the *Essex* Air Group 15 and the *Lexington*'s Air Group 19.

It was a masterly performance under Halsey and Mitscher, with the Japanese losing sixty-six aircraft in the air and Task Force 38 losing six.

Nimitz read Halsey's report following the attack with great satisfaction. "Approach to Luzon apparently undetected and surprise complete thanks to convenient weather front." Halsey said, "Weather over target good but foul in launching area. Operations possible only because of superb judgment, skill and determination of Task Force 38 and its commander Vice Admiral Mitscher."

Carrier aviation had destroyed over 1,000 Japanese planes and 150 ships since August 31, while their combat losses were only fifty-four planes.

These raids proved the Philippines were susceptible to approach by invasion forces. The Joint and Combined Chiefs felt vindicated now that the separate drives by Southwest Pacific and Central Pacific forces had achieved such overwhelming successes. They ordered that these separate drives should now be joined in the Philippines.

The opening air strikes prior to the landings on Leyte were targeted by Mitscher's Task Force 38 carriers on October 10 against airfields on Okinawa and the Ryukyus. Built around seventeen carriers, this force hit airfields on northern Luzon and Formosa October 11 through 14, and on the 15th it attacked the Manila area. A total of 438 Japanese aircraft were destroyed on the ground during five days of intensive carrier operations. Attacks were concentrated on reinforcement areas to clear the air for landings on the 20th by MacArthur's Southwest Pacific forces.

The main attack force of 130,000 troops remained off the entrance to Leyte Gulf on October 19 while bombardment ships moved closer to the shore. At dawn, the big ships belched steel at enemy positions while Rear Admiral Sprague's eighteen escort carriers—organized in three elements under Task Group 77.4—sent in their planes to bomb and strafe the invasion front.

General MacArthur watched his troops wade ashore at 10:00 A.M. the next day from the deck of the cruiser *Nashville*.

The general later waded ashore at the same spot he had stood forty-one years before when he had participated in the first Philippine mission as a second lieutenant in the U.S. Army.

The Japanese had retired to more entrenched positions, so the landings were made against only light opposition.

The amphibious operation was executed with masterly precision but then the Japanese planes appeared on the scene and not only attacked the ships but destroyed themselves in savage kamikaze suicide attacks.

Before the landings were made on Leyte, Admiral Halsey had transferred his big ships north of the Philippines in case the Japanese left the home islands to strike at the invasion forces. His carrier planes had attacked Okinawa on October 10, but only a few Zeroes rose to attack the bombers as they hit airfields and harbors. Three hundred and forty planes participated in these attacks, and extensive damage was done, particularly to the port city of Naha. In conjunction with these bombing attacks, photographic planes covered the island extensively with their cameras.

Halsey moved south, feinting toward the Philippines, but heading instead for Formosa to engage in a three-day attack to help reduce enemy air strength, which might be thrown into the battle for Leyte. The Japanese Second Air Fleet had hundreds of airplanes on Formosa, so when Halsey's planes attacked on October 12 they received a brisk, fiery reception. Over a thousand land-based airplanes rose to fight it out with Halsey's fliers.

Air Force B-29s made three successive strikes on the 14th, 16th, and 17th from forward bases in China, so the Japanese reeled under attacks from all sides.

The Third Fleet did not complete the action unscathed. The *Canberra,* the *Houston,* and the *Reno* were all damaged.

Halsey had a good laugh when Tokyo Rose gave the Japanese version of the action. "All of Admiral Mitscher's carriers have been sunk tonight instantly!" Halsey recommended issuance of a press release. "All Third Fleet ships reported by Tokyo radio as sunk have been salvaged and are retreating in the direction of the enemy."

It was estimated that the Japanese had lost over 500 planes both in the air and on the ground, whereas Third Fleet losses totaled seventy-nine planes and sixty-four pilots and crewmen.

The Third Fleet next headed for Leyte Gulf to protect the invasion forces from intruders. Despite the fact that they had been fighting continuously for two weeks, there was no tine for replenishment of Halsey's ships or to give crews a much-needed rest.

Halsey did send McCain's Task Group 38.1 to Ulithi, because their supplies were depleted to the point where they could not engage in sustained operations. The *Bunker Hill,* which needed a replacement group after heavy losses, was sent to the Admiralties, escorted by two destroyers. The three remaining groups speeded toward the Philippines the night of October 23.

Admiral Sherman's carriers, with Mitscher on board the *Lexington,* went north to search an area near the island of Polillo, 150 miles from Manila.

Admiral Toyoda was at his Tokyo headquarters when he was informed that American troops were expected to invade Leyte. He had a difficult decision to make. Loss of the Philippines would cut lines of communication between the main islands of Japan and her remaining occupied positions in the Southwest Pacific. Although he hated to sacrifice his fleet, he could see no sense in saving it if the Philippines were taken.

American submarine attacks on Japanese shipping had drastically reduced supplies of precious oil at home. Toyoda knew Admiral Ozawa's carriers in the Inland Sea were short of trained pilots and planes. Toyoda discounted Japanese reports of a dozen American carriers sunk in recent weeks, but he believed many of them had been destroyed by his navy. He heeded the cries from commanders on Formosa who had lost heavily during attacks by carriers and B-29s, and sent 150 planes and pilots from Ozawa's four carriers. They were only partially trained, but Ozawa voiced vigorous protest. He now had only 116 aircraft for his carriers.

Toyoda instructed Vice Admiral Takeo Kurita to maintain his heavy ships at Singapore because of limited oil reserves in the Empire. On October 18 Kurita was

Admiral Soemu Toyoda of the Imperial Navy aboard his flagship *Oyodo.*
(Courtesy of U.S. Navy.)

ordered to put to sea. Heading north to Leyte Gulf, his fleet was impressive, composed of five battleships, including the huge *Yamato* and the *Musashi,* each with nineteen-inch guns and heavier than any American battleship, plus ten heavy and two light cruisers and fifteen destroyers. But he had only scouting seaplanes, so he traveled almost blindly.

Kurita was only part of the Japanese Fleet—which had been identified by the Americans as composed of three elements, including southern, central, and northern forces—which had been ordered by Toyoda to converge on Leyte Gulf from as many directions.

Toyoda was aware that Kurita's fleet would be no match against Halsey's Third Fleet, so he decided to lure Halsey away by dangling Ozawa's almost planeless carriers in front of him. They would probably be sacrificed, he thought, but they were of little value without planes and trained pilots.

American submarines dogged Kurita's fleet. After leaving Brunei Bay, the submarine *Darter* torpedoed his flagship *Atago*. The Japanese tried to avoid further torpedo attacks, but next the *Maya* was sunk and the *Takao* was hit. Three of Kurita's best cruisers were now out of action, and Kurita himself had to shift flagships twice, ending up on the *Yamato*. He grinned wryly to himself when he read the latest orders from Fleet Admiral Toyoda in Tokyo. "Probable enemy are aware of concentration of forces."

Toyoda informed Kurita that the Americans would probably concentrate in the San Bernardino and Surigao straits, and instructed him to proceed east of San Bernardino Strait and Tacloban and destroy the transports. Toyoda warned him to be in position to attack the invasion fleet by the afternoon of October 24. "Carry through our original plans," Toyoda concluded.

Kurita was advised to increase his alertness against the submarine menace and to utilize particular caution in penetrating the narrow strait. Shore-based planes, Toyoda assured him, would destroy the American carriers. The Japanese had 700 aircraft, but most were obsolete and their pilots were young and unseasoned. Admiral Shigeru Fukudome's Second Air Fleet had returned to the Philippines from Formosa, but it was reduced to 350 planes of all types. Vice Admiral Takijiro Ohnishi's First Air Fleet—he had helped to plan the Pearl Harbor raid—was down to sixty pilots, who were totally demoralized by recent heavy losses and inadequate planes and therefore of little use. The Fourth Air Army had about 150 airplanes of all types at bases all over the Philippines. Their pilots had a low record and a total lack of aggressiveness.

Ozawa also received orders to leave the Inland Sea to attract Halsey's attention. He told his subordinates sadly, "Our fleet is so weakened, I expect total destruction. If Kurita can destroy the invasion fleet it will be worthwhile."

Ozawa led his decoy ships through Bungo Channel on October 20. His flagship, the *Zuikaku,* had been in on the attack at Pearl Harbor. Three light carriers, the *Chitose,* the *Chiyoda,* and the *Zuiho,* and the converted battleships *Hyuga* and the *Ise,* with only aft flight decks, made up the main part of his fleet, although three light cruisers and eight destroyers completed it.

The converted battleships had no planes. The others had only eighty fighters, thirty-one torpedo planes and seven dive-bombers, for a total of 118 planes, but the crews were green, hardly able to land and take off from the carriers.

After less than two years of war, this was all that remained of Japan's carrier aviation. Unfortunately, the Americans had no way of knowing this. Halsey considered the Japanese carriers the main threat to a successful invasion of the Philippines. If they had been up to full strength, with trained crews and airplanes, he would have been correct in the decision he was about to make.

13

Combustible, Vulnerable, and Expendable

Toyoda had devised a clever plan. He split his available naval forces into three parts. One under Vice Admiral S. Nishimura and another under Vice Admiral Shima would come in upon Leyte through the Surigao Strait, while the largest force under Kurita would come through San Bernardino Strait to hit Leyte from the north. Under the circumstances it was a good plan, but its success hinged upon timing, and also upon luring Halsey away from the beachhead with Ozawa's "bait" carriers, which would approach Luzon from the north.

It was evident to the American commanders that the Japanese were up to something, but they were not sure just what. A submarine gave the first intimation of possible action when it reported fifteen to twenty warships west of Mindoro on an easterly heading. This proved to be Kurita's central force.

Halsey ordered his carrier groups to launch at dawn, searching every square mile of the western approaches to the Philippines. A search plane from the *Cabot* spotted Kurita's attack force and radioed back to flagplot (the room where the ship's course is plotted) aboard the battleship *New Jersey:* "Four battleships, eight cruisers, and thirteen destroyers are off the southern tip of Mindoro."

Orders went out to fast carrier group commanders from Halsey. Rear Admiral Ralph E. Davison had his group to the south. Rear Admiral Frederick C. Sherman was off Luzon, and Rear Admiral Gerald F. Bogan's group was in the center. "Concentrate your forces with Bogan," Halsey said.

Halsey ordered McCain, en route to Ulithi, to return immediately, even though he was 635 miles from the coast of Samar.

Halsey worried about enemy carriers because none had been reported. He sent Mitscher top secret instructions to keep the area north of Sherman's Task Group 38.3 under close surveillance.

Captain William H. Buracker's carrier *Princeton,* with Sherman's group, ordered his men to battle stations at dawn on October 24 because radar had spotted a large number of unidentified airplanes approaching from the direction of Manila. Then, radar plot called. "There's a second group the same size fifteen miles behind the first one." He immediately ordered fighters into the air, and Sherman ordered more fighters to be launched while he maneuvered his ships into rain squalls for better protection.

The *Princeton's* Fighting 27 ran into the Japanese attackers first, and—with only eight Hellcats—tore into the formation of eighty planes. Twenty-eight attackers fell, streaming flames into the ocean, but one Japanese bomber got through and dropped a bomb among loaded torpedo planes on the hangar deck. Damage control parties went to work to put out raging fires.

Sherman now received orders to take his task group to San Bernardino Strait, so he reluctantly had to leave the flaming *Princeton* behind. Just when it appeared the flames were coming under control, a tremendous explosion killed many of the carrier's men as well as some on rescue ships alongside.

Captain Buracker gave the inevitable order, "Abandon Ship!" American destroyers, after rescuing the survivors, sent a spread of torpedoes into the doomed *Princeton.*

Meanwhile, Kurita continued toward Leyte Gulf, heading in a zigzag course through the islands of the Philippine Archipelago. He had divided his force by the time he passed the southern tip of Mindoro, heading north through the Tablas Strait.

Kurita, who was engaged in a desperate gamble to reach the American invasion fleet, felt a sense of frustration because of his lack of airpower. At 10:15 A.M. his radar picked up a large group of American planes heading his way. He knew he was in a desperate situation, and became even more convinced of it after the first torpedo attack badly damaged the huge *Musashi* and the heavy cruiser *Myoko.* The *Myoko* was in such bad shape that Kurita ordered her to proceed along to Brunei.

Twenty-five planes now hastened to the battle area from the carriers *Intrepid,* the *Cabot,* and the *Independence.* At a quarter to one they attacked, seeking out the damaged *Musashi.* The mighty ship, built with blister-protected sides to reduce the havoc caused by torpedoes, was struck three times. The American fliers watched with awe as she slowed down, evidently badly hurt because she drifted in circles.

Kurita suspected he faced Halsey's fast carriers, which by now should have been lured north by Ozawa's tempting carrier targets. He informed Ozawa, "Subject to repeated enemy carrier-based air attacks. Advise immediately of contacts and attacks made by you on enemy." Then he asked for fighter protection from Manila, but they had none to spare. They had all they could do at the time to stand off Sherman's group off Luzon.

Admiral Davison's carriers of Task Group 38.4 found no ships in their area, so he dispatched planes from the *Enterprise,* the *Franklin,* the *San Jacinto,* and the *Belleau Wood* against Kurita's ships. Commander Dan Smith of the *Enterprise's* Air Group 20 arrived on the scene to find Kurita's ships proceeding at twenty knots eastward despite the damage they had suffered.

Kurita's guns opened up before they came within range, and Smith watched the colored bursts carefully. When he saw that they trailed his formation, he stopped worrying about the flak and ordered an attack against the *Yamato* and the *Musashi.*

Just then Air Group 13 arrived from the *Franklin.* When the planes entered a cloud bank, Japanese guns stopped firing, but they opened up again with renewed intensity once the group emerged in the clear. Smith led his torpedo planes in, and all eight torpedoes seemed to score direct hits on the ship's bow. Bombers came in next, making five direct hits and three near misses.

The *Musashi* lay still in the water, her entire forecastle awash. Kurita, noting her condition—down at the bow—ordered her to reverse course and head back through San Bernardino Strait.

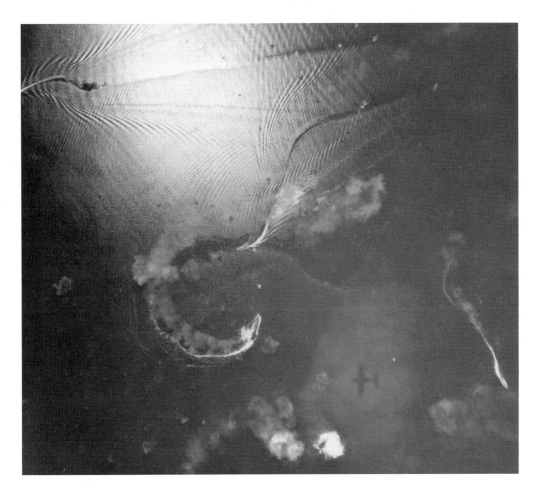

The Battle for Leyte Gulf. Japanese ships try to avoid American air attacks. (Courtesy of U.S. Navy.)

Kurita was almost in a state of shock. His attack force had undergone five air attacks in the last six hours. He was behind schedule because he had intended to steam through San Bernardino Strait after sundown. The fierceness of the American attacks had increased each time, whereas he was completely without air support. He decided to retire temporarily beyond the range of these relentless Navy planes and proceed with the attack later.

Toyoda's carefully laid plans seemed to have failed, because it appeared to Kurita that Ozawa's carriers had not diverted Halsey from the invasion area. Kurita's force had been savagely reduced.

Kurita felt he was fighting the battle alone with nothing to show for his sacrifices. He retreated through the Sibuyan Sea, but fliers from Bogan's 38.2 carriers made a final attack just before 6:00 P.M.

The *Musashi* had received nineteen torpedo hits and seventeen bomb hits and American Navy fliers were astounded that she remained afloat. She didn't for long, and attempts to beach her on the north coast of Sibuyan Island failed. She capsized at twilight, dragging 2,200 members of her crew with her to the bottom.

The *Musashi* under attack. Air Group 20 from the *Enterprise* makes simultaneous bombing and torpedo runs on the huge Japanese battleship in the Sibuyan Sea October 24, 1944. (Courtesy of U.S. Navy.)

In Tokyo, when Toyoda learned that Kurita was in retreat, he was beside himself. He had informed Kurita previously that the assault on the Philippines had to be stopped even at the sacrifice of the entire Japanese fleet.

He wired Kurita, "With confidence in heavenly guidance, the entire force will attack!" Kurita, who had planned only a temporary withdrawal, had already turned his force around and headed back toward San Bernardino Strait before he received this message from his Commander in Chief.

Kurita radioed messages to Toyoda and Admiral Nishimura, whose southern attack force was also en route to Leyte Gulf. "My force of four battleships, six cruisers, two light cruisers, and eleven destroyers will pass through the San Bernardino Strait at 1:00 A.M. tonight. Will proceed southward down east coast of Samar and arrive Leyte Gulf 11:00 A.M. October 25."

Halsey was still obsessed by the absence of the Japanese carriers. He told his staff, "It is unthinkable the Japs would undertake a major operation without carrier support."

Admiral Mitscher, with Sherman's Task Group 38.3, remained under constant attack from bases in the Philippines. Rain squalls gave them some respite from the attackers, but his pilots reported many of the 150 planes they shot down were carrier types. (The latter had flown off Ozawa's carriers en route to land bases in the Philippines.) Mitscher's suspicion that carriers were in the vicinity was strengthened by these disclosures, so he dispatched scouts to patrol to the northeast.

While Kurita's main striking force was retiring at 4:40 on the afternoon of October 24, Mitscher's planes had finally contacted an enemy force 200 miles off northern Luzon. The first contact reported three battleships, half a dozen carriers, and six destroyers. Ranging farther afield, the planes found three carriers, three cruisers, and three destroyers on a westerly heading.

When the information reached Halsey he felt a sense of relief. The reports of a southern attack force were confirmed at noon, apparently coming through Surigao. At last the Japanese fleet had been located.

Halsey paced the bridge of his flagship. The news called for decisive action by his Third Fleet. It appeared that concerted action was planned by the Japanese for the following day—October 25.

Halsey decided upon a course of action. He would take Task Force 34 under Vice Admiral W.A. Lee Jr. with six new battleships and his fast carriers to attack the northern carrier force. Task Force 34 had been integrated into Task Force 38 at Ulithi.

Halsey briefed Admiral Lee on his estimate of the situation and his decision to go after the carriers. Lee objected. He suspected the Japanese would be trying to divert attention from a main striking force, which could spell disaster for the invasion fleet off Leyte. "I don't trust the sons-of-bitches," he said forcefully.

Halsey was adamant, "Destruction of the enemy carrier force offers the best possibility of defending the invasion front."

Lee protested again. "It will leave the San Bernardino Strait unguarded."

"The center force [Kurita] has been struck severe blows," Halsey replied. "They may inflict more damage, but any advantage the Japs may gain can be overcome by prompt action of the fast carriers."

If Halsey had known Ozawa's carriers had almost no planes, his decision might have been different. But he made the biggest mistake in judgment of his long career, and compounded it later by indecisiveness.

He now directed Davison's Task Group 38.4 and Bogan's 38.2 to head north with him. Toyoda's plan was finally working out the way he had envisioned.

At 11:30 that night Sherman's Task Group 38.3 joined Halsey and the other groups for the run north with Admiral Mitscher in tactical command under Halsey.

At 7:25 P.M. this same evening Vice Admiral Thomas C. Kinkaid, Commander of the Seventh Fleet under General MacArthur, received an urgent dispatch that an enemy force was sighted off Sibuyan Island to the south. He was not unduly alarmed at first because he assumed Task Force 34's heavy ships would engage such a force and turn it back.

Kinkaid had received word from Halsey shortly after 3:00 P.M. that Task Force 34 would be formed, although Halsey failed to tell him he was taking it against the northern Japanese carrier force. The San Bernardino Strait, therefore, was completely undefended by modern warships.

Rear Admiral J.B. Oldendorf, commanding the Leyte bombardment and fire support ships, received a message: "Prepare for a night engagement." He dashed off a message to his commanders. "Consider surface attack tonight via Surigao Strait imminent. Make all preparations." Admiral Oldendorf hoped to plug the Surigao Strait as he flanked both sides with cruisers and destroyers and sent five picket ships deep into the strait.

Japanese air attacks persisted from land bases because Admiral Fukudome's Second Air Fleet in the Philippines had been strengthened to 350 planes by flying them in from the homeland and China.

As Admiral Nishimura's seven warships headed north for Leyte Gulf they were repeatedly attacked by PT boats and destroyers. Kinkaid now received reports that this southern force of Japanese warships was in two groups, one forty miles behind the other. Vice Admiral S. Nishimura was ahead with the battleships *Fuso* and the *Yamashiro,* a heavy cruiser, and four destroyers.

In his rear, a second force under Vice Admiral K. Shima had two heavy cruisers, a light cruiser, and four destroyers. Shima had left Japan's Inland Sea under orders to assist Nishimura. But the latter had no report of Shima's whereabouts, and the only word Kurita had received from Nishimura was that he had located the Americans.

Oldendorf had positioned his battleships across the entrance to Leyte Gulf and sent his PT boats in first. Then destroyers were sent against Nishimura's ships, crippling several with their torpedoes and sinking the destroyer *Yamagamo.*

Nishimura's force was now reduced to the *Yamashiro,* the cruiser *Mogami,* and the destroyer *Shigura.* They proceeded north to continue the battle.

Nishimura blithely sailed into a well laid trap. Moving in a single column, the Japanese ships allowed Oldendorf's older battleships to "cross the T" ahead of them, so that only the forward batteries of Nishimura's ships could be brought to bear in battle. It was an event of which admirals had long dreamed but few had ever seen, as the amassed firepower of Oldenberg's old ships sent full broadsides into Nishimura's ships. The exploding hell killed hundreds, with Nishimura one of the first to die at 4:19 A.M.

Vice Admiral Shima entered the Surigao Strait just as the holocaust ended. His ship flashed a challenge to a ship heading south. "I am the *Shigura,*" the ship replied. Then it sped on, the only survivor of Nishimura's task force. "I am the *Nachi,*" Shima's ship answered. Neither ship bothered to question the other further, but went their separate ways.

Shima passed the burning cruiser *Mogami* and quickly realized that the ships ahead were not Japanese. He ordered a turn, which was executed in such haste that his flagship collided with the *Mogami.* Although his cruiser was badly damaged he steamed south to get away from the battle zone. The *Mogami* got under way later and escaped Surigao Strait, but planes from escort carriers found her. She was sunk along with the *Bukomo* of Shima's force.

Rear Admiral Clifton Sprague stood on the bridge of the escort carrier *Fanshaw Bay* at 6:45 A.M. watching his planes take off for another search for the enemy. Combat Information Center gave its Captain Douglas F. Johnson a message from one of the pilots. Johnson immediately showed it to Sprague. "Enemy surface force of four battleships, seven cruisers, and eleven destroyers sighted twenty miles northwest of task group and closing at thirty knots!"

"Tell that pilot to check his identification." The admiral said. This was the last place he expected to find enemy ships. Johnson spoke into the squawk box. " Confirm report, " he told the center. "Identification confirmed," air reported. "Ensign Brooks says ships have pagoda masts."

Sprague, who had been looking intently toward the northwest, now saw antiaircraft bursts in the sky. "Course 090. Flank speed," he ordered. "Launch all aircraft."

None of them had expected Kurita to return, particularly after the drubbing he had taken. But there he was, and Sprague found himself with only six escort carriers, three destroyers, and four destroyer escorts to bar Kurita's path to Leyte Gulf.

The Seventh Fleet's escort carriers (CVEs) were designed to support amphibious operations. They were never intended to fight against heavy warships because they were thin-skinned, their attack planes were equipped with light bombs and weapons, and the CVEs were, as many bluejackets (enlisted men) humorously said, "combustible, vulnerable, and expendable."

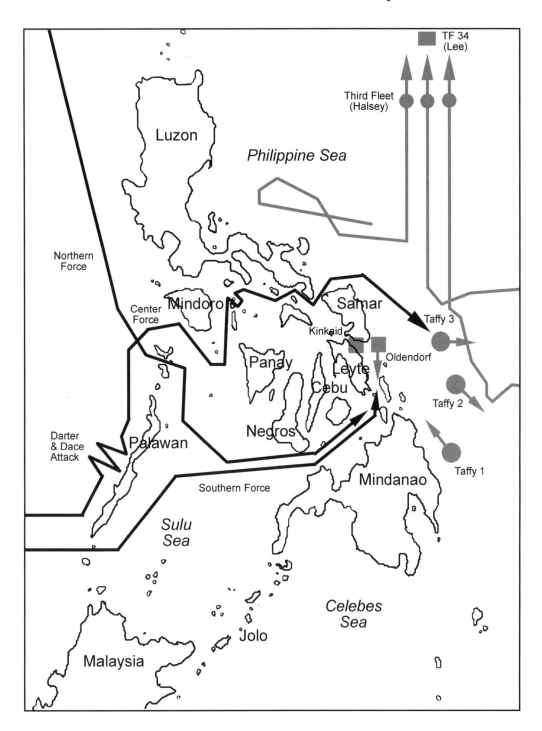

The Battle for Leyte Gulf in the Philippines. (Drawn by Stewart Slack.)

With a deck half the length of a standard carrier, rigged over a hull designed for a tanker or merchantman, the baby flattops were used primarily for air support of ground troops and antisubmarine patrols. Now they faced some of the mightiest warships in the Japanese Navy.

Sprague's group was one called "Taffy One," "Taffy Two," and "Taffy Three" that had arrived October 18 to support the invasion. As Kurita headed directly for them, they were fifty miles off the southern coast of Samar. Rear Admiral Felix Stump headed another group thirty miles to the southeast, and Rear Admiral Thomas L. Sprague had a third group 120 miles away.

The American and Japanese groups spotted one another simultaneously.

Kurita quickly changed course to the east, and when eighteen miles away he ordered the *Yamato* to open fire. This was the first time these big eighteen-inch guns had been fired in combat.

It was a minute before 7:00 A.M. on October 25 when salvos crept closer to the Seventh Fleet's "jeep" carriers. There had been no time to change the loading of the planes, and even the torpedo planes had only small, general purpose bombs and depth charges.

Sprague ordered his destroyers to attack first. "Lay a smokescreen between the Japs and the carriers," he ordered.

The slow CVEs formed a large circle with the *St. Lo* in the north, then the *Kalinin Bay,* the *Gambier Bay,* the *Kitkun Bay,* the *White Plains,* and Sprague's flagship *Fanshaw Bay.*

The *Gambier Bay* and the *Kalinin Bay* received heavy punishment as the Japanese cruisers poured their fire into them. Thin-skinned destroyer escorts attacked the cruisers with their five-inch guns against the cruisers' heavy guns.

It was an unequal contest, but the "jeep" carriers made wildly evasive turns at full speed, barely missing most of the heavy shells from the cruisers. Japanese cruisers closed in, in spite of air and torpedo attacks, until hits were scored on the *Fanshaw,* the *Kalinin,* and the *Gambier.*

Sprague barked an order over the radio. "Small boys [destroyers] on my starboard quarter. Intercept enemy cruiser coming down on my port quarter."

Destroyers and destroyer escorts fought violently, but there was little they could do. Soon the destroyer *Hoel* sank after heavy fire and the *Johnston* received mortal wounds. Sprague's force was now completely out of torpedoes. It was pressed closer and closer toward Samar, only fifteen miles away, by battleship attacks from astern.

At 8:30 A.M. Sprague spoke over the TBS system, "Small boys on my starboard quarter. Interpose with smoke between carriers and enemy cruisers." Rolling clouds of smoke created momentary protection, but the torpedo planes of the task group were forced to make dummy runs on the enemy ships—they were completely out of bombs and torpedoes. Sprague believed it was just a matter of minutes before his ships and MacArthur's transports would be destroyed.

While Kurita's main striking force had been retiring temporarily the previous afternoon of October 24, Mitscher's planes had reported an enemy fleet 200 miles off northern Luzon, including carriers on a westerly heading.

While Halsey's fast carriers and the modern battleships of Task Force 34 had continued to search for the Japanese carriers, Captain James Russell, Davison's chief of staff, monitored the tactical radio network. He listened intently when Halsey's watch officer came on and, using Halsey's voice call identification, asked Mitscher what he made of the situation. Mitscher's chief of staff, Commodore Arleigh Burke, used Mitscher's voice call to respond immediately. "I recommend you form Leo." "Leo" was Task Force 34's voice call.

After a pause, Halsey ordered, "Form Leo."

Now some of Davison's ships designated for Task Force 34 began to break away from the formation and report to Vice Admiral Willis A. Lee. Davison's Task Group 38.4 was thus reduced to two-thirds of its normal size.

Commodore Burke had been pressing Mitscher to urge Halsey to get his battleships down to Leyte Gulf to plug the strait because there were insufficient carriers and other ships to take on the Japanese carriers. Mitscher refused, saying that Halsey must have information they did not. He added, "But this doesn't look right to me."

Halsey was convinced that night that the Japanese planned on concerted action on the following day—the 25th. He told his staff that the situation called for decisive action by his Third Fleet. Despite some of their protests, he insisted that Lee's Task Force 34 remain with his fleet, claiming he would need its six new battleships as well as the fast carriers.

When Lee continued to dispute Halsey's decision, the Third Fleet Commander replied, "Destruction of the enemy carrier force offers the best possibility of defending the invasion front."

"It will leave the San Bernardino Strait unguarded," Lee said in protest.

"The center force has been struck severe blows," Halsey said. "They may inflict more damage, but any advantage the Japs may gain can be overcome by prompt action of the fast carriers." In his opinion, he said, if Kurita did return he was confident that Kinkaid's battleships could repulse him.

At 11:30 P.M. on October 24, Sherman's Task Group 38.3 joined Halsey and the other two task groups for the run north to attack the Japanese carriers. Mitscher was in command of the fast carriers, and Halsey's flagship the *New Jersey* was integrated tactically with Task Group 38.3 after Task Force 34, or Leo, was formed.

As they all steamed north, Davison turned to his chief of staff, Captain Russell, and said "Jim, we're playing a helluva dirty trick on the transports at Leyte Gulf."

At 7:25 P.M. the night of October 24, Vice Admiral Thomas C. Kinkaid, Seventh Fleet Commander, received an urgent message that an enemy force had been sighted off Sibuyan Island. Although he did not know it, this was Nishimura's task force. Kinkaid had no doubt it was headed for Leyte Gulf, but he was not concerned because he assumed that Task Force 34 was there to protect the invasion ships. He had received word from Halsey at 3:00 P.M. that afternoon that Task Force 34 had been formed, and he believed that it was now covering the San Bernardino Strait. Instead, however, it was with Halsey's fleet heading north. Halsey had told Kinkaid that he was taking three groups to attack the enemy carrier force at dawn on the 25th. The fourth, McCain's Task Group 38.1, was still en route back from Ulithi under emergency orders from Halsey, but Kinkaid believed it was guarding the San Bernardino Strait. Kinkaid also assumed that Task Force 34 would be available if Kurita returned, but Halsey had had no such intention.

During the early morning hours of October 25, while Kurita's ships moved along the coast of Samar to engage Oldendorf's fleet near Leyte Gulf, Mitscher's scout planes lost contact with Ozawa's fleet. Mitscher was concerned that the Japanese would locate them first and launch an attack, so he ordered some of his bombers into the air to fly fifty miles ahead of his carriers, as search planes continued to roam widely over the surrounding area.

Halsey received a message from Kinkaid off Leyte Gulf: "Is TF 34 guarding San Bernardino Strait?" He radioed back that TF 34 was with him. This prompted an immediate message from Kinkaid that Kurita's battleships were shelling his escort carriers off Samar.

Mitscher's concern about Ozawa's carriers grew in direct proportion to the lengthening period of time during which his scout planes continued to fail to locate them again. When a staff member suggested that the enemy carriers might be further east, Mitscher responded, "You may be right. Send four planes from the *Essex* combat air patrol to search the northeast."

A pilot reported at 7:30 A.M. that he had found enemy carriers on a northeasterly course, and radioed back their longitude and latitude. This report indicated that the Japanese carriers were 130 miles away.

In the clear, Mitscher radioed David McCampbell, target coordinator for his attack planes, "Ninety-nine Rebel. Take charge of incoming strike and get the carriers."

McCampbell ordered his attack groups, "Follow me in," after he sighted the Japanese fleet. He quickly noted that there were four carriers, two or more battleships, and a number of smaller ships all within a radius of five miles. His wingman, Lieutenant Wayne Morris, followed McCampbell in an attack on the Japanese fighters. McCampbell soon increased his score of enemy planes. By war's end he had shot down thirty-four Japanese planes—more than any other Navy pilot. McCampbell's group as a whole was quite proficient as well, boasting twenty-seven aces.

Rear Admiral "Ziggy" Sprague called in plain English from his flagship in Leyte Gulf while Mitscher's pilots were going after the Japanese carriers: "I am being shelled by a Japanese battleship."

Seventh Fleet Commander Kinkaid on board the *Wasatch* in Leyte Gulf heard that transmission. Now that he knew that Task Force 34 was not guarding the San Bernardino Strait, but rather with Halsey far to the north, he anticipated that Sprague's support fleet would soon be destroyed.

While McCampbell's fighters afforded protection to the strike groups, Commander Daniel Smith called his air group: "Pick one out, boys, and let him have it."

When McCampbell saw fifteen to eighteen Oscar fighters dead ahead, he ordered an immediate attack, and they were quickly shot down. To his surprise, not another enemy plane rose to challenge the attacks by the strike groups, although flak was heavy around them as they approached the carriers.

Ozawa and his carriers had headed for Luzon on October 23, and the next day he sent out patrols with orders to attack the American ships. It was a hollow gesture, and he was perfectly aware that his few remaining airplanes would be sacrificed in vain. Inasmuch as it was part of Toyoda's plan for the Americans to locate his four carriers, three light cruisers, and eight destroyers, their failure to do so had caused Ozawa increasing concern.

On October 24, only 150 miles from Luzon, he ordered his attack units to take off at 11:45 A.M. He gave their commanders precise instructions to find the American carriers, attack, and then proceed to airfields in the Philippines. He watched them assemble and fly away. There were seventy-six planes in all. This was part of the attacking force that had hit Sherman's task group and sunk the *Princeton*.

When Ozawa received reports that Kurita's ships were under heavy air attacks in the Sibuyan Sea, he felt his mission was a failure. Obviously, he believed, Halsey failed to fall for the ruse of his decoy ships. He decided to send his battleship-carriers the *Ise* and the *Hyuga* on ahead with four of his large destroyers in the hope they could engage the Americans the night of October 24.

But it was not until the next morning that Mitscher's planes located Ozawa's fleet. With Commander David McCampbell of the *Essex*'s Air Group 15 in charge of the planes from Sherman's group, Commander William Ellis of the *Intrepid*'s Air Group 18 with Bogan's planes under him, and Commander Daniel Smith of the *Enterprise*'s Air Group 20 in charge of Davison's fliers, target coordinator McCampbell signaled for the attack to begin against Ozawa's carriers.

McCampbell watched one carrier as it turned out of formation as if preparing to launch airplanes. (There were none to launch.) Bombs plummeted down on the hapless carrier, so McCampbell ordered torpedo planes to head for a battleship.

After Smith's group finished their attack, every ship trailed oil, and the carrier *Chitose,* smothered by bombs from the first strike, sank. The largest carrier, the *Zuikaku,* was hit with a torpedo, the light carriers *Zuiho* and the *Choyoda* were heavily bombed, and the *Choyoda* had to leave the protection of the formation. Other ships also took a drubbing: the destroyer *Akitsuka* blew up and sank, and the light cruiser *Tama* was slowed drastically by torpedo hits.

While the first strike was under way, Halsey received another urgent dispatch from Kinkaid: "Jap battleships and cruisers are fifteen miles astern of my escort carriers off Samar."

Halsey was furious with Kinkaid, believing he had panicked in a tight situation. He could not understand his frantic calls for help, knowing Kinkaid had been in every major Pacific action and had always demonstrated a superb coolness under fire.

Halsey listened to the calls from his strike groups with satisfaction, noting that Commander Malcom T. Wordell reported that the carrier *Zuiho* was under attack. This was the Japanese carrier that had sunk the old *Hornet.*

But Halsey became increasingly angry as Kinkaid continued to send frantic messages. At 8:02 he reported that enemy ships were retiring through the Surigao Strait. These were the surviving ships from Nishimura's task force. Halsey believed that their retreat indicated the Leyte Gulf situation was under control. He therefore failed to appreciate the significance of another message from Kinkaid at 8:22 A.M., saying, "Enemy battleships and cruisers reported firing on TU 77.4.3." This was a reference to Sprague's escort carriers.

Kinkaid advised at 9:00 A.M., "Our CVEs being attacked by four BBs, eight cruisers plus others. Request Lee cover Leyte and at top speed." This message was followed at 9:22 by, "CTU 77.4.3 under attack by cruisers and BBs. Request immediate air strike. Also request support by heavy ships. My OBBs"—a reference to his old battleships—"low on ammunition."

Despite Kinkaid's increasingly urgent messages, Halsey refused to change his plans for continuing the attack with all his ships against Ozawa's fleet. This failure prompted Kinkaid to dispatch a desperate message at 10:00 A.M. in the clear (not coded), saying, "Where is Leo? Send Leo."

This message was picked up by Nimitz's radio monitors on Guam. After it was referred to the Commander in Chief, Pacific, he sent off a message with the customary padding wordage at the beginning and the end to confuse Japanese decoders: "Turkey trots to water GG, From CincPac, Action Com Third Fleet, Info CTF 77, where is rept where is Task Force 34? RR the world wonders."

All padding was normally deleted before a message was delivered, but despite the RR the last three words were kept in the message that was sent to Halsey. Halsey flushed angrily when he read the message, assuming that "the world wonders" was a rebuke. His hands shook, and he snatched off his cap, hurled it to the deck, and swore.

Rear Admiral Robert B. Carney grabbed Halsey's arm. "Stop it! What the hell's the matter with you? Pull yourself together." Halsey thrust the message at Carney and rushed off. He was bitter and for an hour angrily refused to give the order to Admiral Lee to return.

In Washington, Admiral King had been following the messages. Halsey's actions angered him, and he told his chief of staff, "He has left the Strait of San Bernardino open for the Japanese to strike the transports at Leyte!"

Halsey had lost control of his better judgment. The Third Fleet's battle lines—with Admiral Lee's new battleships—was only forty miles from the damaged enemy carriers far to the north of Kurita's surface forces. Throughout the Pacific War battleships never had a chance to prove themselves against other capital ships. It was an opportunity that Halsey did not want to miss.

Meanwhile, at Leyte Gulf, Admiral Sprague was convinced that his fleet of escort carriers would soon be overwhelmed, leaving the invasion fleet open to quick annihilation by Kurita's ships. Kurita's ships moved within fifteen miles of his carriers at 8:30 A.M.

Rolling clouds of smoke gave the carriers momentary protection. Their torpedo planes, unable to rearm, were forced to make dummy runs on Kurita's ships.

Sprague moved his carriers to a southwesterly direction at 9:10 A.M. so he could head for Leyte Gulf. He expected total destruction because his ships were almost defenseless. Japanese cruisers bore down on them only 10,000 yards away—point-blank range for their eight-inch guns. Sprague resigned himself to the inevitable. He had lost the *Gambier Bay,* three destroyers were sunk or sinking, and other ships were badly hit.

Sprague, lost in thought, looked up as someone shouted, "they're turning back!" Kurita, on the verge of total victory, had ordered his ships to reverse course and abandon the battle. After learning what had happened to Nishimura he had decided to retreat at 9:11 A.M. after his force had been cut in half with the loss of one of his five battleships and half of his ten cruisers. He headed back for San Bernardino Strait, en route to Brunei.

Two American escort carrier groups in the south were now in position to launch attacks. Admiral Stump, in the center, had found time to equip his planes with torpedoes and semi-armor-piercing bombs, as well as rockets. He launched six attacks, but a lone kamikaze plane got through to the escort carrier *St. Lo* and sank it.

Now McCain's task group, returning at high speed from his aborted trip to Ulithi, harassed Kurita's ships from the air while the Thirteenth Air Force's B-24s bombed him en route to Brunei.

Halsey finally ordered Lee to return to the San Bernardino Strait at approximately 11:00 A.M. at maximum speed. Halsey was furious because he believed that he had been called off the biggest sea battle of the century. But by this time Kurita was in full retreat and all danger to Kinkaid's fleet had passed, although Halsey was unaware of this turn of events. Halsey ordered Bogan, despite his protests, to take his task group with Lee to provide air cover, particularly after Lee's battleship task force came within range of Japanese bombers in the Philippines.

Mitscher, with his two remaining task groups, carried on the attack against Ozawa, whose force was in a deplorable state after the second wave of American planes finished their bombing at 11:00 A.M. His flagship, the *Zuikaku,* proved difficult to steer, so Ozawa transferred to the light cruiser *Oyodo.*

He also decided to draw the American carriers farther north, but then a third wave of 150 planes, directed by Commander Richard L. Kibbe of the *Franklin's* Air Group 13 and Commander Hugh Winters of the *Lexington's* Air Group 19, appeared on the scene. Mitscher had told them, "Stay over the target as long as your gas holds out."

Winters looked down on the Japanese fleet and counted three carriers. They lacked formation and there were cripples among them. He called his formation's attention to the carriers heading north.

Winters ordered an attack on the *Zuikaku.* Their bombs covered the carrier's deck, then their torpedoes hit and plumes of water shot high in the air. The *Zuikaku* caught fire, but burned for two hours before she sank. The *Zuiho* was tougher to hit. Smaller, and extremely maneuverable at high speed, she stoutly took several hits.

Winters wrote rapidly on his knee pad, not trusting his memory at the height of the air battle, that a cruiser and a destroyer were sneaking off to the northwest. He called another air group commander: "Follow that oil slick to the northwest. You'll find a big Jap cruiser."

Mitscher, apprised of the situation, sent the cruisers and destroyers forward to pick off the stragglers and cripples, and the *Choyoda* became the first victim.

Air Group commanders, in the excitement of ordering one attack after another, failed to satisfy the curiosity of carrier commanders as to details of the action. The calls repeated over and over, "Get the carriers!" Finally, in exasperation, Winters called back to the *Lexington,* "They're all going under the water." He heard a whoop on his radio, then silence. At last an excited voice said, "That's all we wanted to know."

Commander Wordell of the *Langley's* Air Group 44 coordinated the last two strikes of the day. His planes were equipped with armor-piercing bombs, which finally sank the stubborn *Zuiho* after three hours of action.

Ozawa had fulfilled his mission and turned hastily toward Japan with the remnants of his fleet, including the battleship-carriers *Ise* and the *Hyuga.*

Planes from the *Hancock,* the *Hornet,* and the *Wasp* in McCain's Group in the south found Kurita's force along the coast of Samar at 1:10 P.M. and swooped down to attack the fleeing ships. The planes were lightly loaded because of the 335 miles they had had to travel to reach the coast of Samar. But McCain's carriers had closed the gap, so their return flight would not be so long. A second strike of fifty-three planes went after the Japanese with bombs and rockets at 3:00 P.M.

By now Halsey was aware that Kurita's fleet was in full retreat and he ordered his ships to block the San Bernardino Strait—but Kurita had slipped through three hours earlier. The large destroyer *Nowake* failed to make it and was destroyed.

Planes from Bogan's group joined with McCain's at dawn on October 26 to attack enemy ships in Tablas Strait along the east coast of Mindoro. The light cruiser *Nashiro* succumbed to their bombs. Fifth Air Force B-24 bombers also made an attack on Kurita's remaining ships, but only managed to straddle the *Yamato*. Kurita's First Division Attack Force at last reached the Sulu Sea beyond the range of Third Fleet carriers.

It had been a nerve-shattering experience for all participants, and Halsey was warm in his praise of their handling of a tight situation. "For brilliance, courage, and tireless fighting heart the all-hands performance since early October will never be surpassed. It has been an honor to be your commander. Well done. Halsey." Halsey's superiors were not quite as pleased with his actions in the battle.

14

Operation Gratitude

Under cover of the great sea battles, Japan managed to get 2,000 men ashore at Ormoc, with protection provided only by the destroyer *Uranami* and the light cruiser *Kino*. After Japanese troops were safely on the beaches, Stump's escort carriers sank the two warships and two of the four transports.

One final action completed the historic battle: the *Shiranuhui*—one of Shima's destroyers—got caught off Panay by carrier planes and was sent to the bottom.

The Japanese almost succeeded in destroying the Leyte beachhead and its support ships. Without carrier aviation in numbers sufficient to be decisive, land-based airplanes found it impossible to protect the surface ships.

Kurita could have dealt considerable damage if he had not broken off the action. He was acting almost blindly (his cruiser scout planes over Leyte Gulf all disappeared without a trace) so he was without adequate air coverage ahead of his fleet. The confined waters of Leyte Gulf were no place to fight without air support.

Japanese reports after the war cleared up several things that puzzled American commanders at the time. The whole Japanese operation had been so inept that it defied classification or explanation, but one major cause was the fact that there never was any coordination between Kurita, Nishimura, or Shima. They each acted independently of the other, with disastrous consequences. Kurita said after the war that once the Americans knew he was coming, he did not believe the Americans would leave their transports almost undefended in Leyte Gulf. He assumed, he said, that the transports were no longer there. His views at the time gave Halsey credit for more common sense than he actually demonstrated. He said he also believed that four *Essex*-class carriers, possibly six, had been sunk earlier. In effect, he believed his own country's propaganda. He also confused the escort carriers with fast carriers. Kurita said he believed there were two American airstrips on Leyte—(there were none)—so his ships would be exposed to land-based airpower.

Kurita tried to explain his precipitous retreat from Leyte Gulf by saying that he decided to aid Ozawa by taking action against Halsey but that this action receded so

A Japanese *Yamato*-class battleship under attack in Kure Bay by Navy fliers October 25, 1944. (Courtesy of U.S. Navy.)

quickly to the north that it would require more fuel than his destroyers possessed. He retired instead through the San Bernardino Strait.

If he had known that all of Mitscher's fast carriers were attacking Ozawa on the morning of October 25, Kurita could have continued into Leyte Gulf with little opposition.

Kinkaid's Seventh Fleet, whose primary mission was to land troops and support them, had fought aggressively and with remarkable courage in a battle they were not prepared to fight.

The Japanese suffered a staggering blow. In effect, their fleet had ceased to exist. They lost the battleships *Musashi, Yamashiro,* and *Fuso,* the large carrier *Zuikaku,* and the light carriers *Chitose, Choyoda,* and *Zuiho.* Their other losses included six heavy cruisers, four light cruisers, and eleven destroyers. In all, twenty-six major combat ships totaling 300,000 tons were sunk.

Total American losses were the light carrier *Princeton,* the escort carriers *Gambier Bay* and the *St. Lo,* plus two destroyers and a destroyer escort. The *St. Lo* fell victim to the first planned suicide attack of the war, while six other escort carriers were damaged.

After the invasion of the Marianas was followed successfully by the Battle of Leyte Gulf, the Imperial Navy found it had lost an additional 1,046 aircraft in support operations during October. It was a blow from which the Japanese never recovered.

The Japanese never understood how American carriers could remain on station so long. The answer was not revealed until after the war. Eight to twelve oilers had been stationed 400 miles east of Leyte Gulf—protected by two escort carriers—which routinely refueled ships on a specific schedule.

After four divisions of Lieutenant General Krueger's Sixth Army were put ashore by Kinkaid's Seventh Fleet, Halsey wired Nimitz on Guam and King in Washington: "In the course of protecting our Leyte landings, the back of the Japanese fleet has been broken."

Halsey wired MacArthur on October 26, "After seventeen days of steady fighting, the fast carriers are virtually out of bombs, torpedoes and provisions. The pilots are exhausted. Unable to provide any intensive direct air support. When will your shore-based air take over air defense of the Philippines?" MacArthur replied the next day. "Land-based air will take over responsibility for direct support to the troops in Leyte–Samar area."

After many difficulties the Army established airstrips for P-38s, but defense of the area was beyond their limited capability. Army long-range B-24s and P-38s found it impossible to stop the landing of Japan's crack 1st Division despite heavy attacks.

When Lieutenant General Sosaku Suzuki and his Thirty-fifth Imperial Army landed the same day, it was evident that Allied air had lost control of the airspace above Leyte after most of the carriers had been withdrawn. In this difficult period, the Japanese increased their Leyte forces by 22,400 men because their main source of air strength was on Luzon, out of reach of Army planes.

When units of the Third Fleet were ordered to Ulithi, Halsey had to leave two groups of seven carriers to aid the deteriorating air situation. Every ship in the fleet was stripped of ammunition for the carriers remaining behind. Although carrier crews were exhausted, and hardly fit for daily operations, they fanned out attacks over Leyte, the Visayas, and Luzon.

While the situation on Leyte remained tense and insecure, Halsey retired with the rest of his fleet to Ulithi, postponing with regret his plans to strike at the heart of Japan to test her strength.

In the Southwest Pacific—a forgotten battlefront as far as headlines were concerned—men still died, as attacks continued against Japan's isolated outposts. Submarines continued their relentless hunts, and the *Sealion* found big game when it sank the veteran battleship *Kongo*.

Japanese aircraft from Luzon posed a threat that had to be liquidated. Although Halsey's attack groups badly needed recuperation at Ulithi, the situation became so serious that he ordered his fast carriers to return to the battlefront.

McCain replaced Mitscher in command of Task Force 38, so Task Group 38.1 was placed under the command of Rear Admiral Alfred E. Montgomery. Task Group 38.3 remained under Sherman as they sortied from Ulithi to meet Bogan's group halfway between Ulithi and Luzon. Davison's group returned to Ulithi as part of Halsey's replenishment plan to permit continuous carrier operations.

The fast carriers arrived before dawn on November 5, 160 miles from Manila and only eighty miles from the coast after a high-speed dash from the rendezvous. Montgomery's carriers were assigned the Clark Field complex on Luzon's central plain where six Japanese airfields and fourteen runways were in constant use. Bogan's carriers were assigned the area comprising southern Luzon. The fields in the middle, plus shipping in the Manila area, were assigned to Sherman's group, comprised of the *Essex,* the *Ticonderoga,* and the *Lexington.*

Fighters swept the fields at dawn and caught the Japanese by surprise. Hellcats found good hunting, and fifty-eight Japanese planes succumbed to their skilled attacks. They reported back to the flagship that hundreds of Japanese aircraft, obviously many flown from Japan and Formosa, were camouflaged on the ground.

Sherman's pilots pounded shipping in Manila Bay, sinking the heavy cruiser *Nachi* in shallow water as it attempted to escape. Top secret war plans, detailing Japanese strategy and operational procedures, were recovered by divers from the ship after Manila was captured.

Back on the carriers, two kamikazes attacked, but were destroyed before they could do any damage. A third, braving a tornado of fire from the 20- and 40-millimeter guns and the five-inchers, slammed into the bridge of the *Lexington.* Captain Ernest W. Litch winced as the blinding flash quickly turned into smoke and flames. Surgeons and hospital corpsmen immediately took care of the wounded and the chaplain hurried to attend to his sad duties.

The *Ticonderoga*'s guns roared at four planes intent on crash dives from 3,000 feet, managing to destroy them just before they hit the ship. The days were past when ships could destroy most of an attacking force. Now, even one kamikaze could inflict serious damage.

On the following day all three carrier groups went after shore targets. During the two-day assault, nine enemy ships were sunk. Damage was caused to other Japanese ships as well.

Men fighting the desperate battle on Leyte would now have 400 fewer Japanese planes to contend with. Air opposition declined quickly at Leyte, so the strikes were worthwhile. The *Lexington* and the *Reno* had been damaged, and twenty-five planes were lost in the strikes, along with eleven attributed to operational accidents.

The Third Fleet hoped to return to Luzon, but a storm changed their plans. Refueling of the ships was a necessity, so the fleet returned to an area 400 miles west of Saipan. Halsey recalled Bogan's group to Ulithi for a welcome respite from operations, returning Davison's group to replace them.

Sherman assumed tactical command of the force on his flagship, the *Essex.* After midnight on November 10 he was handed a dispatch from Halsey. "Cancel fueling. Reverse course and proceed at best speed to Central Philippines to support our ground forces on Leyte Island." The sudden change of plans was brought about because Japanese heavy warships were reported heading for Balabac Strait.

The more Halsey learned of this new threat, the more concerned he became, because it had the huge *Yamato* with its 18-inch guns, and three older battleships, along with a cruiser and four destroyers. He was also advised that a reinforcing convoy could be expected to arrive at Ormoc Bay the following day. The Third Fleet was a thousand miles away, but it steamed at twenty-six knots and headed back.

After a spectacular high-speed run, carriers arrived 200 miles off the San Bernardino Strait early on November 11. Search planes covered the area but failed to find a single enemy ship. At last the convoy was found, but there was no trace of the warships. Four cargo ships, five destroyers, and a destroyer escort were heading south between Leyte and the northern tip of Cebu.

Sherman launched 347 planes, attacking the convoy as it rounded Apali Point at 11:00 A.M. Bombers from the *Essex,* the *Ticonderoga,* and the *Langley* destroyed the four transports in a few minutes.

Limited airfield facilities on Leyte and the lack of strong Air Force concentrations of fighters and bombers forced MacArthur to insist on continued use of carriers for his Philippine operations. By agreement with MacArthur, Sherman launched his main strikes at Luzon, where shipping in Subic Bay–Lingayen Gulf and Manila Bay offered the greatest concentration of targets. In a closely timed operation, fighters swept Clark Field clear of enemy planes at dawn on November 13, while bombers headed for shipping.

It was a wise decision, because little opposition was found over the ship-crowded Manila Bay. During the attacks, flak caused heavy losses to American planes, with twenty-five shot down during the day. The following day reconnaissance photographs revealed that the docks at Manila and Cavite Naval Yard had been hard hit. Japanese ships in the bay also suffered, with a light cruiser, four destroyers, three tankers, and three cargo ships resting on the bottom. These losses were serious, but the loss of the tankers began to have an adverse effect on all Japanese operations.

Task Force 38 withdrew for refueling and replenishment of military supplies.

Army engineers struggled valiantly to complete airfields on Leyte but heavy rains caused one delay after another. It was inevitable, therefore, that MacArthur should request of Nimitz that Halsey's fast carriers be brought back for the Mindoro operation scheduled for early December. During November, fast carrier forces had sunk 134,000 tons of shipping and destroyed over 770 aircraft, while losing 117 planes but no ships.

In Washington the Bureau of Aeronautics continued to think about the future of Navy airpower after the war. It was now recognized that turbojet and turboprop power plants would ultimately replace reciprocating engines. Bauer, on November 6, requested that the Naval Air Material Center begin to study requirements for a laboratory to develop and test gas-turbine power plants. This action led to the establishment of the Naval Air Turbine Test Station at Trenton, New Jersey.

On November 28, Commander Joseph F. Enright led his submarine *Archer Fish* in an attack against the largest aircraft carrier afloat. While the *Shinano,* a converted battleship, tried to sneak to safety into Japan's Inland Sea, it was sent to the bottom by a spread of torpedoes. It had never launched a plane, and in fact was still not fully completed, and therefore was not watertight. It sank quickly as a result.

MacArthur's troops found recapture of the Philippines more difficult than he had thought. Heavy rains made jungles a quagmire, and advances were measured in yards instead of miles.

The Japanese Air Force, with a continual flow of replacement crews, made the job much more difficult, and, despite all efforts, the Japanese Army managed to reinforce its beleaguered troops.

Despite these efforts, the Japanese were fighting a losing battle, and by the first week in December they found themselves jammed into the northwestern part of

Leyte. It was not until December 26, however, that MacArthur could announce that all organized resistance on Leyte was at an end.

Fifty-six thousand Japanese had died, and the American Army had lost slightly more than 3,000 on Leyte. It had been the toughest fighting the veteran Army had encountered since Guadalcanal. Before all Japanese troops were eliminated in the next six months, MacArthur's army lost another 24,000 men in the Philippines.

General T. Yamashita had wanted to fight only a delaying action on Leyte while he prepared for the decisive battle on Luzon. But he was overruled, so his army needlessly lost a vast number of troops, along with thousands of tons of vital supplies.

After a brief period for refueling, Task Force 38 under Admiral McCain made strikes against Luzon's airfields and shipping in support of the Mindoro landings between December 14 and 16. MacArthur had insisted on taking Mindoro—an island slightly larger than Leyte—to build airfields for his coming invasion of Luzon. Mindoro, south of Luzon, is separated from it by a narrow channel. With seven heavy and six light carriers, McCain's fliers flew successive combat air patrols to spread an aerial blanket over Luzon, effectively pinning down all enemy aircraft on the island. They accounted for a major share of the 341 enemy aircraft destroyed in this brief campaign.

In more direct support of the Mindoro invasion, planes from six escort carriers of Task Unit 77.12.1 under Rear Admiral Stump, along with Marine Corps shore-based aircraft, flew cover for the passage of transport and assault shipping through the Visayas. The escort carriers also provided direct support for landings by Army troops on December 15, and in the assault area on December 16 and 17. On the night of D-Day, Navy seaplanes joined with operations from Mangarin Bay.

With a hundred enemy airfields to cover on Luzon, McCain and his staff devised a plan to keep Japanese aircraft on the ground night and day for the first three critical days of the Mindoro invasion. McCain assigned each of his task groups a specific area of responsibility, and he divided the airfields equally among them. He ordered that first priority be given to the destruction of enemy aircraft. Fuel tanks, gas trucks, and storage facilities that might interfere with air operations were given a second priority. He advised his pilots that if they became temporarily out of assigned targets, they should attack ships.

The carriers were positioned 200 miles northeast of Manila on December 14, and for the next three days they stood guard over Japan's airfields, attacking immediately any time a Japanese plane appeared. Kamikazes flew out to strike the carriers, but the combat patrol dispatched every one of them.

The Mindoro operation proceeded without enemy air opposition as McCain's fliers ranged far and wide of the battlefront, shooting down sixty-four Japanese planes and destroying 200 on the ground. Of greatest satisfaction was the fact that not one American plane was lost to enemy fighters, although Japanese antiaircraft guns proved brutal, destroying fifty-four American planes.

Vice Admiral Shigero Fukudome, commander of the Japanese Second Air Fleet on Luzon, had been soundly defeated. He had been able to reinforce his air fleet to almost 700 planes until the middle of November, but McCain's pilots made every one of them ineffectual.

McCain ordered his carriers to withdraw again for refueling once American troops were dug in on Mindoro, promising to return on December 19 if they were needed.

Halsey's Third Fleet, refueling east of the Philippines, blundered into a killer typhoon on December 18 that capsized three destroyers and damaged several ships, including four light carriers of Task Force 38 and four escort carriers of the replenishment group. Loss of life was heavy, totaling 700 men. One hundred-fifty carrier planes were also lost in the storm.

In Washington, Admiral King called for a private investigation of the incident, assigning Vice Admiral John Hoover to head the inquiry. Hoover blamed Halsey for the disaster despite Halsey's insistence that he had not been negligent. Halsey blamed the weather. But the investigating committee exonerated Halsey, claiming that his mistakes were the result of errors in judgment brought on by the war's stress. With utter inconsistency, the committee commended Halsey for his desire to keep hitting the Japanese even if he had to fight his way through a typhoon.

King privately agreed with Hoover, but he permitted Halsey to retain his command. Halsey was a big hero in the eyes of the American people, and King uncharacteristically backed off by not censuring him for his gross negligence.

Although McCain had promised MacArthur to return on December 19, Halsey was forced to advise the general that the fast carriers would not return on that date. "Unable to strike Luzon before 21st at dawn."

The Mindoro invasion proceeded smoothly until Admiral Kinkaid received word on December 26 that Japanese surface forces were heading for Mindoro. Four cruisers and several destroyers were immediately dispatched to meet this threat. While Japanese warships headed for Mindoro Strait, Army planes attacked, and then PT boats went into action and repulsed the threat. Ships off the beachhead, however, were damaged by Japanese bombers.

After the first of the year, with the Mindoro operation proceeding according to plan, Kinkaid went to see MacArthur about the invasion of Luzon. "I believe a landing in the Manila area would be less costly," Kinkaid said. "Convoys would be less exposed to kamikaze attacks." MacArthur shook his head. "For Army purposes, the plains of Luzon are needed as a place to maneuver." They discussed plans for several minutes, then MacArthur said thoughtfully, "I believe the Japs are off balance. We should hit them quickly with a knockout punch—a great assault on Luzon."

After extensive arguments with his staff, MacArthur agreed to postpone the Lingayen Gulf operation until January 9, 1945, because he needed close cooperation of the Third Fleet. Halsey, meanwhile, laid plans for his carriers to attack Japanese airpower on Formosa and the northern tip of Luzon prior to and during the landing phases.

In coordination with Southwest Pacific Forces, Halsey divided Luzon with an imaginary line that crossed inland, north of the Lingayen Gulf. "All Japanese airfields north of that line are fair game for us," Halsey told his commanders. "General Kenney's Fifth Air Force has all hunting rights south of it."

The Japanese now learned a lesson about airpower they themselves had tried to teach earlier in the war. But without the industrial capacity of the Allied nations, particularly the United States, they never found it possible to attain the heights of air employment now unleashed upon them.

Bombers of the Fourteenth Air Force and the 20th Bomber Command in China continued to strike telling blows at Formosa, while Marianas-based B-29s of the 21st Bomber Command laid the foundations for the eventual destruction of the homeland.

Despite almost daily attacks against the Japanese air force in the Luzon–Visaya area, by scraping the bottom of the barrel of their air reserves throughout the Far East, the Japanese still had 450 planes on Luzon; 300 were north of Manila in the complex of fields around Clark.

The Third Fleet, after leaving Ulithi on December 30, arrived on the scene with two fighter squadrons of Marines. Their gull-winged Corsairs were placed on the *Essex,* the first time Marine pilots had operated from fast carriers since a few days before Pearl Harbor. At that time the *Saratoga* delivered twelve Marine fighters to Wake Island.

While the Japanese hoarded their planes on Luzon, they kept another 400 in reserve on Formosa. This prompted Task Force 38 to pay another visit to the strategic island, and despite bad weather their attack destroyed planes on the ground and sank a number of ships.

Admiral Oldendorf's heavy ships and minesweepers preceded the invasion fleet into Lingayen Gulf on January 3, while seventeen escort carriers of Task Group 77.4 under Rear Admiral C.T. Durgin stood by to protect them. Oldendorf sent spotter planes over land prior to starting ship-to-shore preinvasion bombardment. After the first pilot circled the area, he reported, "I can't see a single military installation. If there are any, they're camouflaged."

"Cancel the bombardment," Oldendorf said. Then the minesweepers went to work. Not one mine was found.

Oldendorf found it hard to believe. He looked up as antiaircraft fire broke out through the fleet. Japanese Army fighters swept down, and it was obvious they were out for big game. The kamikazes shifted quickly to the battleships, and the *New Mexico* reeled under the impact of a hit.

Oldendorf decided to withdraw his bombardment ships as the raid intensified. He dictated an urgent message to Kinkaid: "Consider need of additional airpower urgent and vital. Our CVEs entirely inadequate providing air cover. Japanese suicide bombers seem able to attack without much interference owing to radar difficulties.

"Believe in addition all fields small as well as large near Lingayen must be continuously bombed and kept neutralized."

When Kinkaid received this message on January 5 he was disturbed. The situation was far worse than he realized. He looked again at the losses listed with the memorandum. They included the sinking of the *Ommaney Bay* on January 4, and on the following day several other ships had been hit, including the escort carriers *Manila Bay* and the *Savo Island.* A total of twenty-one ships had been either hit or sunk in the past twenty-five hours.

Kinkaid requested that Halsey return his fast carriers immediately to protect his invasion forces. Halsey had just refueled his ships and was about to move into the China Sea, but abruptly changed his plans. Although the weather was foul, Halsey sent fliers back over Luzon. The men of the Seventh Fleet had to be protected at all costs.

When the kamikaze attacks had been reduced, Oldendorf sent his bombardment ships back into the Lingayen Gulf to continue the work they had started. When troops went ashore on January 9, the XIV Corps landed near the town of Lingayen and the I Corps at San Fabian.

Planes from the escort carriers rushed to attack ahead of the advancing troops, lashing Japanese defenses with bombs and rockets. The section northeast of San Fabian developed the enemy's greatest resistance, so planes concentrated there.

After Sixth Army troops seemed safely ashore at Lingayen, Halsey felt Japanese airpower was sufficiently neutralized on Luzon and Formosa to undertake a pet project that he called "Operation Gratitude." In other words, he had not forgotten his original intention of sweeping the South China Sea.

Halsey's fliers attacked airfields along the coast of China, destroying fourteen enemy planes in the air and another ninety-five on the ground. Four Japanese convoys, hugging the coast for protection, also fell prey to the attackers, and the light cruiser *Kashii* went down.

On Luzon, Admiral Fukudome, commander of all naval planes in the archipelago, was ordered to Singapore with his remaining thirty flyable airplanes. His two air fleets—originally totaling almost 700 aircraft—had been almost entirely wiped out.

Task Force 38—with seven heavy and four light carriers in three groups, and one heavy and one light carrier in a night group, accompanied by a replenishment group and seven escort carriers—concentrated on the destruction of enemy airpower and air installations in surrounding areas. In spite of almost continuous bad weather that hampered flight operations during the entire month of January, Task Force 38 launched offensive strikes on Formosa and the Ryukyus on January 3 and 4, made a two-day attack on Luzon on January 6 and 7, and struck airfields in the Formosa–Pescadores–Ryukyus area on January 9. In one week of preliminary action, over one hundred enemy aircraft were destroyed and 40,000 tons of merchant and combatant ships were sunk.

Halsey was particularly intrigued by a report that the battleship–carriers *Ise* and the *Hyuga* were at Singapore. When he learned they had left on January 9, possibly to interfere with the Lingayen supply routes, he ordered his Third Fleet that night to head out of Luzon Strait. Halsey anticipated that he would find the *Ise* and the *Hyuga*—along with their cruisers and destroyers—near Cam-ranh Bay on the Indochina coast. What Halsey did not know was that American submarines had destroyed the Japanese tankers, forcing the battleship carriers to return to Singapore.

During the night of January 9 and 10 Task Force 38 made a high-speed run through Luzon Strait, followed by the replenishment group, which passed through Balintang Channel for operations in the South China Sea. The task force struck targets along the Indochina coast on January 12 for 420 miles, as far south as Saigon, where they caught ships in the harbor and in coastal convoys. The results of these raids proved devastating to the Japanese, as twelve tankers were sunk, and twenty passenger and cargo ships and numerous small combatant ships totaling 149,000 tons were destroyed.

Halsey ordered his task force now to move north to evade a typhoon, and on January 15 it struck targets at Hong Kong, along the China coast, and on Formosa. The next day he again ordered attacks on the Hong Kong area, where additional shore installations were destroyed, and another 62,000 tons of shipping were sunk.

Bad weather forced Halsey to leave the South China Sea, and his force made an after-dark run through Balintang Channel on January 20 to hit targets on Formosa, the Pescadores, and Okinawa. Enemy air action caused damage to the *Ticonderoga* and the *Langley*, but Halsey repeated an attack on the Ryukyus the next day. This ended three weeks of action, with an aerial score of 600 enemy aircraft destroyed and 325,000 tons of shipping sunk.

Opposition was heavy at only one place during this period. Returning pilots from Canton described the flak in awed terms. For the first time in months American losses exceeded those of the Japanese. With forty-seven enemy planes down, the carriers lost sixty-one planes, thirty-one due to operational accidents.

Halsey was exuberant. He wired Nimitz: "We have completed a 3,800 mile cruise in the China Sea," he said, "and not one Japanese plane got within twenty miles of our ships. That's gratitude."

MacArthur, recalling vividly how ticklish the situation had become when Japanese warships almost reached his beachhead at Leyte, still worried about a repeat

An SB2C approaches the *Hornet* after an attack against Japanese shipping in the China Sea in January 1945. (Courtesy of U.S. Navy.)

performance in the Lingayen Gulf. Nimitz advised him not to be concerned about the fact that four enemy battleships were definitely in empire waters and other powerful Japanese fleet units were at Singapore.

"Best naval protection for the Philippines is to proceed with offensive operations against Japan," he said.

While the Third Fleet was busy at sea, the fighting on Luzon continued against serious organized resistance. Manila did not fall until February 23, and it was not until April 16 that the Philippines were officially liberated. Some Japanese resistance groups fought for another three months.

Even though Japanese snipers were still on Corregidor on March 2, General MacArthur returned in a PT boat the same way that he had left "the Rock" thirty-six months before. As MacArthur walked up to the assault commander, Colonel George M. Jones, the men stood stiffly at attention and saluted smartly. With a new flag in his hand, MacArthur approached the color guard at the base of the battered flagpole. "I see the old flagstaff still stands. Hoist the colors to its peak and let no enemy ever haul them down."

Iwo Jima, a Japanese-held volcanic island between the Marianas and Japan, now stood almost alone in an American-controlled Central Pacific. Now that B-29s were participating in strikes against Japan from bases in the Marianas, its importance as a haven for badly shot-up Superfortresses grew acute. The Japanese were aware of this, and planned to contest every foot of ground.

Its occupation would also permit fighter escorts for the big bombers. Search planes covering the southern coast of Japan, and Japanese patrol vessels flashing early warnings of B-29 raids to the homeland, could be eliminated. Almost all Pacific commanders agreed that Iwo Jima had to fall into American hands. It was a job ideally suited to the Navy and Marines, so preparations had been under way for months.

Spruance, Commander of the Central Pacific Task Forces of the Fifth Fleet, was placed in charge to assault this small island only four and a half miles long. He was promised the fast carriers, but these ships were only a small part of the 900 ships scheduled for the invasion.

Rear Admiral Calvin T. Durgin's support carrier group of eleven escort carriers headed for Iwo to attack on February 16. In a two-pronged offensive, the fast carrier forces stood off Honshu, Japan, to help isolate Iwo from reinforcements.

It was the largest carrier task force ever assembled. Task Force 58, the advance units of the Fifth Fleet, had seven carriers, including two carriers and five light carriers. It also had eight fast battleships, a battle cruiser, five heavy cruisers, eleven light cruisers, eighty-one destroyers, and more than 1,200 aircraft. Such was the resurgence of American military might since the disaster at Pearl Harbor that only four of these ships had been with the Fleet on December 7, 1941.

These same ships, with different designations, again replaced top commanders. In other words, the Third Fleet became the Fifth Fleet, with Spruance and Mitscher instead of Halsey and McCain.

After they left Ulithi it was obvious to the men that something big was up. There was much speculation, but the truth did not come out until they were out of sight of land. Officers on each ship reported to their wardrooms to bear an identical message from their captain. "Gentlemen, it gives me great pleasure to announce you will strike Tokyo." Later the public address systems reverberated with the excitement of their commanders' voices, "Attention all hands! This is the captain speaking. Our target is Tokyo. We'll strike at dawn on the 16th."

While escort carriers were neutralizing defenses at Iwo Jima, it had been agreed that the fast carriers could perform a more useful function if they struck hard at Tokyo and the Kobe–Nagoya area that same day in advance of the invasion on February 19th. Everyone knew it would be difficult. Spruance told his staff, "The fleet will be in a dangerous position off the coast of Japan. If we can't take it, it will be worse off Okinawa later on."

The ever-present picket boats ahead of the Fifth Fleet were replaced by a large screen of destroyers.

One hundred and twenty miles off the coast of Honshu, silent pilots soberly read Mitscher's final memo on the bulletin boards, "The coming raid on Tokyo will produce the greatest air victory of the war for carrier aviation, but only if every air group commander, squadron commander, combat team leader, station leader and individual pilot abides by the fundamental rules of air combat that have been taught since the war started. Those of us who can't get over the target will be doing all we can to get you back safely."

General quarters sounded in predawn darkness. Planes were ready, wings folded to their fuselages, as fighter pilots hurried on deck from the final briefing. The air officer's voice brought them to a standstill. "On the mark it will be 6:32. Mark!" Pilots shivered as the cold wind penetrated their flying clothes as they synchronized their watches.

"Standby to start engines...stand clear of propellers!" came the call. "Start engines." A rippling roar broke the stillness on deck and blue flames stabbed backward from the planes.

The plane director, identified only by the dim light emanating from the palms of his hands, signaled the first plane to unlock wings. Handlers hurriedly pushed the wings forward and into flying position. Waving lights, intelligible to the pilots, brought each plane forward on the dark flight deck.

"Launch planes!" the air officer yelled. Bombers from the carriers waited until fighters finished their sweeps. The sky over Tokyo was so overcast, however, that fighters had to circle back to the coast.

It was over China that they ran into their stiffest opposition. The *Hancock's* Fighting 80 downed seventy-one Japanese planes, surpassing the previous record by the *Lexington* during the Marianas "Turkey Shoot." Over Honshu, fighters bore the brunt, while bombers from the *Lexington,* the *Bunker Hill,* and the *San Jacinto* attacked the Ota airframe plant. B-29s had struck it a week before, but Navy bombers caused extensive new damage.

In all, 700 interceptors rose from the Tokyo plain to stubbornly resist this heavy attack. Twenty-four American planes fell victim to Japanese fighters, but the Americans, in turn, destroyed 281 enemy planes in the air and another 200 on the ground.

That night the *Saratoga* and the *Enterprise* launched their night fighters for a sweep of Tokyo's airfields, ranging as far west as Hamamatsu.

Bad weather limited sorties the next day, but on the 17th 200 planes—half of them bombers—attacked airfields and aircraft and engine plants on the northwest outskirts of Tokyo. Enemy fighters met them at the coast, and they had a running fight to and from the targets.

Planes from the *Essex,* the *Bunker Hill,* and the *Cowpens* singled out an important plant for significant damage when they attacked the Tama-Musashino Aircraft Engine Factory.

The weather grew progressively worse, so by noon Spruance and Mitscher called off the operation and proceeded to Iwo Jima.

Although attack groups had found the Japanese fighters alert and aggressive—losing thirty-four fighters and two torpedo bombers—not a single Japanese bomber penetrated to the carriers offshore. The Japanese lost 416 planes in the air and another 354 on the ground—planes they desperately needed.

In the dusky dawn of a Pacific morning two days later, blacked-out ships approached Iwo Jima, or "Sulphur Island." It was part of a larger group including the Bonins, extending more than 700 miles from the entrance to Tokyo Bay and south to within 300 miles of the northern Marianas Islands. Thus, Iwo Jima was the key to a strong inner ring of Japanese defenses.

The northern half of Iwo Jima is roughly circular in shape, forming a broad dome with a maximum elevation of about 380 feet and two and a half miles in width. But the island tapers to half a mile in width at the base of the 554-foot Mount Suribachi near the southern tip.

The north coast is rocky and dominated by cliffs. Therefore the only available landing beaches were on the southeastern and western shores. The southeastern shoreline extends northward from the base of Suribachi for two miles and ends in high, rocky cliffs. Almost at the water's edge, terraces of varying height and width, created by heavy storms, fringed the beaches.

American intelligence sources knew that Iwo Jima was defended by Lieutenant General Tadamichi Kuribayashi of the Japanese Army's 109th Division. He had lived in Canada and had visited the United States before the war, and he had been impressed by the Western world's mass production techniques.

After the Marianas fell that summer, Kuribayashi had told Imperial Army headquarters what he would need to properly defend the strategic island. He said he must have all the guns and ships they could send him and far more troops. "Give me these things," he radioed his superiors in Tokyo, "and I will hold Iwo."

After August of 1944 American planes and submarines choked off his supply routes to the mainland, and his garrison was reduced to a 30-day supply of rice and a 15-day supply of other food. He also warned that he faced a continuous water shortage, alleviated only by trapping as much rainwater as possible during the frequent rainstorms.

Lieutenant General Holland M. Smith commanded the expeditionary troops while Major General Harry Schmidt was in charge of the Amphibious Corps. They agreed that the landing on the island's southeastern beaches should be spearheaded by the 4th and 5th Marine Divisions. They would land abreast at H-Hour on February 19, while the 3rd Marine Division was held at sea as a floating reserve. The 5th Division was ordered to cut off the southern part of the island at its narrowest point at the foot of Mount Suribachi, and then move north by advancing along Iwo's long axis on the island's western side.

The Marines had trained on the Hawaiian Islands' largest island. Hawaii had been selected because its terrain was somewhat similar to that of Iwo Jima, with its volcanic soil.

Captain John Tanner, commanding officer of B Battery, 1st Battalion of the 5th Marine Division's 13th Regiment, had left Hawaii on board an LST with his men on February 22. LSTs were poor sailers in rough weather, and they encountered plenty of that. Approximately 200 Marines were jammed into the 180-foot vessel, so there was little privacy. Some of Tanner's men had been in on the invasions of Guadalcanal and Tarawa, but most, like Tanner, were new to combat. There had been constant shipboard drills en route, including extensive map reading to acquaint themselves with the island's topography.

On the morning of the invasion—set for 9:00 A.M.—Tanner's LST and dozens of others arrived on the west side of the island and circled to the south before turning east to stand offshore at Red Beach 2.

Preinvasion bombing and shelling had been going on for days. B-29s had used the beach as a practice range for new crews. The bombing was not as effective as usual, because the Japanese were dug deeply into the volcanic soil, often to a depth of twenty feet, where bombs and shells could not penetrate. Even worse, this type of soil cushioned the impact of an explosion more than other types of soil.

After the first troops moved on shore (the Japanese had elected not to contest the landings, but to rather wait until the troops got on shore), wicked-looking red flashes erupted on the purplish-colored slopes of Mount Suribachi and casualties began to mount. Mortar fire was particularly heavy, and it took a great toll. Soon the beaches were littered with equipment and shattered landing craft that were hit by the vicious mortar fire while machine guns chattered incessantly whenever a head emerged from a temporary foxhole carved in the soft soil.

Orders went out to delay further infantry landings until the beaches could be cleared. But Tanner's battalion was ordered on shore earlier than planned to get its artillery into action against the withering fire of Japanese guns and mortars. They boarded their amphibious DUKWs and moved toward shore at noon. Underwater demolition teams had cleared the entrance of mines, but Red Beach 2 was so jammed they did not try to land. The control ship ordered them to land on Red Beach 1 instead. This time, despite the clogged beach, they landed without casualties despite increasingly heavy mortar fire. Tanner jumped off the rear of his DUKW and found himself in chest-deep water. Almost immediately they ran into the terraces formed by storm action through the years, where mortars fell unmercifully on them. They were pinned down, unable to move inland.

Tanner asked a man whose unit had preceded them, "Where are the front lines?" The man replied that they were right over the knoll in front of them. Their situation was intolerable: the position they had been assigned was still in Japanese hands. Until it was taken, Tanner knew they had no place to go.

Inching forward over steep terrain, each movement was impeded by heavy incoming fire. Mortars were deadly, and their firing was constant, but no one could spot their location. All they knew was that it was coming from their left from the slopes of Mount Suribachi, and to their right as they tried to move the half mile across the island to cut it in two. Close air support proved impossible because the battle lines were too close together and often intermixed with enemy positions. Somehow they forged ahead, and by nightfall advance troops reached the other shore.

That night the 3rd Division's 27th Marine Regiment in front of them called on Tanner's battery to support their 700-man attack against Motoyama Airfield 1. (There were three airfields on the island.) Tanner established a fire-direction center in a small Japanese-built tunnel. They fired all night long, but they took a beating from counterbattery fire. They lost 350 rounds of 75-mm ammunition when enemy rounds blew them up, and two of Tanner's howitzers suffered direct hits. As casualties mounted, Tanner tried to report to battalion headquarters, but he found that his communication system was out.

He discussed the situation with his executive officer, and they agreed that the first sergeant should be ordered to pull everyone back a hundred yards on signal. They were convinced there was no point in remaining where they were and just taking it.

Once communications were restored, and headquarters learned that his battery was largely ineffective due to losses of ammunition and guns, one of Tanner's remaining howitzers was assigned to another battery, and his executive officer took the other to the front. Thus the battalion ended up with two five-howitzer batteries instead of three four-howitzer units. In effect, Tanner's outfit was disbanded, and he and his men were assigned to infantry units or other artillery units.

On D-Day plus four a patrol from the 2nd Battalion, 28th Regiment moved up Suribachi's gutted slopes. It was a straight head-on attack over land defended by Japanese soldiers. The battalion's leader, Sergeant Sherman Watson, reported to the rear that the Japanese were still holed up.

The battalion commander, Lieutenant Colonel Chandler W. Johnson, put together a forty-man patrol under Lieutenant Harold G. Schrier. He told the lieutenant, "If you reach the top, secure and hold it. And take this along." He handed Schrier a small American flag.

They filed through reeking battle debris and blasted gun pits and started up the northern face of Suribachi, often on their hands and knees as the slope steepened. Despite heavily mined trails, they reached the top and spread out along the crater's rim. The Japanese had fled. Suribachi had fallen, although there were days of mopping-up operations. Schrier ordered that the small flag, given him by his commanding officer, should be raised on the summit. It was attached to a hollow pipe. Japanese snipers tried to kill the flag raisers, but the snipers were quickly dispatched by Marine marksmen. Four hours later the small flag was replaced by a larger one that could be seen by many Marines below them in the surrounding area. It was a cheering sight in an otherwise cheerless situation.

Meanwhile, as the 28th Marines were taking Suribachi, the rest of the 5th Division, the 4th Division, and the 3rd turned north. Now Tanner was operating almost independently with a radio operator, PFC Henry Walter, to spot artillery for anyone who needed help. They made only minor advances during the next three days—only fifty yards in one difficult day—because of increasing opposition. The cold, rainy weather added to their misery. They encountered Japanese mines everywhere, and the roads to the north were thick with them.

On D-Day plus nine the 5th Division reached Hill 362, the highest point on the western side of the island, and the backbone of the Japanese defenses in this sector. Sharp, barren edges of land were chopped up by caves, concrete pillboxes, and blockhouses on its almost vertical cliffs. At its base, jagged, rocky outcrops commanded every approach to the hill. Japanese mortars proved deadly, causing the majority of the casualties. They would hit without warning, unlike artillery shells. (If you could hear an artillery shell you knew it had gone by you.)

There were extensive underground tunnels, built prior to the invasion by Korean conscript laborers. When the Americans fired, the Japanese would duck into their tunnels, relatively safe from even heavy shells. The island was honeycombed by these tunnels. Unique Japanese blockhouses puzzled the Americans until they learned their secrets. Their double-walled construction left hiding places for Japanese soldiers, who removed a brick in the inside wall to fire at the advancing Americans. At first, when blockhouses were overrun, there would seem to be no one in them, but then fire would issue from them. Throughout the drive north, furious hand-to-hand fighting occurred all along the front.

For Tanner and everyone else the nights were particularly bad, because the Japanese would wander through the American positions dressed in American uniforms they had stripped from dead Marines. All such persons were suspect, but in that first moment of indecision, the Japanese would fire at the Americans. Confusion became the order of the day as the Marines continued their advances.

The Navy had misjudged the amount of time needed to take the island, and after several days the frontline units ran short of flares. They were reduced to two flares an hour for each regiment, and the Japanese found it easy to infiltrate in pitch-dark conditions as they crawled through the American lines trying to bayonet Americans in their foxholes. It was always an eerie feeling for Tanner and his men to hear someone groan with pain, knowing the Japanese had found another victim.

At first "spider traps" also baffled the advancing Marines. Tanner and his radio operator came under rifle fire one day from behind them, but they couldn't figure out where it was coming from. Then they realized that a Japanese sniper was hidden below ground with a manhole-type cover. After they passed, he would shoot at them and then pop back into his hole, pulling the cover over him. Once the secret was disclosed, a grenade would finish off the sniper.

During the hard drive north, Tanner and his radio operator received messages from one battalion after another to direct artillery fire. One day, while it was raining hard, and they were under constant rifle fire, they tried to figure out its source. Then they noticed that dead Marines were being brought to the rear on stretchers covered by ponchos. While they watched, a Japanese soldier lifted a poncho, took quick aim at them, then quickly disappeared back under the poncho. They were reminded again how ingenious, and cagey, the Japanese could be.

Dobermans, a breed of hunting dog, were used effectively to sniff out Japanese. Once these "devil dogs" located Japanese in a tunnel, the Marines would throw in a grenade, or use a flame-thrower.

Four days after the 3rd Division landed on Iwo Jima its troops reached the west side of the Japanese airfield Motoyama 2. Two battalion scouts, corporals Bradley Spencer and John Perigo, were alerted by a battalion runner to report to the command post. There they were assigned to escort a stretcher team from the front lines to the rear aid station. Bradley Spencer turned to Perigo. "It should take us less than two hours, so we can get back before dark." Perigo agreed.

Unfortunately it didn't work out that way. It was almost dark by the time they arrived at the pickup point. The terrain was treacherous in daylight, but at night with wounded the task took on ominous overtones.

They started back with two men on stretchers—one with a serious head wound—and ten others still able to walk. It wasn't long before mortar shells started to drop around them and they hit the deck. Once the shells stopped coming, they moved on, but soon the mortars began to fall in greater numbers. With the group's rapid pace, and dropping to the ground each time mortars fell among them, the wounded on the stretchers suffered through their ordeal with unavoidable bumps over the rough terrain.

Spencer saw a partially destroyed bunker up ahead and waved them to it. It was not an ideal protective shelter, but Spencer knew it would protect them from anything but a direct hit. Perigo whispered to Spencer, "The Japanese can either see us or hear us. We've got to move on."

Spencer nodded. They were all exhausted. Those who were able, including Spencer and Perigo, spelled the stretcher-bearers, who were almost out on their feet. The medication the wounded had received was wearing off, and they were hurting more and more with each step. The man with the head wound begged them to leave him or shoot him. Spencer replied, "We're staying together. No one will be left behind." He turned to Perigo, "Should we stay here until daylight or one of us leave to seek help?" "Let's stay together."

They both studied the terrain, noting that if they went along a depression in the opposite direction (away from their lines), they might free themselves from the mortar fire. Perigo said, "Let's go. It's now or never." They moved away from the bunker and mercifully remained free of the shelling.

Later they turned in the direction of their lines, and the rest of the trip was uneventful. When they reached the aid station, a corpsman said, "You're back from the dead. We'd given you up for lost." They were relieved of their charges and Perigo and Spencer crawled into a foxhole and slept undisturbed.

The bubbling pools of sulphur on Iwo Jima were a godsend for the foot soldier. When he had time to eat a can of C-Rations, it could be eaten hot. One merely scooped a small depression in the volcanic soil, placed the can of ham and lima beans or whatever in the hole, and a few minutes later he had a hot meal. Care had to be taken not to leave the can too long, however, because it would explode.

A shallow foxhole or a depression could be a plus. But the part of one's body next to the ground roasted, while the top part of one's body was cold and wet. As a result, sleep was interrupted each night by continual tossing and turning.

During the mopping-up phase, Corporal Bradley Spencer was on patrol in the northern part of Iwo Jima. Rocky outcrops made the ground uneven and treacherous. After an hour they received rifle fire from a cave complex. The patrol took cover until the rifle fire stopped. A Marine interpreter, who spoke Japanese, used a bullhorn to try to persuade the Japanese to surrender. There was no answer, so the interpreter made another appeal. "Come out with your hands up, and you will not be killed." Still no answer. The interpreter tried again, warning the Japanese that smoke would be used against them, and the cave sealed by explosives. Part of the patrol watched the cave's rear in case it had another exit, but there was only one. Finally a small white flag was seen moving back and forth at the cave's entrance. The interpreter again assured the Japanese of safe conduct.

The Japanese soldier holding the flag said that most of the men inside were wounded and hungry, but that the senior noncommissioned officer refused to let them surrender. The American interpreter warned him of the consequences, and the man's head disappeared inside the cave. A few minutes later the flag reappeared and a few heads were seen above the rubble. Now the soldier with the flag stepped into the open. Tension mounted among the Americans, fearful of a trick to get them to reveal their position. The American patrol leader instructed the interpreter to move toward them and away from the cave's entrance. As the Japanese soldier stepped up on a small rock, he looked around him and said in perfect English, "War certainly changes things, doesn't it?" Spencer joined the patrol leader and walked over to him while the remainder of the American patrol helped to treat the Japanese wounded.

Spencer learned later that the flag bearer was equivalent to the grade of an American corporal. A clerk, he soon made it clear that he wasn't happy with the Japanese Army. He had been a medical student in Japan, and the Army had forced him to leave school and drafted him. During their conversation, a B-29 Superfortress coming in for a landing on the airfield flew overhead. The Japanese soldier hit the ground, petrified. He was all too familiar with the bomber, having seen them in their destructive raids against mainland targets.

While the patrol was tending to the wounded, grenades exploded inside the cave. It was obvious the remaining Japanese soldiers had chosen suicide rather than surrender. The Americans used satchel charges to seal off the cave's entrance.

At sea, kamikaze pilots struck the ships off Iwo Jima at twilight on February 21. The *Saratoga* should have gone down after savage attacks by suicide planes, but through the tenacity and courage of her crew she remained afloat. A light escort carrier, the *Bismarck Sea,* succumbed to the fiery onslaught.

Losses on both sides during the invasion of Iwo Jima were enormous for the scale of the operation. By the time organized resistance ended on March 16, Japanese losses were 21,300 (almost all dead), and total American casualties numbered 22,083, including almost 5,000 dead.

The 2nd Battalion, 26th Marine Infantry Regiment probably had the greatest number of casualties on a percentage basis. They lost ninety-eight percent of officers and men, as new Marines were continually brought in to fill the ranks of the fallen. These replacements, in the heat of the battle, posed serious problems, because few officers and noncommissioned officers even knew their names.

The butchery on both sides rose to such heights that veteran combat officers were sickened by the slaughter. The smell of death was everywhere as the Japanese died in their tunnels and foxholes, refusing to retreat. It was a smell that lingered for months after the fighting ended.

15

Victory

The fast carriers had provided four days of support for the Iwo Jima invasion, and on the 23rd they refueled prior to a return trip to Tokyo. Two days later, high seas made operations difficult as the rolling carriers pitched violently, but 400 fighters headed for the snow-covered airfields of Tokyo. To their surprise, the pilots found them deserted. The Japanese had flown their planes to safer havens, fully expecting that the American planes would return after their previous raids. The enemy did contest the operation, losing fifty more fighters, while the U.S. Navy losses were limited to three planes.

Further strikes proved impossible because of deteriorating weather over Honshu. Radar-equipped Superfortresses from the Marianas—more capable of riding through bad weather over Japan—dropped hundreds of tons of bombs as 200 B-29s made the largest daylight raid of the war.

Japanese radio broadcasts, monitored constantly, assailed the Americans for their "arrogance and lawlessness." Premier Kuniaki Koiso said, "I have requested an audience with the Emperor to offer the apologies of myself and my cabinet for this unforgivable negligence."

After refueling again, Task Force 58 sniped at Okinawa and the Ryukyus before heading for Ulithi. In its wake 648 enemy aircraft were destroyed and 30,000 tons of merchant shipping were sunk.

Seabees (*Construction Battalions*) performed their usual miracles, stretching Iwo Jima's main runway to 4,000 feet by March 2. Two days later they leaned on their tools and watched with satisfaction as a crippled B-29 made a safe landing. It had to be refueled by hand before it could proceed to Saipan.

Task Group 50.5, under Commodore D. Ketcham, was based in the Marianas, but the group's shore-based aircraft conducted shipping reconnaissance and air-sea rescue between Japan and Iwo Jima. They also flew offensive screens for

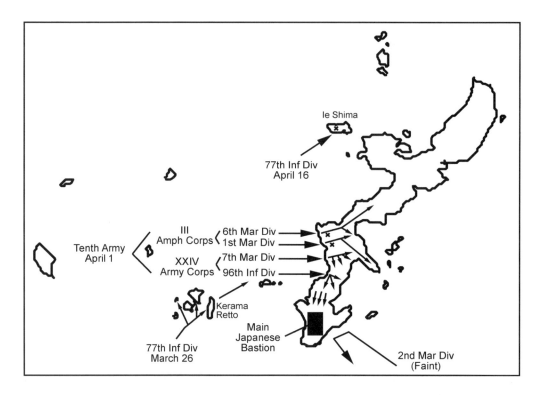

The invasion of Okinawa. (Drawn by Stewart Slack.)

carrier raids and expeditionary forces. Similar operations were carried out by planes of Fleet Air Wing 1 from tenders anchored in Iwo Jima's lee during the battle for the island. Marine Corps Observation Squadrons 4 and 5 arrived on escort carriers and LSTs and began operations from Iwo on February 27. Army Air Force fighters were flown from Saipan on March 6 and Marine Corps Torpedo Squadron 242 arrived two days later. They flew day and night air patrols and provided all air support upon the departure of the last escort carriers on March 11.

While preparations for the April 1 invasion of Okinawa were under way, the 21st Bomber Command in the Marianas was assigned to photograph the island while it continued its attacks on Honshu's principal cities. B-29s were also ordered to carry out diversionary attacks on southern Kyushu while the invasion was in the critical stages.

Spruance, who would direct the invasion, had broad responsibilities beyond the initial amphibious operation. No one doubted that the invasion would be difficult, and it was expected to take three months to secure the strategic island. Ships were assembled seven months prior to D-Day. Excluding personnel landing craft, the total rose to 1,457 ships.

This ambitious campaign was supported by three separately operating carrier forces, by tender-based patrol squadrons, by Marine and Army air units based in the immediate area, and by Army and Navy units based in other areas.

Mitscher's fast carriers of Task Force 58 began the attack on March 18. With an original strength of ten heavy and six light carriers, they launched neutralization

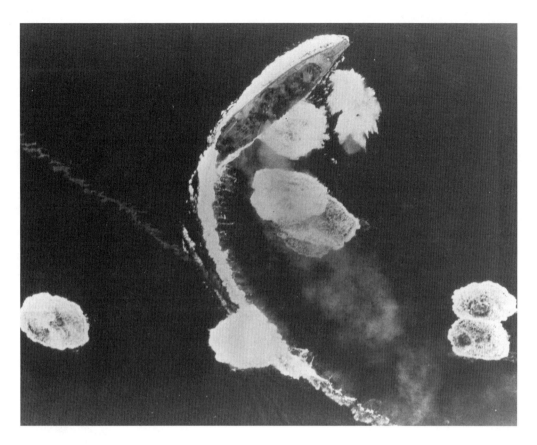

A *Yamato*-class battleship tries to avoid American bombs March 19, 1945, in Kure Bay. (Courtesy of U.S. Navy.)

strikes on Kyushu. They destroyed 482 enemy aircraft in combat while the ships' guns shot down another forty-six.

Preassault strikes were made on Japanese positions on Okinawa on March 23. In retaliation, kamikaze pilots—using conventional aircraft, bombs, and Baka flying bombs (first observed on March 21)—attacked American ships. They seriously damaged the carrier *Franklin,* and made hits on four other ships. They set a pattern that would last for weeks.

Task Group 52.1 under Rear Admiral Durgin started out with eighteen escort carriers. They made preassault strikes and supported the occupation of Kerama Retto on March 26. From positions southeast of Okinawa, they prepared to support the landings on April 1.

British Task Force 57 under Vice Admiral H.B. Rawlings was built around four carriers and was positioned south of Okinawa. From this position it was assigned to neutralize airfields on Sakishima Gunto and Formosa.

Patrol squadrons of Fleet Air Wing 1, based on seaplane tenders at Kerama Retto, conducted long-range anti-shipping searches over the East China Sea to protect the assault forces from interference by enemy surface ships. They also flew antisubmarine patrols in the immediate area, and provided air-sea rescue services for carrier operations.

The invasion of Okinawa. Ships stand off the island on D plus two. The airfield
of Yontan, left center, fell to the American forces on the first day.
(Courtesy of U.S. Navy.)

Five hundred and forty-eight thousand Army and Marine Corps troops landed
on the western shore of Okinawa on Easter Sunday April 1 against light opposition.
They immediately established a firm beachhead, and captured Yontan Airfield the
same day.

Mitscher's fast carriers operated continuously in a sixty-mile-square area northeast of Okinawa—within 350 miles of Japan—to help neutralize Japan's airfields on
Honshu and Kyushu, destroying 900 planes on the ground. Then they joined Durgin's
Task Group 52.1 and British Task Force 57 to provide close air support for the troops
on Okinawa.

After ground opposition stiffened, Marine Corps elements of the Tactical Air
Force began local air defense patrols on April 7, and started their close air support.
On May 13, Army Air Force fighter squadrons began operations from Ie Shima.

The first of a series of mass suicide attacks began on April 6 with 400 Japanese
planes. In seven mass raids, interspersed with smaller scattered ones—during the
critical period from April 6 to May 28—the Japanese lost 1,500 aircraft in attacks
against American and British ships supporting the invasion. During a three-month
period, while the human guided missiles of the kamikaze force sought to overwhelm

the Allied navies, the U.S. Navy took the heaviest punishment in its history. Although Task Force 58 lost no ships during the campaign, eight heavy and one light carriers were hit, along with three escort carriers. All four of the British carriers were hit, but none were sunk.

The Fifth Fleet continued to fight off waves of suicide planes as the relatively unopposed land forces moved inland during the first five days. Due to the kamikaze attacks, Spruance spread his fleet east of Okinawa as much as one hundred miles. Admiral Turner's amphibious force almost circled the island.

Attacks by suiciders were bad enough—especially on April 6 when 355 one-way attackers made a mass attack—but now remnants of the Japanese fleet decided on their own kamikaze mission. Two-thirds of Japan's remaining ships sailed the same day from the Inland Sea.

Day after day fast carriers sent their fighters aloft to destroy the kamikazes before they could crash into the ships. Fliers from the *Essex* broke the carrier record by shooting down sixty-five planes, and the *Belleau Wood*'s Fighting 30 destroyed forty-seven more, but the enemy planes came in an endless stream from Kyushu's airfields.

While crews and pilots were at the exhaustion point, humor relieved the grimness of the period when Captain William G. Thompson of the *Belleau Wood* wired Mitscher, "Forty-seven bogies splashed. Does that exceed game limit?"

"Negative!" Mitscher replied. "This is open season. Well done."

Although hundreds of attackers were downed, only two American planes were lost in the April 6 attempt to destroy the invasion fleet. Twenty-four managed to hurl their planes and bombs onto ships. Three destroyers were sunk, two ammunition ships exploded, and one LST was lost. Attacks abated for two days, but on April 11 the Japanese turned their full fury on the fleet.

"Abandon all close support operations," Mitscher ordered.

There were times when it seemed as if the Japanese would sacrifice every plane in their air arsenal. Throngs of suiciders were splashed flaming into the sea. The battleship *Missouri* suffered damage, as did the carrier *Enterprise*.

Kamikazes ignored the carriers the next day and concentrated on the invasion ships. Carrier pilots lashed out at the deadly attackers, scoring 151 down, but the kamikazes caused more destruction than on any previous day.

Spruance alerted patrol submarines to keep an eye on Japanese surface ships, which had left the Inland Sea on April 6. When submarines spotted the ships, their skippers wired him immediately. "At least one battleship, supporting destroyers, course one nine zero.

Vice Admiral Seichi Ito actually had nine ships. His flagship *Yamato,* the light cruiser *Yahagi,* and seven destroyers were under his command as he left Tokuyama Harbor. An *Essex* Hellcat reported their exact position.

"Pilots, man your planes!" The strident cry echoed off the flattops while pilots rushed to their planes. Avengers were loaded with torpedoes, Helldivers with semi-armor-piercing bombs, and fighters with 500-pound bombs slung underneath their fuselages.

Every task group contributed, as 386 planes headed north while Navy PBMs kept the enemy under constant observation. During the night, pilots found the waiting monotonous, lighting one cigarette after another to while away the hours until the enemy task force came within striking distance.

The approaching carrier planes saw the antiaircraft fire explode dead ahead, bringing them to anxious attention. They dropped through the clouds from 10,000 feet as rain pelted their windshields.

Lieutenant Thaddeus T. Coleman broke through at 5,000 feet and saw the ships squirming nervously on the ocean's surface as he began his dive. Torpedo planes rushed to attack the mighty *Yamato,* scoring hits that promptly made her list. Then planes dropped torpedoes from the other side, momentarily slamming the battleship back on an even keel. The first wave pulled away, noting the *Yamato* and the *Yahagi* were badly damaged, and positive that two destroyers were going down.

Radford's Task Group 58.4 arrived with forty-eight Hellcats, Helldivers, and fifty-three Avengers. They found the *Yamato* still able to fight, but the *Yahagi* was dead in the water. They concentrated on the *Yahagi* and she succumbed to their fierce attacks.

Commander John J. Hyland of the *Intrepid*'s Air Group 10 led his planes against the *Yamato,* and the big ship reeled with explosions from one torpedo and eight bombs. Lieutenant Thomas H. Stetson, skipper of Torpedo 9, led the attack as the battleship listed heavily to port. He could see her massive armor belt on the starboard side as he directed their torpedoes to her vulnerable underbelly. "Lower depth setting from 10 to 20 feet," Stetson called.

They made perfect runs and not a shot was fired at them. Sweeping over the shattered hulk, Stetson saw the bottom rip out of the ship as she rolled on her beam-ends and went down, taking Vice Admiral Ito and most of her crew with her.

It had been a magnificently executed attack, and only four dive-bombers, three torpedo planes, and three fighters were shot down. All but four pilots and eight crewmen were promptly rescued.

The *Hancock* was bombed shortly before noon, but fires were brought under control and she landed her planes safely.

Japan had lost six more of her fleet. Only four Japanese destroyers survived.

Major General Curtis E. LeMay's 20th Air Force had caused enormous destruction to Japan's principal cities during March, but when the Navy's position off Okinawa became highly critical shortly after the invasion began, Superfortresses were diverted again from the strategic bombing of Japan to attack Kyushu's airfields. LeMay had agreed to preinvasion support by his B-29s, but now he strongly objected to using his heavy bombers in a ground support role. He firmly believed that the Japanese would be forced to surrender by October 1, 1945, if his command kept up its massive attacks. He later officially went on record when General Arnold visited the Pacific in June. But his protests were overruled, and he was ordered by a reluctant Arnold in Washington to continue such attacks until they were no longer needed. Long-delay fuses were used extensively to hamper clearance problems, and they seriously disrupted kamikaze missions.

When the 10th Army came up against Lieutenant General Mitsuru Ushijima's first defense line on Okinawa on April 11, the troops found it impossible to penetrate elaborately prepared positions. Ushijima had devised this strategy to force the Americans out from the cover of the devastating guns of the Fifth Fleet.

The fleet, meanwhile, continued to suffer daily attacks, climaxed by another major suicide attack on April 16 involving 165 Japanese planes.

Then the air attacks diminished in strength. The Japanese boasted about the tremendous destruction, but their claims far exceeded the number of American ships that actually were sunk.

Lieutenant General S.B. Buckner's ground forces ran into stubborn resistance the farther they fought inland. The Japanese contested every foot of ground, only permitting gains over their dead bodies.

Carriers provided close support every hour as ground forces fought for small advances. It was an exhausting task for the fliers, flying constantly into deadly ground fire while they sought out pinpoint targets ahead of the troops.

Twelve Avenger crews from Sherman's 58.3 task group were briefed on May 20 about a particularly stubborn ridge surmounting a natural amphitheater to the south. "This small ridge," the briefing officer said—pointing to a map of the battlefield— "has held up our ground forces advancing toward Shuri for a week. Three hundred infantrymen have died trying to capture it. It must be pulverized."

Avenger pilots headed for the target, disturbed that they had to drop bombs only fifty yards from their own troops, but determined to destroy the Japanese stronghold. In pairs they dove to within 200 feet of the ground and dropped their bombs, feeling the concussion slap at them as they pulled away.

Artillery opened a barrage and troops swarmed over the ridge, capturing the gun positions from their stunned Japanese defenders. The Americans lost only two men in this well-coordinated attack between ground and air. This opened up the whole Shuri defense line and American troops charged forward, gaining thousands of feet where formerly they had been lucky to gain a yard.

During May, while American ground forces continued to make gains on Okinawa, Japanese kamikaze pilots were still making life aboard American and British Navy ships a hellish nightmare, while Japanese fighters and bombers harassed Okinawa's airfields. American Marine pilots ranged widely over Kyushu to destroy Japanese planes on the ground. One such mission was flown by Captain Kenneth Walsh and his wingman Lieutenant Steve Furimsky. Walsh had already shot down twenty Japanese planes.

As usual, this day was cloudy with rain squalls as they flew in and out of the clouds at 1,200 feet hoping to catch a Japanese plane on the ground, but none materialized. They had flown up to Kyushu with their usual three belly tanks. Their fuel was used up on the flight to Kyushu and the tanks dropped. Now they had a full load of gasoline to extend their stay over enemy territory to a full hour. This trip they spotted only one Japanese plane on their return trip. A four-engine Japanese flying boat was flying at 300 feet above the waves. It was using the overcast to pop in and out of the clouds to prevent American planes from attacking. Walsh decided that their fuel levels would not permit them to play this cat and mouse game successfully, so they continued on to Okinawa at 12,000 feet.

The weather improved as they listened to the radio chatter from other planes in the vicinity. Then they saw two Corsairs just off Okinawa who had a Zeke bracketed as they flew a thousand feet on each side of it. Walsh and Furimsky listened to the conversation between the two pilots, noting they were not from their squadron. One pilot said, "If he turns left, you roll over and pop him. If he turns right, I'll get him."

While the Zero streaked away without deviating from its straight course, Walsh told Furimsky, "You're always bitching that you don't get a chance to shoot down an airplane." (Furimsky's responsibility as Walsh's wingman was to protect his rear.) "Go down and splash that Zeke." Furimsky took a quick look at his fuel gauge. Noting he had only twenty gallons left, and knowing he still had twenty-five miles to reach the field, he replied, "I've got twenty gallons of gas. I'm not going anywhere but sit down on the ground."

"You had your chance," Walsh said. He nosed over while Furimsky remained above, calling "Red One, you're clear." Walsh flew between the two pursuing Corsairs, firing all his guns as he neared the Zeke. It disappeared in a spectacular fireball. One of the Corsair pilots said on the radio, "Who is that son-of-a-bitch!"

Furimsky called his field to tell the tower operator about his fuel situation, saying he would make a straight-in approach and glide home. The operator responded immediately. "We're operating out of a bunker. Our airfield is under attack. And you're on your own. If the airfield is not being bombed, come on in."

Furimsky was down to 3,000 feet before he spotted the runway. The field appeared to be momentarily free of attack, so he continued straight in. Then he heard "ding, ding, ding" on the armor plate behind his seat. He glanced at his rearview mirror and saw with horror that there was a Zero on his tail, noting the orange flickers of light on its wing as the Japanese pilot opened fire on him. Then another Zero moved in to continue the attack. This pilot was not as proficient, Furimsky noted with relief, as the tracers passed under his wing and streaked far ahead of him. While glancing back he noticed the Zero was disintegrating in an orange fireball. He learned later that his friend, Lieutenant Robert Rouse, had seen his predicament and moved behind the attacking Zero.

While on his final approach Furimsky kept his eyes glued to the rearview mirror as much as possible as he watched Rouse's Corsair do a high wing-over and end up with a three-quarter nose-angle shot at the other Zero. He scored a hit as the Zero pulled away with smoke pouring out of its engine, but it didn't get very far. Another Corsair joined the attack and destroyed the second Zero. Later Rouse got credit for one and a half planes. The other pilot was credited with half a kill.

Throughout this action Furimsky never deviated from his straight approach to the runway. He touched down normally, but at the end of the runway his engine started to cough and then quit. His tanks were so dry he could not even taxi off the runway. He pushed his plane into the grass and crawled into a drainage ditch while the swirling dogfight raged above him.

The two Corsair pilots who had been deprived of a possible kill refused to confirm Walsh's shooting down of the Zeke off Okinawa. They were furious that Walsh had taken their prey away from them, but they should have acted more quickly. After some pressure was put on them, they admitted they had witnessed Walsh's victory, and he was given credit for it.

Organized resistance on Okinawa ended in the latter part of June, but a colossal mopping-up operation remained. In victory, the Army was saddened by the loss of General Buckner, who was killed just a week before the end.

Lieutenant General Mitsuru Ushijima thrust a hara-kiri knife into his stomach three days later, and the rest of his staff followed suit.

Carrier air support was on a larger and more extensive scale than on any previous amphibious operation. Fast and escort carriers flew over 40,000 sorties, destroyed 2,516 enemy aircraft, and blasted enemy positions with 8,500 tons of bombs and 50,000 rockets.

The Marine Corps squadrons on the shore destroyed another 506 Japanese aircraft and expended 1,800 tons of bombs and 15,865 rockets on close air support missions.

Task Force 58's time on the line from March 18 to June 10 was surpassed by the escort carriers. The most outstanding record by carriers was logged by the *Essex,* with seventy-nine consecutive days on station.

Okinawa had been a deadly battle, with a loss of 12,250 Americans killed. The Navy lost the largest number, 4,907.

Plane losses were also high, with 269 destroyed in combat, 299 on board damaged carriers, and 292 from all other causes. Air combat took the lives of 205 men.

Despite intelligence estimates that the Japanese were short of airplanes, American fast and escort carrier planes destroyed 2,576 Japanese aircraft, while Marine Corps squadrons on shore destroyed another 500 planes. Uncounted were at least hundreds more destroyed by B-29s on the ground.

Admiral McCain summarized his feelings about using fast carriers in such operations in a letter to Nimitz. "It is wasteful to use fast carriers in direct support of ground troops over a long period. It fails to exploit the assets of mobility, surprise and concentration and is undesirable. Such use invites danger to the carriers while diverting them from more worthwhile targets which only they can attack."

General LeMay had said much the same thing about using B-29 Superfortresses in a support role.

Halsey added his own objections. "Fast carriers can support landing operations far better by going into offensive action. I recommend defensive deployment only when offensive action is impossible or when such defense use is the only means of preventing defeat of, or disastrous losses to the invasion forces."

One lesson emerged. There was no doubt in anyone's mind that carriers were too vulnerable to suicide attacks so close to Japanese bases.

The British Navy, meanwhile, disposed of the last ships the Japanese Navy had based at Singapore, including four heavy cruisers, minesweepers, and luggers.

The last stepping stones were in Allied hands by the spring of 1945, so now an important strategic decision was needed for the next move. Some of that decision making was accomplished on April 3 when the Joint Chiefs of Staff—despite objections from several people in the War and Navy Departments—decided that General MacArthur would remain in command of Army ground and air forces except for the 20th Air Force, which would still report to General Arnold. Various units of the South Pacific's inactive areas were removed from MacArthur's jurisdiction. Nimitz remained in charge of all naval forces in the Pacific. Functionally there was no change, and Nimitz and MacArthur were told to complete their current campaigns under the old divided command structure.

The war in Europe came to an end on May 8, so the Joint Chiefs asked MacArthur and Nimitz for their views on future operations in the Pacific. Nimitz said the home islands should be encircled and bombing stepped up to a great crescendo before Kyushu was invaded.

There were many in top military positions who favored a direct assault on the main islands. They pressed for an invasion of Kyushu, the southernmost island, to establish naval and air facilities before a mass invasion of the Tokyo plain. General

Arnold was one of those who believed that airpower could force Japan to surrender, but General MacArthur and General Marshall did not agree, and their views prevailed. Admiral Spruance agreed with Arnold, believing the islands should be strangled by submarine and fleet action.

Emperor Hirohito and his government officials desperately desired peace, but army militarists refused to concede. Devastating B-29 raids were now pulverizing Japan's cities, and both Japanese Navy and civilian leaders realized that Japan had lost the war.

High surf precluded any invasion of Kyushu after November 1, so that date became the target for planners. MacArthur favored a November invasion of Kyushu and said he could get the men and supplies by cleaning out bases throughout the Pacific. This plan was adopted.

Halsey relieved Spruance of command of the Fifth Fleet, and it became the Third Fleet on May 27 off Okinawa. Incredibly, he led it for the second time into a vicious typhoon. Extensive damage was caused to his ships and there was a large loss of life.

Admiral King ordered Vice Admiral Hoover to conduct another investigation, again Halsey was found to blame, along with his carrier commander McCain. Hoover told King that Halsey deserved to be court-martialed. He added that if he was not such a war hero he certainly would be. King agreed, but was reluctant to order such a court-martial so close to the end of the war, knowing it would be explosive. Secretary of the Navy Forrestal reluctantly agreed.

MacArthur planned the invasion of Japan as the biggest of any he had led in the Pacific. The Navy offered full fleet support. Halsey's Third Fleet, with the fast carrier groups, would be augmented by units of the British Pacific Fleet to provide support for landings through strong attacks against Honshu and Hokkaido. They would be based at Eniwetok and the Marianas.

Land, sea, and air forces congregated in the Philippines, Okinawa, and the Marianas for the huge undertaking. Okinawa would be the staging point for half a million men, and 212 attack transports and 555 LSTs would transport them to the shores of Kyushu.

MacArthur planned for fourteen divisions—three of which would be Marine—holding three divisions in reserve to meet an expected Japanese defense force of ten divisions.

Every planner and commander knew that it would be a hazardous undertaking. When they looked at the expected 50,000 casualties in the first thirty days—and perhaps a million American casualties to defeat Japan by invasion—they sadly shook their heads and wondered if there was not a less costly way to bring about Japan's final surrender.

Howls of anger and distress emanated from Japan as the air war was stepped up. The U.S. Far East Air Force, the 20th Air Force, and carrier planes dominated the skies over Japan's prostrate empire. While B-29 strikes leveled her cities and factories, carrier planes sought to establish air supremacy in the skies over Kyushu in anticipation of the coming invasion.

One enigma remained: What would Russia do? Imperial General Headquarters in Tokyo expected American landings on their southern island. Their only hope to avoid total annihilation lay in their Kamikaze Corps. Despite a fleet of suiciders numbering 5,350, destruction of the U.S. Fleet and its troop transports would take all the determination of its pilots. Japan did not expect to rely solely on her suiciders, because she had an equal number of planes for normal attack purposes.

The Japanese now consolidated their forces, and Lieutenant General Albert F. Wedemeyer reported they were withdrawing from Shanghai and Hong Kong, permitting a revitalized Chinese Army to concentrate in the Canton–Hong Kong area.

The Japanese government was desperate, looking hopefully for a face-saving formula to get out of its present predicament. The devastating attacks at home and the loss of Japan's bitterly won possessions to the south had created a chaotic situation that was not fully comprehended by the Allied powers.

While B-29s saturated Honshu, Kyushu, and Shikoku with fire bombs and devastated her vital factories, the Emperor initiated discussions to end the war. He received support from every group except the Army, and after Okinawa capitulated, premier Kunikai Koiso's cabinet resigned. Russia's intentions became clearer on April 5 when she renounced her neutrality pact with Japan.

Hirohito ignored his senior statesmen and appointed the reluctant 79-year-old Admiral Kantaro Suzuki as premier. His appointment was viewed with great interest by officials of the American government, because when President Roosevelt died on April 12, he had said, "I must admit Roosevelt's leadership has been very effective. It has been responsible for the Americans' advantageous position today. I can understand the great loss his passing means to the American people, and my profound sympathy goes to them."

Suzuki, with the Emperor's support, decided that Japan could not continue the war because of factory damage, ship losses, a deteriorating food situation, and the attitude of the Japanese people. He revealed his thoughts to the Emperor, and they both agreed that steps should be taken immediately to establish peace. Suzuki started conversations with the Russian ambassador in Tokyo, whom he hoped to use as an intermediary, but the Army's reaction was violent.

Hirohito personally called a meeting of the prime minister and five others of his cabinet on June 20. "I think it is necessary," he said, "to have a plan to end the war at once as well as to defend the home islands." Suzuki called his entire cabinet together. "Today the Emperor has said what everyone has wanted to say but yet was afraid to say," he told them. "The war must be ended."

★ ★ ★

Task Force 38 returned to Leyte Gulf for provisioning on June 13, and after a long rest while ships were refueled and reprovisioned, they headed north on July 1. Crews were wild with excitement when bullhorns announced, "We're headed again for Japan."

Many an eyebrow lifted in wonder when the Joint Chiefs of Staff sent a top secret message to MacArthur, Nimitz, and General LeMay on July 3: "The Japanese cities of Kukura, Niigata, Kyoto, and Hiroshima will not be attacked by elements under your command."

Halsey's Third Fleet operated on Japan's slim lifeline with surgical smoothness, halting all shipping from the Asiatic mainland to the home islands of Japan. Even shipping between the main islands came under attack.

McCain's Task Force 38, part of the Third Fleet under Halsey's overall command, was initially composed of fourteen carriers, but it was augmented later by one other carrier. Halsey had received specific instructions from Nimitz: "Launch air attacks on northern Honshu and Hokkaido, and bombard coastal cities."

Fliers from McCain's carriers took off for Tokyo at 4:00 A.M. on July 10, after waiting out a bad weather front. Fighters swept low, then torpedo and dive-bombers plunged at their targets.

McCain's three carrier groups were headed by Admirals Thomas L. Sprague, Bogan, and Radford. They were later augmented by British Carrier Task Force 57 under Vice Admiral Rawlings with four carriers.

While attackers struck vicious blows, not one Japanese fighter rose to do battle. Even flak was of minor concern. It was evident Japan had decided to hoard its airpower. Attackers found a hundred planes on the ground to destroy, and damaged an additional 231 more. It was a two-pronged offensive because the 5th Air Force and the Tactical Air Force on Okinawa had devastated fields the day before.

Army P-51 Mustang fighters, based on Iwo Jima, added to the devastation by sweeping the Tokyo area following a massive 550-plane B-29 strike at industrial and refinery centers.

If Japanese defenders were confused by the variety of attacks, they had every right to be. Marine Corsair fighters and Avenger torpedo planes struck Kyushu fields on July 11 while carriers were off Tokyo Bay. As if the strike attacks were not enough, Rear Admiral John F. Shafroth Jr. used naval gunfire to bombard the steel-producing city of Kamaishi in what must have been a humiliating attack for top Japanese Navy officers.

Halsey's Third Fleet moved north, and on July 14 and 15 he sent his carrier planes to bomb the ferry route between the northern tip of Honshu and southern Hokkaido. Railroad ferryboats, carrying coal for the industrial furnaces of Honshu, received a devastating blow. Five were sunk, one was left burning, and three others were damaged.

Navy fliers had a field day all over northern Japan. They were given no particular targets. "Choose a target of opportunity," they were told.

Big guns of the warships created a hell on earth for the town of Muroran on the southern coast on July 15, when the Nihon Steel Company and a synthetic oil refinery were devastated by ship bombardment. The Japanese were unable to contest these actions.

For all practical purposes, the Japanese Navy had ceased to exist. There were still ships that might pose a threat—although limited—so Nimitz wired Halsey, "Continued existence of remaining Japanese naval forces make it necessary to use more defensive missions against Japanese naval forces than would otherwise be necessary. Remaining heavy ships must be eliminated. This is the responsibility of the Pacific Fleet."

Halsey put his intelligence experts to work, and they found where the ships were hidden. Some were in coves of the Inland Sea near Kure, close to Hiroshima. They included the battleship *Haruna,* carrier–battleships *Ise* and the *Hyuga,* three carriers, including the *Amagi*—which had never been in combat—and the *Katsuragi* and the *Ryuho.* Six cruisers, the heavy *Tone* and the *Aoba,* and the fast light cruisers *Iwato,* the *Izume,* the *Setsu,* and the *Oyodo* completed the list of major ships in this area. The latter was the flagship of Admiral Ozawa, Commander in Chief, Combined Fleets.

The battleship *Nagato* was tied up and camouflaged at the naval base at Yokosuka on Tokyo Bay. Halsey glanced at the antiaircraft guns plotted around the area and shuddered.

British Task Force 57 joined the Third Fleet on July 17 for two more strikes at Tokyo's airfields. Heavy cloud cover reduced the scope of operations, and only four planes were destroyed on the ground.

On this date the cruiser *Indianapolis* started its fateful journey across the Pacific with two atomic bombs.

While the 34,000-ton battleship *Nagato* remained at Yokosuka, it was a challenge to Halsey's pilots. Despite dozens of flak batteries, fighters armed with 1,000-pound armor-piercing bombs dove on the *Nagato* while torpedo planes tried to neutralize antiaircraft positions. The ship was badly damaged but refused to sink, although an adjacent destroyer went down.

All across the Tokyo plain carrier planes struck again at airfields without opposition except for the deadly antiaircraft fire. American planes got thirty-nine enemy planes, and the British thirteen, but American losses numbered twelve planes and twenty-two men.

July 24 dawned bright and clear. "A good day to strike Kure," Halsey told his staff. Words were translated into action and bombers headed for the naval base. Ships here could not be attacked by torpedo planes—a lesson learned early in the attack against the *Nagato*—because of the manner in which the ships were anchored, but dive bombers and fighters scored five hits on the *Ise,* and the *Hyuga* was struck such deadly blows that it was grounded.

The veteran *Haruna,* which had often been announced as sunk since the earliest days of the war, received only one hit. The *Amagi* received a hit on its flight deck, and the carrier *Ryuho* was so well camouflaged that pilots failed to find it. The carrier *Katsuragi* was also not touched. Ozawa's cruiser *Oyodo* received four direct hits and some near misses, and the older cruisers *Iwato* and the *Setsu* were hit so severely that they settled into the mud.

The Japanese had taken enough. They struck back, particularly when carrier planes approached Nagoya, Osaka, and Miho.

It had been a good two days for the Third Fleet, with over a quarter of a million tons of enemy warships either sunk or badly damaged. One hundred planes were destroyed on the ground and, with the resurgence of enemy activity in the air, another thirty-one were shot down. This was not without its price. Thirty-two American planes were lost.

Halsey ordered his carriers to strike Kure on July 28, because warships berthed there did not seem to be seriously damaged. "Concentrate on the *Haruna,*" he said. Carrier pilots did as they were instructed and finally the ship settled to the bottom, Far Eastern Air Force B-24s from Okinawa also attacked. They failed to score hits on the *Haruna,* but sank the *Aoba.* The *Ise* also met her doom, as well as the carrier *Amagi,* which was ripped apart from incessant attacks.

When the strike was over the *Oyodo,* the *Tone,* and the *Izuma* were on the bottom, and the carrier *Katsuragi* was badly damaged, but again no one found the *Ryuho.*

Halsey was jubilant. He sent a personal message to all ships. "Mark well this day the 28th of July. To the Dumbos and lifeguards, to combat air patrol and men of the surface team, to the valiant British force on the right flank. Well done. For the flying fighters who fought it out over Japan to a smashing victory, I have no words that can add to the record with their courage, their blood and their lives."

The industrial center of Hamamatsu suffered a night raid as fliers from the *Bon Homme Richard* dealt a deadly blow.

The Third Fleet now steamed back and forth across Japan's doorstep, dealing body blows that weakened and disheartened the distraught enemy.

In this final carrier action of World War II, Navy planes destroyed 1,223 aircraft, of which over 1,000 were on the ground. They sank twenty-three warships and forty-eight merchant ships totaling 285,000 tons.

Emperor Hirohito personally intervened to end the slaughter of his people. He appealed on July 20 to the Soviet government to receive Prince Konoye as special ambassador in an endeavor to improve relations between the Russians and the Japanese. He also asked the Soviet Union to intercede with the United States to stop the war.

Moscow bluntly declined because the Potsdam Conference was about to begin. Hirohito and Prime Minister Suzuki read the Potsdam Declaration with dismay, because it called for unconditional surrender. Despite their feelings about these terms, the cabinet wanted to accept the terms, to do anything to call a halt to a war that threatened their survival as a nation.

But War Minister Anami refused to concede, claiming the Army was as yet undefeated, so the Allied demand was rejected.

In New Mexico, the United States tested its first atomic bomb on July 16. Scientists were awed and speechless as the huge flash lightened the desert sky with a brilliance greater than the sun.

Without hesitation President Truman ordered the bomb used to shorten the war, so the 509th Composite Group of the 20th Air Force's 313th Wing prepared for its first mission from its base on Tinian in the Marianas.

Now top commanders understood why certain targets had been given sanctuary from B-29 and carrier attacks. At first Kyoto, Hiroshima, Kokura, and Niigata were listed as targets. Kyoto was removed from the list because it was the site of a historic shrine. Nagasaki was chosen instead.

The world was shocked by the destruction at Hiroshima on August 6 as the world's first atomic bomb was exploded above the city. Three days later, a second bomb destroyed Nagasaki.

The Japanese War Minister and his chiefs of staff still refused to accept the Potsdam terms, holding out for a negotiated peace.

Hirohito took decisive action. "To stop the war on this occasion is the only way to save the nation from destruction," he told his people. "I decide this war shall be stopped!" He had no constitutional right to make such a decision, but while the cabinet ministers broke down and cried at the humiliation their nation had suffered, the Japanese people stood by their Emperor. They, too, had had enough of war.

The government accepted the unconditional surrender terms on August 10, but with one exception—that the Emperor not be deprived of his throne. Swedish and Swiss emissaries transmitted the agreement through diplomatic channels to Washington.

Nimitz signs the Japanese surrender document aboard the *Missouri* in Tokyo Bay. Looking on, from left, are General Douglas MacArthur, Admiral William F. Halsey, and Rear Admiral Forrest P. Sherman, Deputy Chief of Staff for Nimitz. (Courtesy of U.S. Navy.)

After serious discussion, the Allies agreed that Japan could keep its Emperor, but he must take orders from them. They put their views in these words: "From the moment of surrender, authority of the Emperor and the Japanese Government to rule the state shall be subject to the Supreme Commander of Allied Powers who will take such steps as he deems proper to effectuate the surrender terms.

"The ultimate form of government for Japan shall, in accordance with the Potsdam Declaration, be established by the freely-expressed will of the Japanese people."

Halsey, when advised of what was going on, said, "Have we got enough fuel to turn around and hit the bastards once more before they quit?"

While governments pondered their next move, a raging typhoon stopped further carrier attacks against Japan.

When no final determination for ending the war was received by August 12, Halsey ordered his fleet to resume its strikes. McCain wired his carrier commanders, "The fact we are ordered to strike indicates the enemy may have thrown an unacceptable joker into surrender terms. This war could last months longer. We cannot afford to relax. Now is the time to pour it on."

The Pacific Fleet assembled for the formal surrender ceremony aboard the *Missouri*. Mt. Fujiyama is in the background. (Courtesy of U.S. Navy.)

Fast carrier planes struck Tokyo on August 15 and were met by forty-five enemy fighters, twenty-six of whom they downed. While the second strike headed for their targets, pilots heard, "Cease fire. Cease fire. The war is over. Cease fire!" They jubilantly jettisoned their bombs and headed back to their carriers.

McCain issued sharp orders after they returned. "I want wartime vigilance maintained," he said, "until we're sure the Japanese plan no further attacks." McCain queried Halsey what his ships should do if attacked.

"All snoopers shall be investigated and shot down," Halsey said, "not vindictively, but in a friendly sort of way." Their alertness paid off, because thirty-eight attacking Japanese fighters were shot down on V-J Day.

In the minds of many people the two atomic bombs were given credit for ending the war. In fact, they only precipitated an earlier end to the war that would have come without their use and without an invasion. Actually, Japan was defeated by August 1. The sea-air blockade of her home islands—and the bombing attacks by B-29s against industrial and urban targets—had destroyed Japan's ability and will to fight. The production of civilian goods had dropped so low that the nation faced economic collapse. Munitions production had been reduced to the point where military operations could no longer be sustained.

After the war the Strategic Bombing Survey revealed its findings, claiming that Japan would have surrendered prior to December 31, 1945, and in all probability by November, even if the atomic bombs had not been dropped and even if no invasion had been planned or contemplated.

Japan's island empire was entirely dependent upon sea transportation, and was particularly susceptible to seapower. When Nimitz's fleets cut off Japan's ocean commerce and B-29s mined the approaches to the main islands, it marked the beginning of the end.

It was not realized by many people at the time that Japan had been defeated by the effective use of airpower. Although it was the primary reason for her defeat, it was not the sole reason. The long and costly drive across the Central Pacific to secure the Marianas as air bases was a prerequisite for the final strategic air offensive against Japan. Therefore, her capability of sustaining her armed forces, particularly those outside her home islands, was undermined by the war at sea. The end was inevitable.

Allied fleet units entered the coastal waters of Japan on August 27 while hundreds of planes flew above them. Landings were made at Yokosuka, Yokohama, and along selected points off Segami and Tokyo Bays. Major occupation forces did not land until August 30.

General MacArthur, who had been appointed Supreme Commander of all Allied powers, flew to Atsugi airfield on August 29, and Admiral Nimitz arrived the same day.

At 9:00 A.M. September 2, the surrender ceremony took place aboard the battleship *Missouri,* anchored off Yokohama.

In the bay 258 warships, representing nations that had fought Japan, rode at anchor. The big carriers remained outside to launch planes at the appropriate moment.

The morning had been cloudy as the dignitaries assembled on the deck of the *Missouri.* When everyone had signed the Articles of Surrender, the sun broke through as General MacArthur said, "Let us pray that peace be now restored to the world and that God will preserve it always. These proceedings are closed."

Then 450 carrier planes, and a large force of Air Force bombers, swept majestically over the ship—a triumphant climax to the end of hostilities.

Due in large part to Admiral Towers, the architect of naval aviation, victory had been assured in the Pacific war. (Towers finally got his chance for a Pacific command in the closing months of the war, when he replaced Spruance as head of the Fifth Fleet.) While others gained the recognition, his strong personal efforts behind the scenes developed naval aviation to the point where it became the backbone of the Navy's striking power. This feat was accomplished only thirty years after the Navy had flown its first airplane, and only nineteen years after it had acquired its first aircraft carrier.

The United States was forced to fight a two-ocean war, but the two theaters had little in common. In the Atlantic, air operations were essentially used to protect Allied ships delivering raw materials to run America's factories and returning munitions to the war fronts in Europe, although the Navy did participate in three amphibious operations. In the Pacific, it was a matter of stopping an enemy's advance, which in a few short months had spread over all the Western and parts of the Southern and Central Pacific, and then carrying out the bitterly contested task of driving the Japanese homeward across the broad, island-dotted expanses of the Pacific Ocean.

The United States began the war incapable of waging a two-front war. But the United States' distance from the war fronts saved her from destruction, permitting the buildup of its enormous industrial power to manufacture the ships, planes, and equipment to go on the offensive.

In the course of the war, Navy and Marine pilots destroyed more than 15,000 enemy aircraft in the air and on the ground. They sank 174 Japanese warships—including thirteen submarines—totaling 746,000 tons, and sank 447 Japanese merchant ships totaling 1,600,000 tons. In the Atlantic theater, American Navy and Marine pilots destroyed sixty-three German U-boats. (In combination with other forces, they helped sink another 157,000 tons of warships and 200,000 tons of merchant ships, and another six Japanese and twenty German submarines.)

Many opinions expressed before the war by so-called experts on the effect of airpower on naval operations proved false. Despite claims by some Air Force enthusiasts that the sinking of warships by bombers had made navies obsolete because they could no longer operate within the range of land-based airplanes, fast carrier operations in the Pacific proved that such statements were fundamentally in error. The usefulness of close air support of ground troops was demonstrated time and time again, proving that such operations were not only possible but indispensable.

World War II proved that neither the Army nor the Navy could survive in war situations, or achieve their objectives, without first attaining air superiority. It now became obvious to most strategic planners that neither service could exert as much force by itself as it could with the aid of airpower. Truly, aviation had finally come of age, and the Navy officer most responsible was a quiet, self-effacing man by the name of John H. Towers. While serving under Nimitz, he had with quiet efficiency made the recommendations that resulted in the successful development of the fast carrier forces. Fortunately, Nimitz had appreciated Towers's background in Naval Aviation and had wholeheartedly supported him.

Part IV

The Post-War Years

16

A Course of Folly

In August 1945, the United States had the greatest air strength the world had ever seen. Naval aviation alone had 41,272 planes, 60,747 pilots, 32,827 ground officers, 344,424 enlisted men, with twenty-nine fast carriers and seventy-one escort carriers.

Naval aviation was liquidated at this time because the nation took the peace seriously. Aircraft production was virtually stopped. Fortunately, research and development continued.

Drastic reduction of the nation's airpower was a tragic mistake, understandable because few thought the United States would have to fight another major war in the foreseeable future.

Unfortunately for the free world, its potential enemies kept right on building their airpower. Within a year after the end of hostilities, the onboard figures for the men of naval aviation fell to a mere one-quarter of its World War II peak. Only a skeleton of the wartime force remained to carry out new operational demands made upon it to counter threats to the peace posed by the Soviet Union and the rise of communism in China and throughout much of the Far East.

The aircraft industry reached its lowest production point shortly after the war. From a peak of 93,318 military aircraft of all types built in 1944, production dropped to 1,669 in 1946.

Eventual realization that this was a course of folly reversed the downward trend and a slow buildup began. It resulted, a few years later, in combat aircraft available for the Korean War. These new airplanes, although few in number, gave the United States a superior edge over airplanes produced by the communist countries.

Research and development is often misunderstood by the average person. It is the "seed corn" of airpower, which must be planted early to improve future aircraft design and performance.

The unsettled international situation in the postwar years raised familiar problems for the U.S. Navy. Fleet elements were given the task of supporting the nation's policy in areas on opposite sides of the world. A task force built around one or two carriers cruised the Mediterranean, and as the years passed, became a fixture in

Rocket-packing F4Us warm up on the deck of the USS *Bunker Hill*
after the end of World War II. (Courtesy of U.S. Navy.)

that sea. A similar force in the Western Pacific provided the same tangible symbol of
American might and determination to support the free people of the world.

Within the Navy there were problems of adjusting to a new departmental
organization formed by what was only a compromise agreement. At the bureau
and office level there were problems of reducing staffs and of realignment of the
functional elements of technical and administrative units to meet new requirements. In the Fleet there were problems of transition, partly in size, but particularly in weapons and tactics developed either as a result of combat experience in
World War II or as a result of technological advances. The introduction of jet
aircraft posed special problems for carrier operations. Superimposed were new
concepts based upon guided missiles, which had been introduced in World War
II with only limited success, but which continued to develop design-wise for operational use.

It was a period in which changes occurred at an ever accelerating rate and came
to be accepted as normal. It was a time of constant readjustment of plans, as tactical
doctrines were repeatedly revised. The urgency was due, in large part, to the realization that the atomic bomb had forever changed the perception of war.

After World War II, the aerial torpedo plane designation was dropped from the
Navy's roster. In this era of rockets, missiles, nuclear warfare, and accurate detection systems, the vulnerable torpedo planes retained a torpedo-carrying capability
only for special missions.

During the war, in response to a Navy request for a new single-engine attack plane, Douglas Aircraft Company designed the XBTD-1. It was converted from the SB2D-1 dive-bomber by removing the rear seat gunner and installing extra fuel tanks. Deliveries began during June 1944, and a total of twenty-eight were built before the end of hostilities. Neither aircraft was successful, so the XB2D-1 was designed. It later became the AD-1 Skyraider.

Simplicity was the keynote of this 345-mile-an-hour bomber. The Navy ordered $50 million worth to rearm its postwar fleet in January 1946. Amazingly light, the 19,000-pound airplane was designed for heavy loads on long missions at least fifty miles an hour faster than any wartime-built dive-bomber. The AD-1 also incorporated the first production application of fuselage dive brakes.

A Navy decision to combine the tactical functions of scout, dive, and torpedo bombers into a single airplane resulted in the Skyraider. By integrating these three functions, development costs were limited to a single prototype, and subsequent production concentrated on one basic airplane. Furthermore, multipurpose airplanes added to the tactical effectiveness of a carrier task force without increasing the number of its airplanes. This also simplified and reduced procurement of spare parts and accessories.

The Skyraider's success was immediate, and the airplane provided the background for many additional tactical uses. A total of 3,180 airplanes were produced during the following years.

A contract was issued to Douglas on April 3, 1946, for the design and construction of the XF3D-1. Such a night fighter was needed to give the fleet around-the-clock protection, and to counter Soviet Russia's fleet of night bombers.

A contract was issued to North American Aviation, Inc. on June 24 for the design and construction of three XAJ-1 aircraft. This proved to be the first step in the active development of a long-range carrier-based bomber capable of delivering nuclear weapons. Navy officials were convinced that it was essential to attain a nuclear delivery capability to preserve their aircraft carriers.

Tests to determine the effects of atomic bombs on naval targets began on July 1 in "Operation Crossroads" at Bikini Atoll in the Pacific. A Nagasaki-type bomb was used first. It was dropped from a B-29 Superfortress from 30,000 feet on ships anchored in the lagoon. Five ships were sunk outright, and heavy damage was caused to five others.

On the 25th, a shallow underwater burst raised the total number sunk either directly or indirectly to 32 of the 83 ships of all types assembled in the lagoon. Among them was the carrier *Saratoga*, which sank in shallow water on the 25th. It had served the Navy for 19 years. The carrier *Independence* was so heavily damaged and contaminated that she was no longer fit for service.

These tests emphasized again the importance of naval aviation for controlling the seas during wartime, especially in this new age of nuclear warfare. Much detailed data was accumulated by the tests on the effects of nuclear blasts on ships, which led to the development of tactics and equipment to minimize damage to a naval task force.

The "Truculent Turtle"—a Lockheed P2V Neptune manned by Commanders T.D. Davies, E.P. Rankin, W.S. Reid, and Lieutenant Commander R.A. Tabeling—flew from Perth, Australia, to Columbus, Ohio, in fifty-five hours and seventeen minutes. Their flight broke the world's record for distance without refueling, with a distance of 11,235 miles.

The Truculent Turtle. A P-2V Neptune flew nonstop in 1946 from Perth, Australia, to Columbus, Ohio—a distance of 11,236 statute miles—in 55 hours and 17 minutes. This distance record remained unchallenged for 16 years. The crew included, from left, Commanders Eugene Ranklin, Walter Reid, and Thomas Davies, and Lieutenant Commander Roy Tabeling. (Courtesy of Lockheed.)

Marine Lieutenant Colonel Marion E. Carl made four free takeoffs and five arrested landings in a jet-propelled P-80A aboard the *Franklin D. Roosevelt* on November 11. His flights were part of an extensive investigation of carrier suitability for jet aircraft that had begun in late June of 1945.

Development of low-drag bombs for release at subsonic speeds was now imperative, and on June 26, 1947, the Bureau of Aeronautics authorized the Douglas Aircraft Company to design a bomb release system with smooth flight characteristics. This step became necessary to overcome aircraft buffeting induced by conventional bombs when carried externally at three-quarters of the speed of sound. The Douglas El Segundo Division was charged with designing the shape of an external store, which could contain conventional bombs, machine guns, and rockets, and also be adapted for use as an external fuel tank.

In the late 1940s, the Navy concluded there were enough single-place ADs. It was decided that they should be replaced eventually by the new A2D Skyshark, and that future ADs should be of the special purpose type. The ideal solution was to redesign the basic structure for all configurations and provide equipment for special configurations in kit form. The AD-5 airplane evolved into a multitude of missions, such as night and day bomber, early warning, photographic, litter carrier, and anti-submarine.

The A2D was a conventional aircraft, but it was powered by a turboprop engine. It was hoped that this marriage of jet speed with the lower fuel consumption associated with propeller-driven aircraft could be advantageous. Technical difficulties, however, proved almost insurmountable, and the Skyshark was abandoned in the prototype stage as newer, pure-jet bombers came along faster than anticipated.

The Navy continued to explore several futuristic programs based on wartime experiences. Priority items included angled decks, steam catapults, and a mirror-approach system for carrier pilots.

The Navy also initiated a scientific flight research program in the summer of 1945 because aeronautical engineers faced great unknowns in the field of advanced speeds. In the absence of adequate wind-tunnel facilities to test models at the speed of sound, the D-558-1 Skystreak program was approved while a further extension of the program was authorized to take man to the outer rim of the earth's atmosphere.

The Skystreak, a straight-winged jet that in the course of its research investigation shattered the world's speed record twice within five days in August 1947, was just the start of a program that lasted several years.

The Douglas Aircraft Company at El Segundo, California, designed and built the airplane in cooperation with the National Advisory Committee for Aeronautics and the U.S. Navy.

The Air Force had a parallel program to explore the frontiers of supersonic flight.

For decades after the Wright Brothers made the first powered flight of an airplane, many scientists and engineers were convinced that a "sonic wall" known as the sound barrier would prevent supersonic flight. It was known that the speed of sound varied with altitude and that at 40,000 feet it was about 660 miles per hour. A condition known as "compressibility" seemed impregnable. Fortunately, there were those who believed it was just a matter of sorting out the problems and eliminating them one by one.

During World War II some fighters had come close to the so-called barrier with unpredictable and frightening results. In high-speed dives their controls would freeze, and buffeting became so severe it was almost impossible to maintain control of the aircraft. Some airplanes actually disintegrated.

Engineers at Bell Aircraft in Buffalo, New York believed the sound barrier could be broken if an airplane was shaped like a .50-caliber bullet, which did exceed the speed of sound. The X-1 was the result of their technical skill. Thirty-one feet long, it was powered by four rocket engines that produced an amazing 6,000 pounds of thrust. The wing had a razor-sharp leading edge and was at most only three and a half inches thick. The tail assembly was mounted high above the fuselage to avoid shock waves as they passed over the wing as it entered the sound barrier. The cockpit was small, and a chore for any pilot to get in or out of.

A 24-year-old former World War II P-51 fighter ace was the logical choice to fly the X-1 because he was a test pilot at Edwards Air Force Base, California, and he was familiar with experimental aircraft. He named the orange-colored rocket plane "Glamorous Glennis" in honor of his wife.

The X-1 was attached to the underside of a B-29 Superfortress to permit it to reach high altitudes, where it could be released to make its runs. With only 600 gallons of liquid oxygen and alcohol, a takeoff under its own power was out of the question.

On the night of October 12, 1947, Chuck Yeager and his wife went horseback riding and his horse crashed into a corral gate that was supposed to be open. Two of Yeager's ribs were broken and he went to a civilian doctor who taped them secretly. He knew that if Air Force authorities knew of his injury he would be grounded. His already formidable risks were intensified by his physical condition.

The next day Yeager prepared for his first flight in the X-1. On his seventh flight the X-1 achieved ninety-four percent of the speed of sound. He eased the control column back, expecting to gain altitude, but nothing happened. He had to shut down two engines to regain control of his aircraft.

On the ground it was determined that a shock wave had formed directly above the elevator hinge on the X-1's tail. (The elevator controls the pitch of an aircraft.) It was realized that the shock wave had nullified the elevator's effect.

Captain Jack Ridley, the plane's flight engineer, suggested to Yeager that the elevator be locked down and the pitch controlled by moving the entire horizontal stabilizer. A small motor built into the X-1 could be used to control minute calibrations from the cockpit. Since then the "flying tail" has become standard on all supersonic aircraft because it resolved the X-1's problem.

Yeager released his X-1 from the B-29 "mother plane" at 26,000 feet the following day, October 14, 1947. He fired all four engines and the plane shot ahead. He shut down two and climbed higher. With all four engines on again the X-1 easily accelerated above .93 Mach. With infinite caution, Yeager shut down two engines again and the X-1 performed beautifully.

He relit one of the engines and shock waves rolled off the airframe. People in the chase plane and on the ground for the first time spotted shock waves and heard the sonic boom when the shock waves hit the ground.

Yeager watched the machmeter as it rose to .965 and then to .988. Suddenly the needle jumped off the scale. Yeager was thunderstruck, realizing he had exceeded the speed of sound without a bump. Years later he wrote that "after all the anxiety, breaking the sound barrier turned out to be a perfectly paved speedway...Grandma could be sitting up there sipping lemonade." The adjustment to the tail proved to be a simple solution to what had been considered an almost unsolvable problem.

Yeager's achievement created a revolution in the design and operation of all future air and spacecraft, but even more than fifty years later his success is not fully appreciated.

The Skyrocket takes off with JATO assist. (Courtesy of McDonnell Douglas.)

Paralleling this effort, the D-558-2 Skyrocket was conceived by Douglas engineers. It was designed as a research laboratory to probe the supersonic envelope beyond the speed range of the Skystreak and the X-1.

Twenty months after design began, the first airplane was completed. With Douglas Chief Test Pilot John Martin at the controls on February 4, 1948, the history-making airplane accomplished its maiden flight at Edwards Air Force Base, Muroc, California.

For advanced tests, the rocket craft was refitted for aerial launching from the underside of a specially modified B-29 mother airplane at altitudes between 30,000 and 40,000 feet. During these flights the jet engine was eliminated and three tons of additional rocket fuel were installed. It was air-launched to conserve fuel while making its climb.

During one flight on August 21, 1953, Lieutenant Colonel Marion E. Carl of the Marine Corps exceeded earlier achievements and thrust the needle-nosed research vehicle to 83,235 feet, for a new altitude record.

On October 20, Scott Crossfield, veteran NACA (National Advisory Committee for Aeronautics) research pilot and scientist, made aeronautical history when he reached 1,327 miles an hour, or twice the speed of sound, with the Skyrocket.

"Passage of the exploratory craft into the supersonic range has become routine," he said. "It has repeatedly exceeded the speed of sound in level flight."

Veteran Douglas test pilot Gene May said, "The Rocket is a flight machine refined to the nth degree. It is extremely stable, yet as sensitive and light as an arrow. I thought the Skystreak was a dream airplane but the Rocket's flying qualities are without parallel."

Douglas turned the three Skyrockets over to the National Advisory Committee for Aeronautics on August 27, 1954, for further advanced research.

During demonstration programs the Skyrocket more than doubled its specification requirements. Over 8,500 flight and engineering reports were distributed to the armed forces and to the aircraft industry.

The Skyrocket blazed a trail for practical supersonic and rocket aircraft. It was a cooperative endeavor that paid dividends as the entire national sonic research program headed toward extension of man's aeronautical knowledge in realms far beyond the speed of sound.

While spectacular advancements continued in aircraft design, modernization of existing aircraft carriers was given top priority. This was necessary to maintain an acceptable degree of combat readiness. Also, modernization could be completed in half the time required for construction of new carriers.

The *Essex* class of World War II quickly became obsolescent. The Chief of Naval Operations approved a series of changes on June 4, 1947, so that these carriers could meet operational requirements of the postwar era.

The USS *Oriskany* was first to be modernized, in October, and at regular intervals the *Essex,* the *Wasp,* the *Kearsarge,* the *Lake Champlain,* the *Bennington,* the *Yorktown,* and the *Randolph* followed. The *Hornet* completed the program in October 1953, and was returned to the fleet. Changes were made in catapult systems and decks to permit operations by aircraft weighing up to 40,000 pounds.

A nuclear weapons capability was given the last six of these nine carriers, and special provisions were made for the operation of jet aircraft. Increased flight deck safety and pilot comfort were also provided.

In April 1947, the *Franklin D. Roosevelt* was given a special weapons capability during another ship improvement program, and her sister ships, the battle carriers *Midway* and *Coral Sea,* followed her into the yards. Then the *Oriskany,* the *Essex,* and the *Wasp*—which had not been given a special weapons capability under the original *Essex* class program—were brought up to renewed flight readiness.

As new jet aircraft were ordered into service, fleet qualifications progressed normally until many squadrons became jet-equipped. Initially it was planned to expand this program to nine more carriers, but development of a steam catapult for smoother takeoffs caused a delay. The British had developed the steam catapult, and tests in the United States were so satisfactory that it was decided to include it on the new flush-deck carrier the *United States,* which was later cancelled. The *Hancock* was the first American carrier to use what became known as the "steam slingshot."

While changes were incorporated in older carriers, the number three centerline elevator was replaced with a deck-edge type of greater capacity. These changes were built into the new *Hancock,* the *Intrepid,* and the *Ticonderoga,* and the Bureau of Aeronautics proposed in June 1952 to install a new type of flight deck on the *Antietam.*

The British had also designed an angled deck that was simulated aboard the *Midway* with excellent results. It was agreed that the *Antietam's* deck would be extended outboard on the port side from the normal flight deck to permit aircraft landings angled ten degrees off the ship's centerline. The new deck was completed in mid-December at the New York Naval Shipyard. This arrangement permits aircraft to catapult off and to land simultaneously, thus improving a carrier's combat proficiency.

Pilots liked the angled deck because it gave them an added margin of safety by removing the danger of crashing into planes forward of the landing area. If they did not "hook up" they could easily fly around again for another attempt. The idea proved so successful that attack carriers were changed again to incorporate the new improvements.

Modernization was accomplished in successive steps to permit orderly retirement of aged carriers. World War II carriers, as they outlived their usefulness as attack carriers, were redesignated antisubmarine carriers.

The familiar escort carrier of World War II disappeared. Its role was taken over by older carriers for antisubmarine warfare and helicopter vertical assault operations.

The attack carrier's role now faced serious opposition in Congress, in the other services, and among influential persons in the United States. Its performance in the past was not sufficient to justify its continuance as a weapons system in a world where technologies were expanding at an explosive rate. It had to prove itself all over again.

President Truman tried to establish a national policy that would govern the procurement needs of a sane national defense policy. One of his first steps was to order his Air Policy Commission to hold extensive hearings during a three-month period and report its findings to him. He appointed Thomas K. Finletter as its chairman.

The Commission filed their report—"Survival in the Air Age"—on December 30, 1947, but it was only a broad review of the problems facing the nation's armed forces. The Commission agreed that airpower had proven its effectiveness in World War II, stating that the advent of the atomic bomb only added to its potential for destruction.

The report stressed the need to maintain military forces large enough to make foreign aggression dangerous. It particularly emphasized the urgency of building up strong military aviation supported by industry and the nation's civil air transport. It concluded that progressive research and development programs must be encouraged to maintain the existing margin of superiority held by the United States.

This vacuous report resolved nothing. If anything, it exacerbated the situation. Would the U.S. Navy's traditional mobility and flexibility be valueless in this age of atomic and thermonuclear weapons? Navy airmen fought back in as bitter a struggle as they had ever waged, but the results were inconclusive, often damaging airpower in the eyes of the American people.

At the Potsdam Conference in July 1945, President Truman, Premier Stalin, and Prime Minister Clement Attlee of Great Britain had established the guiding principles of the postwar Allied Control Commission. Among other terms, they called for Germany's complete disarmament and demilitarization, destruction of its war potential, and decentralization of Germany's political and economic structure. Germany was divided into four national occupation zones, and each was headed by a military governor.

Right from the start, the work of the Allied Control Commission was rendered ineffective due to the complete lack of cooperation by the Soviet Union, and it soon became evident that Russian vetoes made the organization worthless. The situation was aggravated by the walkout of the Soviet Union's representative to the United Nations on March 20, 1948.

The United States and Great Britain retaliated by enlarging their zones economically. These two nations, in association with France and the Benelux countries, agreed to set up a German state comprising the three western zones.

The Soviet Union struck back by initiating a blockade of all ground communications between the western zones and Berlin, which had become an island of Western influence within the Soviet zone. The Western allies rose to the challenge by initiating a gigantic airlift to fly supplies into Berlin. At its peak 60,000 men were assigned to the airlift. The best monthly total of the airlift was made in March of 1949, as American aircraft delivered 154,475 tons of cargo to the beleaguered city. In making its contribution to the total, Navy Transport Squadron VR-8 set an all-time airlift record of 155 percent efficiency for the month, and a daily aircraft utilization of 12.2 hours per aircraft.

The Soviet Union was forced to end the airlift on May 12, 1949, to avoid the growing threat of another war. Participation of Navy Transport Squadrons 6 and 8 during their eight months in Germany set records for the period. They flew a total of 45,990 hours, carried 129,989 tons of cargo into Berlin, and established a record for payload efficiency and aircraft utilization at the unparalleled rate of more than ten hours per day for each airplane during the period.

Unification of the armed forces had been under discussion for decades, often resulting in the expression of controversial opinions by military leaders. After World War II the subject gained new adherents, because the Army Air Forces insisted upon coequal status. It had earned the right after its war achievements, and the Truman administration and Congress for the most part agreed that the time had come to resolve the problem once and for all.

Naval leaders expressed serious doubts, remembering the bitter days of General "Billy" Mitchell. Fleet Admiral Ernest J. King expressed the views of many in the Navy when he said on October 9, 1945, "Seapower will not be accorded adequate recognition because the organization contemplated would permit reduction of seapower by individuals who are not familiar with its potentialities."

Despite such views, the National Security Act of 1947 was passed on July 26. It created a National Defense Department Establishment, with the secretaries of the Army, Navy, and Air Force subordinate to a Secretary of Defense.

James Forrestal, a former Secretary of the Navy, and the nation's first Secretary of Defense, spoke realistically when he said, "I would be less than candid if I did not underline the fact there are great areas in which the viewpoints of the services have not come together."

His views understated the controversy. Interservice bickering intensified both in public and private between the Navy and the new Air Force. Emotions often clouded the basic controversy, and in the final analysis caused disruption in the orderly advancement of airpower in both services.

The issue became so intensified that men like General Dwight D. Eisenhower found it necessary to speak up. "These are distinguished Americans," he said, "who are in honest disagreement with one another. We should not question their motives."

The issue reached a climax in 1949 when new Secretary of Defense Louis A. Johnson discontinued work on the Navy's new supercarrier, the USS *United States*. Unfortunately, the manner in which it was done inflamed the Navy more than anything else. Johnson made his decision when Navy Secretary John L. Sullivan was out of Washington, and Sullivan resigned in protest three days later.

Inevitably, wild rumors were started that the Marine Corps was going to be abolished and the Navy reduced to convoy and escort duty. Johnson denied these charges, but tempers were so high that a full-scale Congressional investigation was started. The venerable Carl Vinson, chairman of the House Armed Services Committee, called witnesses to present their cases. General Hoyt Vandenberg, Air Force Chief of Staff, said, "The only war you really win is the war that never starts." It soon became evident to Vinson and the committee that the controversy really was one of strategy, and of the role each service should play.

During this period, when it was clear that Marshal Stalin and the Soviet Union would disregard most of the promises made at the postwar Yalta Conference, the delivery of nuclear weapons against a potential enemy assumed first priority in the armed forces.

Air Force Secretary W. Stuart Symington put it succinctly, "If war comes, we believe the atomic bomb, plus the airpower to deliver it, represents the one means of unloosing prompt, crippling destruction upon the enemy with absolute minimum combat exposure of American lives.

"If it is preferable to engage in a war of attrition," he said, "one American life for one enemy life, then we are wrong. That is not our way. That is not the way in which the mass slaughter of American youths and the invasion of Japan was avoided."

Hearings continued to evade the central issue. Many in the Air Force and Navy regretted that the dispute had been brought into the open. Obviously, they said, the military leaders of the nation should have been able to resolve their differences without a public hassle.

Symington continued to espouse Air Force doctrine, and he was sincere when he said, "We can hope, but no one can promise, that if war comes the impact of our bombing offensive with atomic weapons can force a decision without the use of surface forces.

"Disregarding such an illusory hope, we do know the engagement of surface forces will take place with greater assurance of success and much fewer casualties to the United States and its allies if an immediate, full-scale atomic offensive is launched against the heart of the enemy's war-making power."

Some members of Congress found the Air Force concept appealing because the Army and Navy might be able to be reduced to the point where the draft would become unnecessary.

Unfortunately, serious consideration was not given to the possibility that a nuclear war might be impractical for moral and political reasons.

Symington called for concentration of America's greatest asset, which he identified as quality of products, super weapons capable of development, and mass production under the nation's free economy system. He mentioned specifically the B-36 intercontinental bomber, which appeared now as the source of the bitter debate.

Navy officials had called the B-36 a "billion-dollar blunder." This ridiculous statement contributed nothing to an understanding of the basic problem dividing the two services, and it prolonged the controversy.

"I am in favor of great development of carrier aviation to whatever extent carriers and their aircraft are necessary for fulfillment of a strategic plan against the one possible enemy we have to face," Symington said. "Less than this would be unsound. More than this would be an unjustifiable burden upon the American taxpayer."

General Vandenberg then said, "I am not only willing but insistent that the types of carrier which can help meet the threat of an enemy submarine fleet shall be developed fully and kept in instant readiness. The sealanes must be kept open."

While Navy brass bristled, Symington added, "I do not believe there is justification for maintaining large carrier task forces during peacetime unless they are required by strategic plans of the Joint Chiefs of Staff. In my judgment they are not required by those plans. My opposition to building the *United States* comes from the fact that I can see no necessity for a ship with those capabilities in any strategic plan against the one possible enemy.

"Any war we may have to fight will obviously be unlike the Pacific War against Japan," he said. "It will tend to resemble the war against Germany, though with certain differences. There will be the same problem of killing submarines, there will be the same problem of protecting Atlantic Ocean supplies, although the threat will come almost wholly from the sub.

"The industrial heart of a potential enemy lies not on any seashore, not on any island, but deep inside the Eurasian land mass. It is to that type of war we must adapt all our forces, including carrier aviation."

Although there was much talk of strategy, it became apparent that both services were in agreement, but the basic difference was how to achieve air supremacy.

The Navy argued that the long-range, land-based bomber and the large carrier duplicated each other. They insisted both were designed for strategic air war.

Actually, the Navy wanted a greater role in strategic air warfare, whose mission had been assigned to the Air Force.

General Omar N. Bradley, chairman of the Joint Chiefs of Staff, said, "I do not believe our country should rely solely on strategic bombing or atomic weapons. Properly balanced land, air, and sea forces are required. I seriously doubt any future campaign will be like the war in the Pacific. I also doubt there ever again will be large-scale amphibious operations."

Admiral Arthur W. Radford, Commander in Chief, Pacific, said, "The major issue deals with the kind of war for which this country should be prepared. It is impossible to predetermine a fixed concert for fighting a war." He was one of the few men who had weighed the controversial items carefully, and he spoke with clarity on a difficult subject. "There are issues more important than the B-36," he said. "I firmly believe atomic retaliation has been over-emphasized." He stated at great length that the security of the nation depended on mobile airpower to ensure control of the air in vital areas. "The Navy today must be built not to meet an enemy, but with the idea, after evaluation, of the need for airpower to theaters of war and parts of the world where they can't get airpower any other way."

Captain Arleigh A. Burke, who like Radford had served with distinction in the Pacific fighting, said, "If war develops, one of the first duties will be to gain and hold command of the seas. If we fail to command the seas, we cannot support our war effort overseas."

The Marine Corps, which unfortunately was dragged into the dispute, was represented by its commandant, General Clifton B. Cates. "Without a well-trained landing force," he said, "the fleet is not a balanced implement of warfare. Marine forces are possessed of great utility in augmenting the national defense if they are permitted to do so."

Admiral Louis B. Denfield, Chief of Naval Operations, said "Naval forces could help discourage aggression either on a large or small scale. The pressure of the Fleet in the eastern Mediterranean has effectively contributed to keeping local conflicts from degenerating into global war.

"Operations of carrier task forces, through applications of the principle of mobility and surprise, have repeatedly demonstrated the ability to concentrate aircraft strength at any desired point in such numbers as to overwhelm the defense. No other force and no other nation possesses this capability to a like degree.

"We have a lead of more than a quarter of a century over any probable enemy." (President Truman's disclosure September 23, 1949, that the Soviet Union had set off an atomic explosion later made Denfield's statement academic.) But Denfield was speaking as if the status quo would remain the same forever. "Let us not squander it [the nation's lead in atomic energy] for any false doctrines, any unsound concept of war. That would be the real extravagance," he emphasized. He continued, "The properly balanced Fleet must have as a major component a Fleet Marine Force of combined arms, including its close-support tactical aviation. The inclusion of such a force permits a fleet commander a degree of initiative and flexibility in his operations not otherwise obtainable. He can seize advanced bases as required by the development of the campaign, or, if the situation dictates, be assured of adequate defenses for those bases already in his possession."

The head of Navy Logistics, Vice Admiral Robert B. Carney, put the Navy's viewpoint in clearer perspective. "To settle on a concept of sustained intercontinental bombing, or a program of procuring costly intercontinental bombers, could only be justified by overriding considerations of the greatest urgency because logistically, in terms of treasure and effort, there are better ways of conducting strategic bombing." Unspoken was his belief that the Navy's carriers should be assigned to at least participate in such a strategy.

Carney said the issue was for the nation to decide whether the American Air Force power in its present form was needed to the extent of accepting deterioration and inadequacy of other essential components of the military team.

The debate reached the explosive point, and words were often used that were unbecoming of men of such stature. The only agreement reached by both sides would have astounded Navy and Army airmen of an earlier period, such as John Towers and "Billy" Mitchell. Both sides claimed that the key to eventual victory lay in airpower.

Chairman Vinson, undoubtedly as bored and confused as the average American after this senseless battle of the generals and the admirals, summarized the committee's views. He said they did not want any strategy drafted that would deny the country an efficient and effective arm to play its proper role in the defense of the nation. "We don't want to keep one strong member of the team sitting on the bench too long," he said lamely.

These hearings had one positive effect. On August 10, 1949, the National Security Act of 1947 was amended to provide for a limited increase in the authority of the Secretary of Defense, and the original National Military Establishment's name was changed to Department of Defense. The changes provided that the three military departments would continue to be separately administered, and that Naval Aviation would be "integrated with the naval services...within the Department of the Navy."

While the American military establishment battled over future war roles, Soviet leaders came to the conclusion that the United States would not fight for Korea. The North Koreans were urged to invade the republic to the south.

Representatives of the United States, the United Kingdom, and the Republic of China had pledged at Yalta that Korea, "Land of the Morning Calm," would become a free and independent nation after thirty-five years of Japanese occupation.

The 38th parallel, which caused endless disputes, was chosen initially only as a convenience to accept the surrender of the Japanese. It was never considered a permanent split of the nation by the Allies. Marshal Stalin thought otherwise. He immediately ordered the border closed, and Korea became a divided nation. This action effectively eliminated chances for unifying the country. It was also unrealistic, because it divided the industrial north from the agrarian south.

The United States attempted to set up free elections in the summer of 1947. Soviet Foreign Minister Molotov accused the United States of further dividing Korea by recommending elections. He suggested American and Soviet troops both get out of Korea. "Let the Koreans organize a government themselves," he said. Such an action would have played into communist hands, because the Soviet Union had trained a large army with the latest Soviet weapons, whereas South Korea's army barely existed.

The United States agreed to give the South Korean people assistance and guidance so they might eventually attain freedom and independence. America was most anxious to reduce its costly commitments on a progressive basis as the South Koreans became self-sufficient.

The United Nations General Assembly approved American recommendations that the Korean question was a matter for the Korean people to decide on their own. On November 14, 1947, the assembly passed a resolution that the Korean people should have an opportunity to elect representatives, draft a democratic constitution, and establish a national government. A temporary commission went to South Korea early in January 1949.

Despite Soviet protests, the United Nations instructed the commission to proceed with the program in South Korea, even though they were denied access to North Korea.

The Republic of Korea was established on August 15, 1948. The Soviet Union promptly promoted the Democratic People's Republic of Korea in the north.

Conditions worsened on the peninsula during the next two years as the North Koreans spread disorder and confusion through subversive infiltration. Armed raids across the border gained in frequency.

After the North Koreans felt secure behind their new army, they sought withdrawal of U.S. and Russian forces. American troops were withdrawn on July 1, 1949, except for a provisional military advisory group of 500 men.

Less than a year later North Korean troops crossed the border in force, starting a conflict that eventually included twenty-five nations and five million soldiers from around the world.

There were no American combat troops on the peninsula, and few Air Force combat wings in the Far East. The U.S. Navy had only one cruiser, four destroyers, and a few minesweepers in the Sea of Japan. This was not a primary operating area, however, but there were substantial fleet elements, including an aircraft carrier, in the Western Pacific.

The type of war most military experts said would never be fought again erupted in all its savagery as the unprepared South Koreans gave ground steadily before the communist-trained North Koreans.

It proved to be a familiar type of war, where control of the seas gave ground and air forces the necessary mobility to limit the war, and turn communist military action from victory to failure.

Part V

The Korean Conflict

17

South Korea Invaded

The Korean Peninsula, dividing the Yellow Sea from the Sea of Japan, is roughly 600 miles long on a north-to-south axis and less than 200 miles east-to-west. The northern part is mountainous, and the east coast is isolated by a ridge of mountains from the rest of the peninsula. In the west and south, a series of river basins created the most populous areas. Rail and road routes from the north converged normally toward the South Korean capital of Seoul.

When seven North Korean divisions of infantry and an armored brigade crossed the border at 4:00 A.M. on June 25, 1950, the South Koreans were caught by surprise. The North Korean Peoples' Army advanced rapidly because the Republic of Korea Army was not only unprepared to meet such an assault but was ill-equipped for a major war.

The communists made two amphibious landings with 10,000 troops at Kangnung and Samchok on the east coast as they reached Seoul on June 28.

When the border was violated in force, the U.S. government was informed immediately. The State Department requested a meeting of the United Nations Security Council, which—by a vote of nine to nothing, with one abstention, and the Soviet Union's representative absent—demanded withdrawal of the North Korean forces. The council went on record as saying that North Korea's action constituted a breach of the peace, and asked members to render every assistance to the United Nations in executing the resolution.

President Harry S Truman met with representatives of the State and Defense Departments that evening. The Joint Chiefs of Staff communicated their decision to General MacArthur in Japan. "Assist in evacuating U.S. dependents and noncombatants. You are authorized to take action with the Air Force and Navy to prevent the Inchon–Kimpo–Seoul area from falling into unfriendly hands."

The next day MacArthur, Commander in Chief, Far East, was given permission to use elements of his command to attack North Korean targets south of the 38th parallel.

Truman, unsure whether this presaged an all-out communist offensive in the Far East, ordered the Seventh Fleet to prevent an attack upon Formosa.

F9F Panthers return to the USS *Kearsarge* after a strike over Korea.
(Courtesy of U.S. Navy.)

The UN Security Council adopted a resolution on July 7 that stated, "We welcome the vigorous support which member nations have given to the Republic of Korea. All members are requested to render additional assistance." The council again called for an immediate cessation of hostilities.

Vice Admiral C. Turner Joy, Commander of Naval Forces, Far East, had only twenty-nine officers in his command. Their primary mission had been to promote the rehabilitation of Japan. Naval operations had been of a routine nature.

Joy decided to send the Seventh Fleet to Okinawa, because it would be imperiled at Sasebo, Japan, if Russia entered the war.

The *Valley Forge* was the most combat-ready carrier in the Far East Fleet. It had Air Group 5 on board, which was the most outstanding group in the Pacific Fleet.

President Truman announced on June 30, that—in keeping with the UN Security Council's request for support of the Republic of Korea in repelling the invaders and restoring peace—he had authorized the U.S. Air Force to bomb military targets in North Korea. He also ordered the use of Army troops to support ROK (Republic of Korea) ground forces, and directed that a naval blockade be established of the entire Korean coastline.

MacArthur soon realized the South Korean Army had no recourse but to retreat to a defensible perimeter in the south and hold out until American divisions could

Skyraiders in a ground attack. (Courtesy of U.S. Navy.)

be landed. His efforts in the first eighty-two days were to maintain a Korean bridge-head around the southern port of Pusan.

Carrier aircraft went into action in Korea on July 3, with the *Valley Forge* and its Air Group 5 and the British carrier *Triumph* operating in the Yellow Sea. They launched strikes against airfields, supply lines, and transportation facilities in and around Pyongyang, northwest of Seoul. Railroad yards and bridges were hit first, because they were constantly in use to move enemy troops and supplies into South Korea. This was the first combat test for the Grumman F9F Panther and the Douglas AD Skyraider.

The British carrier *Triumph* sent twelve Fireflies and nine rocket-laden Seafires to attack hangars and installations at Haeju airfield. The *Valley Forge* sent sixteen F4U Corsairs from VF-54 and twelve AD Skyraiders from VA-55 for the attack against Pyongyang. After these attack airplanes were airborne, eight F9F-2 Panthers were catapulted into the air. Their jet engines gave them a superior speed and would get them over the target ahead of the prop airplanes.

This was the first wartime use of the jets, and they destroyed five planes on the ground at the North Korean capital, and two Yak-9s in the air. The war's first combat kills were made by Lieutenant L.H. Plog and Ensign E.W. Brown of VF-51. They swept ground installations, igniting hangars and ammunition dumps, so the area was under a pall of smoke by the time the ADs and the Corsairs arrived.

Lieutenant Commander N.D. Hodson signaled his ADs to push over from 7,000 feet and the Corsairs followed. He ignored light flak that burst hundreds of feet away as he plummeted earthward. His speed climbed to 300 miles an hour. Then he pushed the button on his stick and the bombs salvoed. A glance at his altimeter showed he was down to a thousand feet so he pulled back on the stick. After climb-out, he noted most of the airfield installations showed bomb bursts, with a direct hit on a fuel storage farm.

Carrier planes returned in the afternoon and attacked the railroad yard and rail and road bridges across the Taedong River. The bridges escaped destruction but fifteen locomotives were destroyed and ten others damaged.

Hodson's VA-55 pilots destroyed a span of a vital bridge and ten more locomotives the following day. This time ground gunners found their range and four ADs from the *Happy Valley* were hit—one so seriously it could not lower its landing flaps. Upon return to the carrier, it careened across the deck, leaped the protecting barriers, and smashed into planes on the forward section of the deck. Three were destroyed and many others were damaged.

Reconnaissance photographs proved their attacks had been successful. For the present, the city's rail center was wrecked, a span of a key bridge was dropped into the river, and an enemy air base was demolished. Air opposition had not been severe, but Seventh Fleet pilots destroyed eleven enemy aircraft.

While the defense perimeter shrank before the North Korean Army in the south, the *Valley Forge* pilots daily attacked targets north of the 38th parallel. This was done to relieve the pressure on UN forces, who were waging a delaying action around Pusan. During July, these pilots claimed thirty-eight aircraft destroyed and twenty-seven damaged, but only two of these were in the air. For the remainder of the month they concentrated on strikes against airfields, railroads, and factories in the north. They did particularly heavy damage to the oil refinery at Wonsan.

To obtain maximum effectiveness in the employment of all air resources in his command, and to ensure coordination of air efforts, MacArthur approved a policy on July 8 established earlier by Navy and Air Force officials. The Navy was given control of operations of its carrier aircraft whenever they were on missions assigned to the commander of all naval forces in the Far East, and of its shore-based aircraft whenever they were on naval missions. It was agreed that on all other missions the operations of naval aircraft, both carrier and shore-based, would be under the Air Force. For shore-based Marine air operations this control was direct, but for naval aircraft the control was handled through coordination. Target selection and prioritization, were assigned to a General Headquarters Joint Service Target Analysis Group to ensure that air coordination met overall objectives.

The North Koreans, with their limited airpower, found it impossible to ward off attacks or provide air assistance to their ground troops.

At home, fourteen squadrons of the Organized Reserve were activated on July 20 for duty with naval aviation forces. They included eight fighter and two attack squadrons for carrier service, an antisubmarine squadron, two patrol squadrons, and one Fleet Aircraft Service Squadron.

The escort carrier *Badoeng Strait* arrived at Yokosuka, Japan, on July 22 with elements of the First Marine Aircraft Wing. Four days later, the escort carrier *Sicily* arrived with a load of ammunition and the carrier *Philippine Sea* joined the Seventh Fleet at Okinawa on August 1. The carrier *Boxer* arrived on July 23 with a load of 145 F-51 fighters for the Air Force and urgently needed supplies. In making the trip from Alameda, California, in eight days and sixteen hours, the *Boxer* broke all

The Wonsan oil refinery during an attack by Navy fliers. (Courtesy of U.S. Navy.)

existing records. These steps were taken in an effort to build up the U.S. Navy to eleven attack and one light attack carrier and five escort carriers in the Far East.

The *Valley Forge* and the *Philippine Sea* began their almost three years of continuous fast carrier operations on August 5 with attacks on enemy lines of communication in southwestern Korea, and close support missions on the Pusan perimeter.

The Russians and the Chinese had planned to give their North Korean allies large numbers of older propeller aircraft for ground support, but the appearance of UN jets over North Korea made this inadvisable.

On the ground, 700 men of the U.S. 24th Division went into action on July 7 after they were flown from Japan. Their first objective was to block the communists from advancing along highways to the south. Three days later, it was obvious that the outnumbered American and ROK ground troops could only take up defensive positions before Taejon. The 24th Division had to be fed piecemeal into the front action and they suffered severe casualties as a result.

MacArthur realized by July 18th that the 24th Division could not hope to fight off four enemy divisions. The 25th Division was shuttled from Kyushu to Pusan by the Military Sea Transportation Service. It was also obvious that additional reinforcements would have to be thrown into the battle if American and ROK troops were not to be driven into the sea. Pusan was so clogged with troops and military supplies that another landing site was selected.

Fortunately, Rear Admiral James H. Doyle had ordered his Amphibious Group One ships to get under way from Yokohama, Japan, prior to the start of hostilities for a training exercise.

When Pohang was selected as the alternate landing base, the group was prepared to leave. Pohang, seventy miles north of Pusan, was still held by the Third ROK Division, and it was hoped they could stave off attacks until landings were made on July 18. The 1st Cavalry Division landed unopposed and helped to reduce the enemy's drive down the Taegu-Pusan highway.

Seventh Fleet Commander Admiral Arthur D. Struble ordered his ships northward into the Sea of Japan following the successful landing at Pohang. He wanted to continue attacks against North Korea.

The Chosan Oil Refinery, one of the largest in the nation, was selected for a full-scale attack on July 18. It was estimated to have an annual production of 1,700,000 barrels of oil. Although it was primarily a strategic target, B-29s failed to do extensive damage. By agreement with the Air Force, Navy planes were assigned the task of its destruction. There was no argument about who should destroy it. It was a job that needed to be done, and the *Valley Forge* was in the best position.

Seven Panther jets took off early for a fighter sweep up the northeast coast past the harbor of Wonsan. Commander A.D. Pollack of VF-51 followed the curving shoreline until he spotted the refinery on the south side of the city. It was a large installation, standing out prominently against the countryside.

Twenty-one planes, armed with large bombs and high-velocity rockets, roared off the carrier to make the attack. Hodson spaced his planes so their bombs would cover all buildings of the refinery. ADs dove steeply to the attack and their bombs slammed into the buildings without a miss. When Hodson pulled away, the Corsairs lobbed rockets and fired cannons into the refinery.

While the commander watched, the refinery belched smoke high into the air and long fingers of flame shot skyward as explosions ripped the place apart. Miles away, Hodson continued to watch black smoke pour from the refinery. It burned for four days and reconnaissance photos showed it had been turned into a mass of rubble. For days it was kept under surveillance until it was obvious that operations had ceased.

In South Korea the outnumbered UN forces retreated steadily into a shrinking perimeter around Pusan. Lieutenant General W.H. Walker sent an urgent plea to the Navy for close air support to assist his desperate infantrymen. Admiral Joy immediately ordered the commander of Task Force 77 to provide assistance.

For two months attacks were made daily, helping to provide the necessary margin for saving the perimeter. There were difficulties involved in such support, because air action close to friendly lines needed delicate integration with ground controllers.

The Navy–Marine system, developed initially during the 1920s, required airplanes and ground troops to work as a team. This system was seldom used in Korea, because it differed from the Air Force–Army concept.

The Navy and Marines had proved their system in amphibious assaults during World War II, when aircraft were directed by frontline troops through the use of portable radio equipment.

The Air Force did not believe in tying its planes too closely to ground operations. They felt such a relationship failed to make use of the fullest potentialities of air attack. The result was that the Air Force insisted upon centrally controlled airpower to achieve ultimate air supremacy over the battlefield. Air Force commanders also gave close air support a secondary role. This was unfortunate because this type of war necessitated maximum use of close support with the troops.

Strategic targets and large supply and transportation points between the frontlines and rear areas were few in number because Korea was not an industrial nation. The Tactical Air Force was not equipped, therefore, to conduct a Korean type of war. This was due largely to the fact that for years the highest officials of the U.S. government had given preeminence to strategic airpower.

Despite problems of command and communications, the Navy–Marine close-support system proved effective when it could be used properly.

Carrier commanders readily agreed to commit certain of their aircraft for control by ground commanders. In this way planes constantly orbited the front lines, available to strike once word was flashed from the battalions. With guidance and information from a trained crew on the ground, targets could be struck within 50 to 200 yards of friendly troops.

These were days of deadly monotony for carrier crews. Deadly because ground fire was vicious. Monotonous because they performed the same missions day after day without tangible results.

Pusan was threatened by encirclement in late July. The 6th North Korean Division swept into Moko, a naval base on the southeastern tip, triggering an urgent request for close air support to this area.

The front became so uncertain that Admiral Arthur D. Struble insisted on satisfactory communications when he committed the Seventh Fleet. At midnight on July 24, the Seventh Fleet steamed for the east coast, launching planes from the *Valley Forge* at 8:00 A.M. the following morning. They had instructions to report to the Fifth Air Force's Joint Operations Center at Taegu.

Every available *Valley Forge* plane was sent out, while planes from the British carrier *Triumph* supplied most of the combat air patrol. Carrier planes were frustrated when they arrived over the combat zone because communications were jammed and they had to fly in wide circles waiting for word of targets. The situation became even more senseless when F-80 Air Force jets were called in from Japanese bases. Navy pilots finally chose targets of opportunity and returned to their carriers.

Struble was blunt in his report. "These sweeps are of minor importance because of a lack of targets. My men can see only a few donkey carts and men in rice paddies. It is absolutely mandatory to establish proper communications."

Even though the strikes were disappointing, both Generals MacArthur and Walker were enthusiastic about their continuance. They realized Army and Air Force communications needed improvement, and MacArthur ordered that it be done.

Admiral Joy at Tokyo headquarters decided to do something to alleviate the situation. He offered to assign personnel to train frontline teams of ground controllers. The *Valley Forge* said it would supply technicians and personnel to form a complete tactical control party. Nothing came of this last proposal as the *Valley Forge* continued to make attacks along the battle line. (At this time, the British carrier *Triumph* left to join her own task group.)

The problem came under better control after a fleet representative was assigned to the Joint Operations Center. Direct communications channels between JOC and the Fleet were established, and map problems were finally resolved. It was incredible—but true—that naval aviators were using one kind of map while Air Force pilots were using another kind. It was finally agreed to standardize so ground control coordinates could be integrated by both services.

First priority for the Seventh Fleet continued to be close support under direction of the Fifth Air Force. When they could be released from such duty they were free to operate in other areas.

The carrier *Philippine Sea,* under Captain W.E. Goodney, joined the *Valley Forge* on August 5 in close support for the embattled ground troops.

Liaison pilots from the Navy were also assigned to ground troops to improve communications, since the front lines were divided into four sectors, each with its airborne Air Force and Navy controller.

Now their work began to show results. Lieutenant Commander Hodson led VA-55 in an attack against enemy troops near Korysong while Lieutenant Billy Glen Jackson led another group against Kumchon. They created havoc, but close support was still not as effective as it should have been.

Ground fighting became so desperate by August 8 that even MacArthur feared for the troops' survival. Every effort was made to stem the tide, as waves of fanatical troops gained yard by yard against the American and ROK infantry. The Seventh Fleet kept up the pressure in a maximum all-out effort to provide what assistance it could.

Rear Admiral E.C. Ewen, the new commander of Task Force 77, felt strongly that this type of close support was not only wasteful of naval airpower but also ineffective.

Admiral Joy in Tokyo agreed, and sent a dispatch to MacArthur, "North Korea has many targets better suited to carrier attacks. In South Korea," he said, "the targets are few and well hidden. I strongly recommend Task Force 77 be employed north of the 38th parallel."

While Joy's request was under consideration, the Fifth Air Force asked for additional close air support. The communists were preparing to launch an all-out attack across the Naktong River.

While MacArthur completed plans for a bold amphibious invasion of Inchon west of Seoul in the hope of trapping the North Korean Army in the south, communists launched their last and greatest offensive to crush the beachhead around Pusan.

Task Force 77 was joined by Task Force 96.8 in striking back at this last big push. Rear Admiral R.W. Druble, Commander of Carrier Division 15, had the escort carriers *Sicily* and the *Badoeng Strait* with Marine Squadrons VMF-322 and 214 on board. They gave direct support to the First Provisional Marine Brigade, the 24th and 25th Divisions, the First Cavalry and the Republic of Korea's First and Second Corps and the ROK Marines. With destructive efficiency, they helped the men on the ground turn back the attacks.

MacArthur, who personally conceived the Inchon operation, had to resist opposition by the Joint Chiefs of Staff. They were convinced it would be a disaster.

He had considered all possibilities. He knew there were risks involved in such a daring maneuver, but he also knew it would turn the tide in their favor. He explained to his staff that enemy supply lines were overextended. "If we can seize Inchon by sea assault, these enemy lines will be severed. It will shorten the war," he said emotionally, "and save thousands of lives. The whole course of the war will be reversed, and we won't have to fight a winter campaign."

Inchon, only fifteen miles from Seoul, was deep in enemy-held territory. The harbor tides undoubtedly are the worst in the Far East, and Inchon can be reached only through a narrow, tortuous channel.

MacArthur refused to reconsider his tactical plans, and the Joint Chiefs reluctantly gave their consent. He placed Admiral Struble in command of Task Force Seven. Plans called for initial landings on Wolmi-do Island because it held a commanding position in the harbor.

Three days before the landings, support was provided by two carriers in preliminary strikes in the Inchon area and on highways leading into Seoul. These operations were augmented by two escort carriers the day before the landing and by the arrival of the *Boxer* on D-Day. The HMS *Triumph,* operating with blockade and covering forces, provided air defense for the assault forces en route.

Under heavy support by naval gunfire and aircraft, elements of the 1st Marine Division landed on Wolmi-do on September 15 at 6:30 A.M. An amphibious feint was made on Kunsan in hopes of diverting suspicion from Inchon.

Carrier Division 15 dropped ninety-five tanks of napalm on Wolmi-do and thirty-nine of forty-four buildings were destroyed by pilots of VMF-212 and 323.

Support ships added their deadly fire. Sea and air bombardment proved so successful that the defenders were dazed. Attacking Marines quickly overcame resistance, with only twenty wounded. Enemy dead totaled 120, and 190 surrendered.

With Wolmi-do secured, landing craft regrouped, and after the high tide the Marines then followed up with a successful assault at Inchon's mainland beaches at 5:00 P.M.

As the troops advanced inland, carrier support continued until October 3 with close air support and strikes against enemy lines of communication.

Kimpo airfield was quickly taken, and a line along the Han River west of Seoul was established as fast carriers and escort carriers provided air cover.

For MacArthur the gamble paid off. The invasion was brilliantly conceived and masterfully executed. He wired Struble, "The Navy and the Marines have never shone more brightly than today."

It was a matter of days before the entire half of the peninsula below the 38th parallel was recaptured by the UN forces. The North Korean Army, badly beaten and almost in complete rout, fled north. Their greatest general, Kang Kun, was killed, and his death completed the demoralization of the army.

While the Inchon invasion progressed smoothly, carrier attacks covered areas in the north as well as the south to prevent reinforcement of the communist army. The whole operation was aided materially by the lack of either enemy naval or air opposition.

By the end of September the North Korean Army was in full retreat and entire divisions disintegrated. Only isolated pockets of resistance remained, and these were quickly liquidated. Everywhere in South Korea mountains of enemy military supplies were abandoned as troops fled northward.

MacArthur's imaginative strategy had transformed defeat into an overwhelming victory. The North Korean Army, which had been at the Pusan doorstep, now ceased to exist as an effective force. Its commanders struggled to reassemble its shattered remnants for defense of the territory north of the 38th parallel.

UN forces advanced steadily. The Republic of Korea's First Corps, near the parallel on the east coast, waited impatiently for orders to drive toward the ports of Wonsan and Hungnam. Meanwhile, the South Korean Third, Sixth, Eighth, and Capital Divisions were eager to liberate the entire peninsula. Around Seoul the U.S. Tenth Corps fanned out in hot pursuit.

There appeared to be nothing to stop the victorious UN forces as the 1st Marine Division pushed north to take Uijonbu twelve miles north of Seoul with support of Marine Corsair pilots from Task Group 97.8. The Seventh Division moved eastward from Seoul, capturing the important rail junction of Osan, while the Eighth Army in the south recaptured territory seized by North Koreans in their drive toward Pusan.

Carrier aircraft roamed the skies without opposition, but Task Force 77 lost six planes and one man was killed by ground fire.

Everything that moved on the crowded communications network of North Korea came in for attention, and destruction of bridges, locomotives, and rolling stock was extensive.

Task Force 77 was withdrawn from Korea on October 3 and returned to Sasebo, Japan. The advance of UN forces toward the 38th parallel had reduced the area that could be attacked and there was no further need for its services. Rear Admiral Edward C. Ewen's task force had contributed mightily to the destruction of the North Korean People's Army, but it was now out of targets.

With the North Korean Army defeated, their diplomats tried to stop UN forces at the former dividing line. Most UN delegates believed this was an unrealistic boundary that had ceased to exist after the invasion. President Truman spoke for the United States when he said, "UN forces have a legal basis for crossing the parallel."

Andrei Vishinsky, president of the UN General Assembly, thought otherwise. He said if UN troops crossed the border, they would be the aggressors. Despite his violent denunciation of the United States, the majority of delegates thought MacArthur had sufficient authority to cross the parallel without further orders.

Warren R. Austin, U.S. ambassador to the United Nations, called for the establishment of a free and independent nation to be united under a UN commission supervising general elections throughout the peninsula. He said if the Soviets were sincere in their desire for a termination of the conflict they would agree to an eight-point proposal for unifying the nation.

While politicians fought over the issue, General MacArthur made his own plans to move into North Korea. He was certain neither Russia nor Red China would interfere. The Joint Chiefs of Staff in Washington gave him authority to cross the parallel and complete the destruction of North Korea's armed forces. They forbade use of air or naval action, however, against Manchuria or any part of Russian territory.

Secretary of Defense George C. Marshall wired MacArthur on September 29 that he should feel unhampered tactically and strategically. Before MacArthur moved forward, he called on the North Koreans to surrender.

Indian Ambassador K.M. Pannikar warned the United States that Chinese Foreign Minister Chou En-lai had warned him that his nation would take no action if only ROK troops crossed into North Korea. This was considered just another threat by the communists and MacArthur discounted it. When he issued his second surrender ultimatum on October 9 and it was ignored, he prepared for the final step of moving his UN army into North Korea all the way to the Yalu River on the border of the People's Republic of China.

An extensive minefield off Wonsan on the eastern course held up a UN landing for eight days. Soviet naval experts had sowed 3,000 magnetic and contact mines. Only a few minesweepers were available, so aircraft from the *Philippine Sea* and the *Leyte* dropped two five-mile lanes of bombs 200 yards apart to detonate the mines. Struble refused to continue the bombing because he felt the operation was impractical. Minesweepers completed their deadly job, while troops waited offshore in their ships.

It was not until October 10 that the First Republic of Korea Corps could make a landing. Preinvasion aerial bombing had been carried out, but the landings were routinely completed and the Tenth Corps followed them in. Then carrier attacks shifted to the north to assist the advance of UN forces. The ROK army, against little opposition, headed north to the Yalu River. Hamhung and the port city of Hungnam were taken on October 18. The ROK Capital Division, supported by Task Group 95.2, reached the approaches of the city of Songjin on November 24.

General Walker's Eighth Army ran into stiff resistance in its drive toward Pyongyang, but the capital fell on October 19. As Walker prepared for the final push to the Yalu, UN victory over the North Korean aggressors seemed certain by the end of October and MacArthur jubilantly told his troops they would be home for Christmas.

When advance units of the Eighth Army neared the Yalu, disquieting reports came in that Chinese army units had crossed the river. While Major General E.A. Almond studied the reports, elements of the Tenth Corps swept to the Manchurian border on the east coast.

Peking radio called the army units volunteers, but it was evident that entire Chinese divisions were involved as the Seventh Regiment of the ROK's Sixth Division was surrounded by enemy forces after they arrived at Chosan on the Yalu. The American First Cavalry Division was attacked the first week of November by Chinese horsemen.

Intelligence sources identified four Red Chinese armies, prompting MacArthur to inform the United Nations on November 5 that organized Chinese units were attacking his troops.

When Chinese swarmed across the border, fast carriers were assigned two missions. They were directed to attack targets of opportunity in northeast Korea and help isolate the battlefield by destroying six major bridges across the Yalu River that linked Manchuria and North Korea.

Pinpoint accuracy was needed because the attack had to be made on the Korean side. This ruled out strikes by Air Force B-29s, because long bombing runs would take them over Manchuria. Then, too, B-29s would not have fighter protection, and the slow, World War II bombers were no match for Soviet-made MiG fighters.

Task Force 77 carriers had approximately 150 planes available from the *Valley Forge,* the *Philippine Sea,* and the *Leyte.* It was a difficult assignment, because the Chinese side of the river could not be attacked even to strike antiaircraft emplacements or to pursue attacking interceptors. Dive-bombers would have to fly perpendicular to the spans, increasing the chance of bombing error. Twin railroad and highway bridges between Antung and Sinuiju were given precedence. They were 3,098 feet long and their destruction would be difficult.

The first strike was made on November 9, and they continued until November 21. A total of 595 individual attacks were made on the bridges with 323 tons of bombs. Panther jets were used to protect the bombers from MiGs, while Corsairs destroyed antiaircraft sites on the Korean side.

The power plant at Suiho on the Yalu River after 40,000 pounds of bombs were dropped by eight Navy Skyraiders July 25, 1951. This was part of a coordinated attack by Navy and Air Force planes. (Courtesy of U.S. Navy.)

At first, attacks met limited resistance, primarily because of flak suppression work by the Corsairs. The MiGs rose to the challenge from Manchuria, and three were shot down. The commanding officer of VF-111, Lieutenant Commander W.T. Aman, scored the first kill in his F9F, to become the first Navy pilot in history to shoot down a jet aircraft.

Toward the end of the strikes, flak from Manchuria's antiaircraft sites grew progressively worse. It was obvious that the communists were using this privileged sanctuary for almost all of their antiaircraft batteries. Carriers operated from the east coast and planes had to cross to the west coast over treacherous mountains in bad winter weather. Attacks were successful, but it was frustrating for the pilots to knock out a span only to find it repaired the next day. The highway bridge at Sinuiju and two at Hyesanjin were dropped, and four others were damaged. This did not stop the flow of Chinese, however, because they swarmed across the river once it froze over.

Emergency conditions on the front lines by November 29—created by the deep penetration of a communist offensive—required a shift of emphasis in fast carrier operations from bridge strikes to close air support. As the situation worsened, carrier operations were intensified through December to cover the withdrawal of troops toward east coast ports, where they could be evacuated by ships.

The bitter Korean winter made it difficult to supply the UN forces. For the enemy, it was a relatively simple matter, because the average communist soldier needed only ten pounds of supplies each day, whereas members of the UN command used a daily average of sixty pounds. Attempts to strangle enemy supply lines proved futile, despite constant attacks against the communist railroad system and highway supply points.

The Chinese and North Koreans, with untold millions of laborers to draw upon, used manpower to bring their supplies down the peninsula over trails and paths through the mountains. Animals were also used extensively, so even though highways and rail lines were cut repeatedly, communist supplies could not be choked off.

American Marines retreated to the Chosan Reservoir in an attempt to link up with the Eighth Army. But it was not until November 7 that they met strong resistance. During the latter part of the month a human wave of Chinese attacked the Marines. To the Americans it seemed as if the earth itself was moving toward them as thousands of Chinese swarmed over the hills in their direction. There was only one thing to do, and that was to make a fighting withdrawal to Hungnam, where Navy transports could remove them from their untenable position.

American tanks often found themselves covered with Chinese soldiers, clinging on top as they sought an opening to slip in a hand grenade to kill their occupants. Tank commanders desperately called to other tanks to "scratch my back!"

Machine guns fired long bursts until the Chinese dislodged themselves or were killed. It was a strange, unorthodox war that no one had ever seen before. There was no letup. During the latter part of November, human waves of Chinese soldiers attacked as the American Marines fought their way to the relative safety of Hungnam and its waiting ships. By November 27 the First Marine Division was fighting off six Chinese divisions.

As the situation grew more critical, and airfields were threatened or overrun, Marine pilots operating off escort carriers in the Yellow Sea continued their close support at Yudamn-ni. They attacked the communists with rockets, bombs, and napalm.

Carrier pilots from the *Philippine Sea* and the *Leyte* off the east coast also participated to blunt the communist advances, but communications with ground troops were difficult. The Fifth Air Force gave their pilots priority over the battle zone because the First Marine Division urgently needed every assistance.

On December 2 the Fifth and Seventh Marines got out of their trap at Yudamn-ni. Close air support was a contributing factor. Two days later the Marines returned to Hagaru-ri. They warmly praised the air effort and their commanders emphasized that they could not have made it without air support. During the Marine withdrawal to Hungnam, the Chinese lost heavily due to the effective use of the Navy–Marine system of close support. It cost the carriers two pilots.

The *Leyte*'s Ensign Jesse L. Brown, the Navy's first black pilot, crashed near Hagaru-ri. Brown's plane caught fire on the ground and he was trapped in the cockpit. While comrades circled above him, he sent an anguished appeal over his radio. "Please help me. I can't free my leg."

Lieutenant Thomas J. Hudner watched the plane with horror as it burned slowly, flames licking back to the cockpit. Without hesitation he crashed his plane alongside. He worked frantically to free the trapped pilot, but the cockpit had buckled and Brown's left leg was held securely. Hudner banked snow around the plane to control the smoldering fire and hurried back to his own plane. He was in an agony of suspense as he tried his radio and found it still worked. He put in a frantic call for a helicopter and metal-cutting tools.

Then he hurried to Brown's side, noting with dismay that he seemed to be weaker. Hudner tried everything to free the man, because Brown was in deep shock and begged him to cut off his leg to free him. Hudner himself was almost in a state of shock because, in the freezing weather, he was concerned that Brown could not hold out until a rescue team arrived.

He looked up gratefully as he heard a helicopter. It settled gently in the snow and a medic hurried to Brown's side. They had the tools to get him out, but Brown died before they could free him.

Hudner, who returned safely to the *Leyte,* later was presented with a Medal of Honor by the president of the United States for his act of heroism.

While Marines withdrew from the Chosan entrapment, 200 aircraft flew daily to assist their breakout. The ground situation worsened steadily and the Eighth Army was in critical condition. It was judged so bad that General Almond's Tenth Corps was selected for sealift out of North Korea with redeployment of the Tenth Corps from Wonsan and Hungnam on the east coast and the Eighth Army at Chinnampo and Inchon.

While evacuations were under way the Seventh Fleet provided daylight close air support and air cover of the embarkation crews while its planes disrupted enemy supply lines in the rear. The Hungnam evacuation was completed on December 24, and the last ground units departed Inchon on January 5 for the Pusan area. The communist armies again crossed the 38th parallel, but they were then brought slowly to a halt.

In the United States, as a step in the implementation of a program to provide for early service evaluation of the Terrier and Sparrow I air-defense missiles, together with the development of production engineering information and the establishment of production facilities, an advance order was placed with the Sperry Gyroscope Company for 1,000 Sparrow I air-to-air missiles.

In Korea during the latter part of January 1951, Task Force 77 began a series of air attacks against rail and highway bridges along the east coast of North Korea. When the force was assigned the additional task of bombing highways and lines of communication in northeast Korea, it was made responsible for interdiction, which occupied a major share of its operations until the end of the war.

After a series of engagements in early 1951, a revitalized UN army started a limited advance and by February 10 found the enemy had disappeared from the front. Inchon and Kimpo again fell into UN hands, along with the industrial suburb of Seoul. The battle line then stabilized along the Han River.

Communists next launched an offensive in the central sector, but it was brought to a standstill in savage fighting. UN forces moved ahead on all fronts, and by the end of March they approached the 38th parallel again.

Lieutenant (JG) T.J. Hudner with the Medal of Honor after his attempt to free Ensign Brown on the ground near Hagaru-ri. (Courtesy of U.S. Navy.)

MacArthur proposed to the communists that hostilities cease and a truce be negotiated. When his request was ignored, the Eighth Army pushed closer to the communists' main supply and assembly area. This become known as the "Iron Triangle" between Chorwon, Kumwha, and Pyongyang.

Seventh Fleet carriers were given responsibility for isolating the battlefield as well as routine support. Bridges were given top priority along the east coast.

When Lieutenant Commander Clement M. Craig of Fighting Squadron 193 spotted a key bridge, he studied the area carefully. It was 600 feet long and sixty feet high and appeared to be an ideal target that would be difficult to repair. It had five concrete piers and was supported by six steel spans across a canyon.

"If we can destroy it," he told his superiors, "it will be difficult to bypass. There are two tunnels on each end, but experience has shown they are difficult to collapse," He pointed to a map of the area south of Kilchu. "This is where three rail lines from Manchuria come together. If we can keep this bridge out of service, the flow of traffic over the eastern net from China will be seriously impaired."

They agreed. The bridge was attacked on March 2, but only minor damage was caused. The following day pilots from the *Princeton*'s VA-195 dropped one span, damaged another, and twisted two others. Another span was dropped on March 7, but the communists used interlocking wooden beams to repair the bridge at night.

Now it became a race between the frantic enemy and the fliers to see which side could outwit the other. Napalm fire jelly was used on March 15 in the hope of destroying the wooden beams. Another span of the original bridge was dropped, and four others received serious damage. The carriers concentrated on the area during March, blasting both sides of the railroad tunnel and using delayed-action bombs to harass repair crews at night.

It appeared that regardless of the type of attack, the bridge was repaired. Task Force 77 struck the bridge twice on April 2 until not one of the original spans remained. The communists finally were forced to make a bypass around the bridge on low ground.

Rear Admiral R.A. Ofstie of Task Force 77 reported to headquarters that his fliers had made gaps in all major sections of the northeast coastal rail net. He admitted, however, that supplies still got through by truck. These attacks were certainly disruptive: rail traffic fell from sixty-five percent in February to thirty-two percent in April.

The Seventh Fleet was alerted on April 2 to a possible invasion of Formosa by the People's Republic of China, so it had to leave its interdiction work until the middle of the month. This gave the communists in Korea a chance to repair their bridges and prepare for a big offensive. When the danger at Formosa failed to materialize, carriers returned from the Formosa Strait to their rail and bridge "busting" in the enemy's rear.

18

Persistence Versus Oriental Perseverance

MacArthur was recalled on April 11 because of differences between him and President Truman over future conduct of the war. The general had threatened China with massive retaliation—which he had no right to do—and the president removed him as Commander in Chief of the UN command. He was replaced by General Mathew B. Ridgway.

MacArthur's views on victory in Korea were in conflict not only with the president, but also with the Joint Chiefs of Staff. He said publicly that victory could be achieved only by extending the military conflict beyond Korea, particularly against the Chinese to destroy their capability of waging war.

Meanwhile, the Hwachon Reservoir posed a threat to advance units of the Eighth Army, because the enemy could close the sluice gates, thereby lowering the water level in the Pukhan and Han Rivers so their troops could ford them. The gates could also be opened to prevent a UN advance across the rivers. Carrier planes were assigned the task of destroying the gates.

B-29s were considered initially for the job, but the gates were only twenty feet high, forty feet wide and two and a half feet thick. These targets were too small for high-level bombers. The job was given to the *Princeton,* and Lieutenant Commander Harold G. Carlson was named leader of the formation from VA-195. Less than three hours after Rear Admiral G.R. Henderson received the request on April 30 planes were on their way.

Skyraiders were assigned dive-bombing missions with 2,000-pound bombs, while five Corsairs acted as fire suppressors to protect the six ADs. The mission failed its primary purpose because bombs did not touch the sluice gates, although they knocked a hole in the dam.

The next day Carlson briefed his pilots carefully. "We'll try torpedoes this time, although the terrain around the dam makes such an attack hazardous." He pointed to a photographic blowup. "Notice the hills surrounding the dam? They limit an attack to two-plane sections. When the first two make their attack, the rest of you orbit above until I signal."

Skyraiders drop torpedoes in front of Hwachon Dam. (Courtesy of U.S. Navy.)

Commander R.C. Merrick, CVG-19, led eight ADs to the reservoir. Their torpedoes were set for surface level. Merrick weaved his planes through flak, then dove for his torpedo run over the high hills down to water level in a precise maneuver to make sure his "fish" would not strike bottom. At the right moment he released his torpedo. He pulled up over the dam and climbed steeply for altitude, noting that his men came through safely. Six of the torpedoes ran true: the floodgate in the center was knocked out, and a ten-foot hole was punched in the second gate.

As they circled high above, they watched the impounded waters surge through the holes and down the canyon before they headed back to the *Princeton*. This was the first and only time that aerial torpedoes were used in the Korean War.

Just before the Chinese launched a huge offensive on April 22, General James Van Fleet took command of the Eighth Army. While some of his senior officers felt a retreat might be advisable, he flatly refused to give ground. "Get ready to fight," he said. "The country on the western side of Korea in the Seoul area is flat and open. We know this country. You'll never find a better battlefield for killing communists."

When the Chinese attacked, Van Fleet ordered a counterattack toward the Iron Triangle, using the First Marine Division and the Second Infantry Division to spearhead the attack. "The Chinese aren't flexible," he told his commanders. "They have no conception of fast moves. Their communications system is extremely poor and they have no real logistical support in the front lines."

Rear Admiral G.R. Henderson, who had relieved Ofstie on May 6, now had the USS *Boxer* with the first carrier reserve air group of the war on board, CVG-101, to augment his other groups from the *Princeton* and the *Philippine Sea*. The *Valley Forge* had returned to the States after extensive service in the theater.

Henderson received a request from the Joint Operational Center at Taegu to strike west coast rail lines from Pyongyang to the north. He ordered a strike on May 11 against four railroad bridges. Three were knocked out and the fourth was damaged. His commitments on the east coast did not permit assistance to the Fifth Air Force on a permanent basis, but he readily agreed to send carrier planes across the peninsula for special strikes.

Task Force 77 was called in for close support when the second large communist attack began on May 16. It was stopped and UN forces made a counterattack three days later that cost the Chinese 40,000 men and vast military supplies.

During the latter part of May, Ridgway's headquarters asked for greater efforts to isolate the battlefield by flying interdiction missions behind Chinese lines. Areas were assigned to various air units the first of June for an all-out campaign to destroy bridges, rolling stock, and vehicles of all kinds. Vast damage was done behind the lines, but at night the highways continued to carry supplies to the front.

Carriers were assigned the central routes while the First Marine Air Wing was given the three eastern routes. At night they used searchlights and flares to spot enemy vehicles, resorting to delayed-action bombs to impede the transportation network.

It was a great but futile effort, even though units like Marine Fighter Squadron 513 at Pusan's airfield destroyed 420 vehicles during the first thirty days. "Operation Strangle" failed because communist resistance in the front lines increased, and the number of supply trucks actually remained unchanged. Trucks merely used detours and avoided main routes. If the source of these supplies in Manchuria could have been bombed, the situation would have been reversed. But all areas on the Chinese mainland were off limits to UN fliers.

Admiral Henderson returned his planes to strike against railroads. He told his staff, "At least the Chinese have to expend a much greater effort to repair rail damage. A bomb crater in a road won't stop a truck. Highway bridges are bypassed so easily their destruction is a waste of time."

Again, while carrier planes supported the ground troops, the communists had another chance to repair the damage previously done to their railroad network. On June 2 planes were out in force, continuing attacks for nine days, destroying twenty-four railroad bridges and bypasses on six highway bridges.

It was a frustrating period because it seemed impossible to isolate the communists' supply centers from their front lines. There was no question that quantities of supplies were getting through, because the communists had mounted two large offensives within a month. Henderson told his officers, "It is evident the communists have shifted part of their rail traffic to the western networks." This was the responsibility of the Fifth Air Force but they had no airplanes capable of carrying 2,000-pound bombs. These had proved best for destroying bridges.

The Fifth Air Force had its hands full because the western network of railroads was much larger than those on the east coast. Also, the Chinese were using trucks to a much greater degree on blacked-out highways at night. Actually there was little transport of men and equipment during daylight hours. It was just too dangerous.

Van Fleet was correct in his assumption prior to the May 16 offensive. The Chinese were defeated so badly that by June they were desperate. Morale was so low that 10,000 Chinese soldiers surrendered.

The Chinese had had enough, at least for the time being. They asked for a truce, and the UN command was ordered not to make further advances.

Leaders on both sides realized the war was stalemated. The Chinese could not drive the UN forces out of Korea, and UN forces could not hope to unite Korea by force without extending the war beyond the peninsula. President Truman and the Joint Chiefs of Staff were adamantly opposed to introducing nuclear weapons against China and perhaps broadening the conflict into World War III. All of America's allies in the conflict had similar beliefs.

The Soviet delegate to the United Nations, Jacob A. Malik, proposed immediate discussion for a cease-fire on June 23 between the parties involved in the conflict. Armistice discussions started on July 9, 1951, with Vice Admiral C. Turner Joy, Commander of the Navy in the Far East, as chief of the UN delegation. There were limited attempts to change certain tactical positions, but the front stabilized along a line between Munsan and Kosong. Meanwhile the sea and air war continued.

While truce delegations convened at Panmunjom for the first time, many UN commanders feared the Chinese would stall the negotiations while they prepared for a big offensive.

Rear Admiral Ofstie, now Deputy Commander of the Navy in the Far East, summed up Navy activity. "It has cost the communists very dearly," he said. "They have had to double and even triple their efforts to get supplies to the front lines. In this, the first employment of sustained ground interdiction by Naval forces, I doubt if our forces could have done better on the East Coast." Yet it was obvious that the communists and their repair organizations were matching the interdiction effort.

The naval effort to reduce enemy supply lines during these last critical months had wrought destruction of hundreds of locomotives, railroad cars, trucks, and motor vehicles. The 13,000 rail cuts made by carrier planes, and the 500 bridges and 500 bypasses they destroyed finally began to show results. Movement of supplies to the front slowed down, and the enemy had to devote a tremendous effort to keep adequate supplies en route to frontline soldiers.

Although diversion of manpower and scarce repair materials was an unknown quantity, it was clear that the enemy had to devote a disproportionate share of its total effort to keeping supplies funneled down the peninsula.

Although movement of supplies to the front slowed down perceptibly, they could not be choked off completely. The enemy's volume of supplies at any one time remained adequate for his needs.

Rear Admiral Perry, Commander of Task Force 77, put it eloquently when he said, "Operations resolved themselves into a day-by-day routine where stamina replaced glamour and persistence was pitted against oriental perseverance."

For Task Force 77—working day and night even though no night carrier was available in Korean waters—the burden on the flight deck crews in particular rose to new heights of human endurance.

For the flight crews operating at night over dangerous mountain terrain, their courage reached phenomenal heights. "Heckler" aircraft disrupted night convoys, but these operations—code-named "Moonlight Sonata"—were only partially successful.

The F4U5N Corsairs and the AD-4N Skyraiders served with distinction. The ADs, in particular, were suited to this type of operation because there were two positions for crewmen to operate complicated electronic equipment. They could remain on station longer, and carry a sizeable bomb load.

Although airpower in Korea failed to disrupt ground action due to failure of military supplies, the fact that Manchuria was politically "off limits" must be given the greatest consideration in any assessment of the success or failure of the air mission. The North Koreans and their Chinese allies received all of their military supplies through Manchuria. If these supply lines and depots had been under repeated attacks, supplies to the communist armies would have dried up and made them impotent because UN ships controlled all other ports of entry.

A Korean report about the interdiction of railroads proved that the enemy showed surprising ingenuity in keeping supply lines open. For example, the North Korean Recovery Bureau had 26,000 personnel. They were used constantly to make bypasses and shuttle trains between breaks in the system.

UN forces were surprised initially by how quickly bypasses could be made. This was due to the fact that Korean rivers are shallow and can be forded with ease.

Jet airplanes, used in war for the first time on a large scale, proved effective because of their quiet approach, their speed, and their steadiness during attack.

While jets made headlines, the AD Skyraider proved the war's most outstanding performer. Its versatility was unusual in an attack airplane, and it could carry an ordnance load of 5,000 pounds for an average carrier mission and an overload on shorter missions of up to 10,000 pounds.

Air supremacy was gained early in Korea, and while the ground troops fought magnificently, Navy and Air Force planes destroyed every worthwhile target.

General Van Fleet said, "The Chinese had unlimited manpower to keep highways and rail systems operating. If we had ever put on some pressure and made him fight, we would have given him an insoluble supply problem. Instead, we fought the communists on their own terms, even though we had the advantages of flexibility, mobility and firepower. Interdiction failed because of the primitive nature of the enemy's exposed supply network."

Rashin, located on a projection of land in northeast Korea, is separated from both Manchuria and Russia by the Tumen River. It was an important supply point and a vital transshipment base for the Chinese armies in the south. In August 1951, serious discussions questioned the advisability of bombing it, particularly since there was an obvious buildup of men and materials along the truce line.

A year earlier B-29s had bombed the center, but concern about a possible incident that might bring the Soviet Union into the war had given it the status of a privileged sanctuary. Another B-29 strike was out of the question without fighter support. When approval to bomb Rashin was again received, Admiral Henderson gladly lent a hand with his fighter aircraft to cover the operation.

Air Force crews, aware that a prisoner of war camp was less than a mile from the railroad station, prepared for the mission carefully.

Such was their efficiency that they accomplished one of the most remarkable bombings of the war. Not one bomb fell outside the target area. Photo reconnaissance showed that ninety-seven percent of the bombs fell on the marshalling yards, destroying approximately seventy-five out of 136 railroad freight cars.

An all-Navy attack by planes from the *Bon Homme Richard* and the *Oriskany* struck the place again on December 10, destroying twelve more buildings. These respites from attacks against railroads were few and far between, and came as a welcome diversion for the fliers.

On September 20 carrier crews were relieved of close support for frontline troops and returned to the wearisome task of bombing the northeast coastal railroad system.

There was evidence that tracks were back in limited use, so multiple cuts were made and key bridges attacked. Rail cutting demanded precision flying because tracks were only fifty-six inches wide. Only a direct hit was effective. It was dangerous work because crews flew in all kinds of weather. More than the danger, however, was the morale of the flight crews, which deteriorated after an extended period.

Pilots found it impossible to sustain enthusiasm for the job, because it seemed sense-less to make hundreds of rail cuts one day only to find them repaired the next.

Guerrillas behind communist lines often recommended special targets. During the latter part of October, information reached the carriers that communist com-missars and party officials were meeting on October 29 in the city of Kapsan.

Commander Paul N. Gray, VF-54, assembled eight crews before the attack. "We understand the meeting is scheduled for 9:00 in the morning. We'll strike between 9:15 and 9:20 to make sure they're all assembled."

"How about fighter escort?" one pilot asked. "We don't need it," Gray said. "The city lies in rugged terrain and there aren't any enemy fighters in the area."

They took off early and Gray led his planes to the city, quickly spotting the meeting compound, and dove for a steep attack. After pullout, he noted with satis-faction that the compound was a smoking mass of rubble, but the city itself was undamaged. Guerrillas informed him later that 500 top communists had died in the attack. North Korean radio stridently called the American fliers the "Butchers of Kapsan."

During this month the "Wolfraiders" joined the First Marine Aircraft Wing in Korea. While the war lasted they outshone even other units in this famed wing as they provided close support for the 1st Marine Division. They led in tonnage of bombs dropped on communist lines as they helped to destroy bunkers, observation posts, and strong points in front of their comrades on the ground.

Corporal Louis F. Perleoni of the 224th Infantry Regiment's 2nd Battalion had arrived in Korea in early February 1952. With the war winding down, the 40th Army Division was assigned to Kumwha to take over bunkers vacated by the de-parting 24th Division, along the 38th parallel facing "No-Man's-Land," a mile or so from the Chinese and North Korean armies.

The bunkers were just holes in the ground reinforced by logs with dirt piled on top. They were located along the main line of resistance. There were frequent ha-rassing fire attacks between the combatants, particularly when patrols moved into "No-Man's-Land." Perleoni, a demolition expert—with two twenty-five-pound satchels of explosives on his back—went on one such twelve-man patrol in early February. Each pack contained a time fuse to detonate the cap on the explosives.

In battle reminiscent of World War I, but further hampered by deep snow, these patrols continued their dangerous but necessary work. As they moved forward, Chi-nese mortars zeroed in and they crept along the ground. A mortar blast knocked Perleoni momentarily unconscious, but the deep snow cushioned its impact. He re-covered quickly, noting they were pinned down. He was grateful that Chinese mor-tar shells were not as good as American ones, possibly saving his life. They were made of cast iron and broke into larger pieces. American mortar shells were made of hardened chunks of steel and were much more deadly.

Perleoni's patrol spotted a large stone house, and he and the other men moved toward it. He placed both of his satchels in the house, knowing its destruction would prevent the Chinese from using it as a safe haven.

Now they had a hill to climb on their way back, and Perleoni lagged behind his comrades, struggling through the deep snow. They yelled encouragement: "Come on, Louis. You can make it." With mortars exploding around him he finally made it up the hill and joined his patrol. He was chilled to the bone by his ordeal, but was glad to be still alive.

During the rest of 1952, until June of the following year, ground action reached a stalemate as negotiations continued at Panmunjom, trying to reach a common ground for agreement to end the hostilities.

With the start of the third year of the war on June 25, 1952, opposing armies were dug in, while Navy ships continued their blockade of the coastline. Meanwhile, UN planes roamed North Korea seeking targets to prevent a massive buildup of Chinese troops while negotiators haggled at the truce table.

In the early days of the war, when victory seemed assured, thirteen major electric power plants in North Korea were spared. Commanders knew their destruction would be needless and costly to a unified nation.

While negotiations continued to drag, reconsideration was given to these strategic targets. They were legitimate military targets, providing power for factories in North Korea and Manchuria, so their destruction was ordered. Pilots from the Navy, Marines, and the Fifth Air Force prepared for a two-day series of attacks in "MiG Alley" along the Yalu River separating Manchuria from North Korea.

Commander A.L. Downing from the carrier *Boxer* was appointed strike leader. "This is a heavily defended area of MiG Alley," he told the flight crews. "It will be rough."

At dawn on June 23 Admiral Apollo Soucek, Commander of Task Force 77, aborted the strike because of weather conditions. By noon, with target weather clearing, he ordered attacks to begin at 2:00 P.M.

Soucek had a full task force of four carriers, the largest number since the war began, including the *Boxer,* the *Princeton,* the *Bon Homme Richard,* and the *Philippine Sea.*

"Launch airplanes!" came the familiar cry. It was exactly two o'clock as the planes started on their way. There were thirty-five ADs from VA-65, VA-195, and VA-115. Thirty-one each carried two 2,000-pound bombs and a 1,000-pounder. The other four, in addition to the heavy bombs, carried survival kits. They would be invaluable in case crews were shot down.

Thirty-five fighters joined the bombers as they crossed the North Korean coast. Most were armed with two small bombs, while others carried extra fuel to act as combat air patrols over the Yalu.

Downing led them across the highest mountains in North Korea, then began his descent to remain below radar-detection height. During his careful planning for the strike, Downing had rotated the formation over isolated territory. He hoped this procedure would minimize detection by ground spotters.

He experienced a thrill when he saw eighty-four Air Force F-86s circling above the power plants as they approached.

His earphones crackled as the leader of the Sabres called on the radio, "I can see more than 200 MiGs parked on airfields in the Antung complex. We'll take care of them. Good luck."

Downing felt his throat tighten as he started a high-speed climb to dive-bombing altitude. Still no fire from the ground. As he leveled off he could see the Suiho plant far below him. He relaxed momentarily, feeling certain they had caught the communists by surprise.

Panther jets swung low for flak suppression as the ADs reversed their courses and headed down. Downing nosed over and his AD picked up speed and he knew Commander Neil MacKinnon and his planes were right behind him, followed by planes from the *Philippine Sea.*

Downing forgot everything else as he meticulously prepared to sight on the powerhouse. Pictures of the target had been good and he quickly spotted the aiming point. The powerhouse, with its generators, transformer, and switching equipment, remained in his sight. A favorable crosswind helped to clear the target of smoke. Flak mushroomed around his diving plane. The fire seemed to be erratic, but the large black puffs jarred his nerves.

When he reached 3,000 feet, he salvoed his bombs, and those behind also let loose. Then he pulled up, firing his guns at flak installations, and leveled off at 1,500 feet.

Several bombs made direct hits, and others slammed into the transformer yard.

Flak bursts from the Manchurian side became more numerous and he pulled away rapidly, grateful for the carrier fighters that had done such a thorough job on gun positions around the powerhouse.

It had taken only 180 seconds for planes to complete their dives and drop ninety tons of bombs before they streaked away to the southeast. Glancing back once, a smile of satisfaction appeared on Downing's face as he saw that the powerhouse looked like a volcano erupting smoke and debris high in the air.

Upon their return to the carriers, Downing learned that five Navy planes had been hit by flak. Lieutenant M.K. Lake of VA-115 was the most seriously hit, and his plane momentarily caught fire. His wingman stayed with him as he reached Seoul's Kimpo airfield and made a wheels-up landing.

MacKinnon sent congratulations to Downing on the *Princeton.* "It was a textbook hop," he said. "The timing was perfect and every checkpoint was on schedule and the bombing excellent." Downing replied modestly, "The exercise went off as though we'd been doing it for years."

After Navy planes retired, Air Force F-84 Thunderchiefs made a second strike, which was also successful. This time most MiGs at Antung had retired to more distant airfields, and although eighty remained on the ground, none rose to attack.

It was a magnificent display of precision bombing by the combined elements of the Air Force, Navy, and Marine Corps. The electric power potential of North Korea was virtually eliminated with the destruction of the thirteen power complexes. This two-day attack, which involved over 1,200 sorties, was the largest single air effort since the close of World War II, and the first to employ planes from all U.S. services fighting in Korea.

The power plants smoldered for days, and North Korea's electric power output was reduced to the point that even the capital of Pyongyang was without power. Admiral Soucek was warm in his praise. "I'll bet the lights are out all over Korea and Manchuria," he said.

Carrier flexibility had proven itself again. Attack groups had shown they could hit a heavily defended target and destroy it. The remarkable harmony achieved between the Air Force and the Navy made the difficult strike possible. Fifth Air Force Lieutenant General Glenn O. Barcus wired Soucek, "My hat's off to the Navy for a terrific job. We must get together again some time."

Word also came to Soucek from the chief of naval operations in Washington. Admiral William Fechteler wired, "The excellent performance of duty and high combat effectiveness demonstrated by your forces, particularly the pilots, are deserving of the highest praise and inspiration, and a warning to the enemy of his inevitable defeat. Well done!"

Aside from the extensive damage to the power complexes, the communists had to relocate antiaircraft guns all over Korea to protect other targets.

While armistice talks dragged on, Far Eastern Air Force Headquarters in Tokyo decided to attack the capital of North Korea again. It had been spared for months while the truce talks continued, but commanders hoped to add a further note of persuasion by striking hard at Pyongyang. It was one of the worst flak traps behind enemy lines, so fliers from the *Princeton* and the *Bon Homme Richard* had no doubts about the opposition when they took off on July 11.

It was another joint Navy–Air Force operation, with the *Bon Homme Richard* launching forty-five planes and the *Princeton* forty-six, with Commander William Denton of CVG-19 leading. Denton rendezvoused his planes over the island of Yo-do in Wonsan Harbor, and leveling off at 18,000 feet for the flight, headed toward Pyongyang.

Guns at Yangdok opened up, and when they arrived at the capital, the city stood out prominently at a big bend in the Taedong River. Denton winced as a heavy burst rocked his plane. The accuracy was so good that he suspected they were using radar-controlled mounts.

The strike leader headed in. He swallowed nervously as he saw an AD take a direct hit, blowing part of the tail surface off. He made a mental note of the plane's number as it headed straight down and crashed a mile from the target. He glanced at his strike list and saw the plane had been flown by Lieutenant E.P. Cummings and his observer L.L. Tooker.

The flak got worse and Lieutenant G.G. Jeffries felt his plane rock crazily. He stared pop-eyed at a large hole in the leading edge of his port wing where a shell had gone through—miraculously without exploding.

Bon Homme pilots concentrated on the railroad roundhouse, the locomotive yard, and a large ammunition storage area, and direct hits were made. Planes from the *Princeton* were equally effective, but Lieutenant Commander Dutemple was hit by heavy ground fire and crashed.

When they pulled away the roundhouse was sixty percent destroyed and the repair shop half destroyed. It had been a grueling mission, with the heaviest flak any had seen, but their troubles were not over. En route home weather reports warned of bad conditions in the Sea of Japan. They were appalled when they heard fog had lowered the ceiling to 200 feet.

Jets were instructed to land at Suwon and Kanghong, but the ADs, using homing devices, made it safely back to their carriers. After the weather worsened, additional strikes were cancelled. The Fifth Air Force continued to make attacks because their fields were clear.

If the Chinese and North Koreans needed any inducement to hasten an armistice decision, this strike should have helped. A total of 1,400 tons of bombs along with 23,000 gallons of napalm were dropped on the capital by 1,254 aircraft.

The wailing on North Korean radio stations reached new heights, proving how deadly the attacks had been. They called them "brutal," charging they were ordered in retaliation for the failure of the armistice talks.

Pyongyang radio surprisingly went into detail about the damage. They claimed 1,500 buildings were destroyed and another 900 damaged. "One bomb," the radio said, "made a direct hit on a large air raid shelter causing heavy casualties among high communist party members."

The North Koreans had not seen the last of these bombers. Pyongyang was set for another assault on August 29. An "All United Nations Air Effort," it would be a larger and far more deadly raid, because 1,403 planes were scheduled around-the-clock for the mission, including 216 from Task Force 77's carriers.

In the meantime, the Navy did not remain idle on other fronts. Attacks were continued on industrial targets during the latter period of July. These bombings culminated in missions against the Sindok lead and zinc mill and a plant at Kilchu.

The thunderous roar of revved-up engines reverberated around the deck of the *Essex* as it rode the waves off the coast of North Korea. Carriers of Task Force 77 were beehives of activity early on August 29 as they turned into the wind and blue Skyraiders rolled down the decks, heading westward for another attack on Pyongyang. Some planes carried 2,000-pound blockbusters, and others were hung with strings of smaller bombs.

Planes from the *Essex* rendezvoused with the *Boxer* aircraft and dropped their lethal loads on the most heavily defended city in North Korea. Planes from the *Princeton* followed. Then the first planes returned with new loads to spread destruction on supply concentrations in and around Pyongyang.

Something new was tried when the *Boxer*'s Guided Missile Unit 90 sent radio-controlled World War II pilotless Hellcats against selected bridge targets. These bomb-laden planes, using television guidance systems, were escorted to their targets by Skyraiders. Of the six attempts, only one failed, and the others reached their destination against a railroad bridge at Hungnam.

Vice Admiral J.J. Clark, now commander of the Seventh Fleet, briefed his commanders on another target. "The synthetic oil refinery at Aoji is one of the main communist sources of gasoline," he said. "B-29s can't attack it because they would overfly Russia. I've asked General Mark Clark for permission to knock it out. I know we can do it."

The Commander in Chief of the UN Command decided to refer the admiral's request to the Joint Chiefs of Staff because the target was only eight miles from Russian territory. After due deliberation permission was granted.

Two large strikes departed from the *Essex,* the *Princeton,* and the *Boxer* on September 1, and with no opposition from fighters or flak, they made leisurely runs over the refinery until it was completely destroyed.

When the United Nations and communist truce teams deadlocked on the prisoner exchange issue in October—recessing negotiations for the next seven months—the war was stepped up in the air and on the ground. The Seventh Fleet cooperated with the Air Force on October 8 in a strike at the railroad center of Kowon. B-29s hit hard at extensive flak installations and Navy bombers went in low shortly afterward. After these attacks worthwhile targets of a strategic nature proved difficult to find.

While these targets were under attack, carriers also had resumed close support missions against targets beyond the range of the UN artillery along the battle line. These strikes—more than half of their total efforts during the summer and fall of 1952—were credited with disrupting several major communist buildups.

The Eighth Army developed so much confidence in the accuracy of Navy bombers that, in some instances, the bomb line was moved as close as 300 yards in front of friendly troops.

After winter set in, the war dragged on with little change in the frontlines. It was a period of hopelessness for both ground and air personnel, with negotiations suspended and little hope for a settlement or a victory. During February in 1953 the Seventh Fleet increased its frontline support, but there was only minor activity elsewhere.

Detachments or teams of Composite Squadron 35, under Commander L.E. Burke, had been sent to Korea shortly after the start of the war. Most of its combat missions were night interdiction flights. In a unique operation of eighteen teams they chalked up an impressive record of flights, destroying a large amount of rolling stock in night forays under cover of darkness. They became known as the "night hecklers" or "roadrunners," but the communists had worse names for them. Their greatest contribution to the isolation of the battlefield came in the closing months of the war.

After VC-35 was commissioned in May of 1950 its personnel had increased to more than a hundred officers and more than 650 men. Most pilots were veterans of World War II, averaging better than 1,000 flight hours. They were mature men with a great sense of humor, carefully selected for specialized tasks in night attacks under all-weather conditions. Despite their dangerous jobs they were always ready for a practical joke on one another.

During a miserable day in January 1953, two pilots on the *Valley Forge* were scheduled for an antisubmarine patrol. They awakened at 1:30 A.M. and saw it was snowing heavily, with visibility reduced to a few feet. After briefing, they sauntered to the snow-covered deck fully expecting the "squawk box" to call, "flight cancelled."

Instead, they heard, "Pilots, man your planes!"

They looked at one another in disbelief. They shrugged their shoulders and groped across the deck to reach their planes on the catapults. "This is idiotic," one said. "The snow is piling up on the wings faster than the deck crews can sweep it off." They listened hopefully for word of cancellation as they climbed into their cockpits. "Start engines!" came the call from air ops.

Hands chilled to the bone, they fumbled with switches until engines sputtered indignantly and then broke into a roar that sounded weird in the raging snowstorm.

Near launching time, while the catapult officer stepped to the flight deck swinging his arms to keep the snow out of his eyes, one pilot called the ship: "Uncle George. This is Crock One. For your information," he said bitingly, "I'm on the starboard catapult and in my opinion the weather is unsatisfactory for flying."

He listened carefully for a reply from air ops. Then the officer on duty, struggling to keep from laughing, said, "Crock One, this is Uncle George. Roger your message regarding weather on starboard catapult. All planes will be launched from the port cat!"

A long period of silence followed while the pilots shivered in their freezing cockpits, softly cursing to themselves at the stupidity of operations officers. Finally—in a voice choked with laughter—came the call "Cut your engines. Your flight is cancelled."

VC-35 performed many jobs not directly associated with its primary mission. It was given the task of evaluating the 2.75-inch folding fin rocket nicknamed "Mighty Mouse."

Executive Officer Commander F.G. Edwards was in charge of the rocket's evaluation with Team Mike on the *Philippine Sea*. Lieutenant Commander F.F. Ward, officer-in-charge of Team Mike, was assigned to cover a section west of Wonsan in March of 1953. It was a clear moonless night as he, along with crewmen Aviation Machinist Mates E.B. Willis and R.M. Yenke, throttled back at 5,000 feet to scan the

A gun captain on the USS *Oriskany* inspects ice on the ship.
(Courtesy of U.S. Navy.)

area for the telltale headlight winks of a truck convoy. They were familiar with the tactics employed by enemy drivers. Trucks were well spaced, and every fourth truck periodically blinked its lights as they rumbled along.

Ward looked over the side of his Skyraider at the valley below, its grayness relieved by the outline of a sandy river bottom. He peered intently as two tiny beams of light stabbed at the darkness. Two more pierced the darkness to the west, and then others appeared. Ward swung to the left to make a "dummy" run.

The flickering lights went out as the Skyraider swept down the valley, but Ward could see a string of trucks winding down from the hills through the valley. Just as the lead truck reached the river he pulled the AD into a climb.

He watched the lead driver flash his lights across the narrow bridge and stop abruptly. Ward grinned, appreciating the driver's hesitation about crossing without lights but being afraid to turn them on with his plane overhead.

Ward made another pass and noted that vehicles were piling up behind one another. He dropped a flare and the valley shone almost as brightly as daylight. He climbed quickly to 2,000 feet, leveled off at the head of the valley, checked his rocket-launching switches and nosed into a dive. Thirty trucks were directly ahead. "We can't miss!" he shouted on the intercom.

He watched the altimeter needle swing around as he dove, his hand tightly gripping the stick. When the glowing needle showed he was at the release point, his thumb snapped down on the red button. The plane shuddered as forty-two "Mighty Mouse" rockets streaked toward the ground, and then the trucks seemed to explode as they connected. Flames leaped from their beds and canopies as five trucks disintegrated and the others backed frantically away, slamming into one another in their haste.

In early April, as the truce talks resumed, following communist acceptance of a United Nations invitation to exchange seriously sick and wounded prisoners, the "night hecklers" continued their work.

Haze clung to the Korean coast as Lieutenant T.P. "Teeps" Owens and his crew, J.C. Peckenpaugh and R.M. Rial, stared into the darkness as they followed the roadbed of rail track "Mable." It could be seen only intermittently as they followed the canyons inland. Members of Team George on the *Oriskany* were out again on a train busting job.

Owens lost sight of the track, so he banked sharply and released a flare. The flare blinded him for a moment, then it flooded the countryside in white light. "There's the track," Owens called to his crew. "There's a locomotive and cars almost beneath us," he said excitedly. He alerted another "roadrunner" three miles ahead, piloted by Lieutenant T.G. McClelland. "Hey, Mac, I've got a good one. Fly for the flare." McClelland called his crew, F.B. Georges and E.L. Hazelwood. "Looks like 'Teeps' got us a train. Here we go."

Owens swung around to fire at the train and the first boxcar burst into flames. The engineer then uncoupled the engine and headed for a tunnel. Owens brought his AD down for a second run. This time he made a direct hit on the track 200 yards ahead of the locomotive. He watched the big eight-wheeler plow into the break under a full head of steam. It nosed up and rolled over on its side before it exploded.

McClelland, meanwhile, raked the freight cars with cannon and bombs until Owens could rejoin him. Owens felt almost sorry for the train crew as the cars remained helplessly on the track at their mercy. One by one they blew up the cars and watched them flame brightly.

In a few hours the track would be repaired, but the engine, its freight cars, and most of the valuable cargo were irretrievably lost. Similar night operations were conducted from Task Force 77's carriers by Composite Squadron 3, commanded by Commander R.E. Harmer. These night fighters (F4U-5NX) were used with equally successful results.

After the communists agreed to exchange sick and wounded, truce talks began again on April 6, 1953, and agreement was reached five days later. In "Operation Little Switch," starting on April 20, 6,670 communist and 684 UN prisoners were exchanged.

Now steps were initiated to reopen the main truce talks and on April 26—199 days after negotiations were recessed—they began again. While truce talks bogged down again on the question of the Chinese and North Koreans who refused repatriation, Task Force 77 continued its monotonous rail-cutting, close support strikes, and attacks deep into North Korea.

The AD Skyraiders had proved to be stellar performers throughout the war, but their capability was demonstrated to an incredible degree when one took off from the

Naval Air Station at Dallas this month with a bomb load of 10,500 pounds. Combined with the weight of its guns, ammunition, fuel, and pilot, the AD-4's useful load was 14,491 pounds. This was 3,143 pounds more than the weight of the aircraft!

Just when it appeared that hope for a settlement was futile, the communist representative at Panmunjom asked on May 7 for the establishment of a neutral commission to discuss the problem of the 114,500 Chinese and 34,000 North Korean prisoners of war in Allied hands. He suggested it be called the Neutral Nations Repatriation Commission, to be composed of four nations that already had been agreed upon as members of a Neutral Nations Supervisory Commission.

UN negotiators made a counter proposal on May 13 that all nonrepatriates be released immediately after an armistice.

Syngman Rhee, President of the Republic of Korea, announced that his government would never consent to an armistice that left his country divided. He threatened to continue the war with his own troops.

This further complicated the situation, but on June 4 the communists agreed to interview the Chinese and Koreans who refused repatriation under supervision of a five-power commission. In this way they hoped to induce them to go home.

The demarcation line posed another stumbling block, and heavy enemy action was initiated to gain new ground before the armistice was signed. During this period, with intense activity at the front, the *Boxer,* the *Lake Champlain,* the *Philippine Sea,* and the *Princeton* were engaged in close support to help throw back the communist armies.

At last a truce agreement was reached on June 16 at Panmunjom. President Rhee promptly released the 27,000 anticommunist prisoners in his custody. He said that if the United States signed the armistice his government would consider it an act of betrayal and appeasement.

The communists reacted furiously, and immediately began one of the heaviest attacks of the war. It was directed primarily at the Second ROK Corps. As they fought courageously, giving ground only when it was absolutely necessary, the carriers lent their assistance.

For two weeks the four carriers pounded the communist lines, setting a new record of sorties, with 1,856 flown on July 24 through July 26. During these weeks of fighting, carrier fliers made 7,570 individual strikes, half of them right on the bomb line, in an attempt to stabilize the front.

While operating with the Fifth Air Force during this period, Marine Major John F. Bolt shot down his fifth and sixth MiGs. His victories made him the first naval aviator to become an ace in jet aerial combat.

Such a massive ground offensive could not be continued indefinitely, and the communist onslaught subsided on July 19. Their hard-won miles were quickly reduced by vigorous counterattacks.

Unfortunately, President Rhee's release of the prisoners prolonged the war for five weeks, during which UN troops suffered 46,000 casualties and the communists at least 75,000.

The armistice was signed on July 27, ending a war that had cost the United States 142,000 casualties and $20 billion. It had been one of the most savage wars in American history, and no clear-cut victory was achieved. It was resolved because command of the seas had been maintained, particularly around the Korean peninsula.

The outbreak of the war in Korea caught the U.S. Navy in the midst of a transition. The establishment of the Department of Defense in 1947 and its reorganization in 1949 required readjustments within the services that had not had time to be assimilated. There had been a number of decreases in military budgets, which had caused problems in the integration of new weapons and equipment during a period when Naval Aviation had become a composite force of new jet airplanes and older propeller models.

Combat operations in Korea were unlike the island-hopping campaigns of World War II, with the exception of the amphibious landing at Inchon. The decision by Allied governments to limit the UN action to the peninsula resulted in reducing attacks to support of ground troops.

The seesaw action on the ground as the battle line shifted and as action flared up and quieted again required naval commanders to demonstrate greater flexibility in their operations than before.

In comparison to Navy actions in World War II, the Korean Conflict was a small war, with never more than four carriers involved at the same time. But Navy and Marine aircraft flew 276,000 combat sorties, dropped 177,000 tons of bombs, and fired 272,000 rockets. This was within 7,000 sorties of their World War II totals in all theaters, and bettered the bomb tonnage by 74,000 tons.

In the midst of the war, the carrier modernization program continued, and was revised to incorporate the steam catapult and the angled deck—the most significant advances in aircraft carrier operations since World War II.

In a period when naval aviation was called upon to demonstrate its continuing usefulness in war—and its particular versatility in adapting to new combat requirements—it also moved forward to meet the requirements of a new navy that was totally unlike anything it had been in the past.

Whether the United States won a victory or suffered a military setback in Korea, the war was a remarkable demonstration of collective security by the new United Nations. Expansion of the war had been prevented, and the horror of a nuclear holocaust averted.

The Korean War taught military men a familiar lesson in strategy. It was more apparent than ever that a balance between the services, plus a strong air arm, must be preserved.

The conclusion of the war resulted in endless arguments. Vice Admiral C. Turner Joy, who had been the Navy's theater commander during the first two years of the war, and later chief of the UN Command's Truce Team, gave the most realistic appraisal: "If Korea has taught us that in unity lies the strength that will preserve our freedom, then Korea has not been in vain."

Part VI

The New Navy

19

A New Threat to Peace

In spite of the truce in Korea, peace in the world remained elusive. Within months the worsening situation in the Far East, a series of crises in the Middle East, and a general deterioration in international relations gave new importance to the traditional policy of deploying naval forces to the world's potential trouble spots.

Many military men had revised their viewpoints about attack carriers after the Korean War began, prompting the Navy to renew requests for heavy duty carriers. The Navy Department announced a contract for a new carrier on July 12, 1951. A joint resolution from Capitol Hill proclaimed, "Be it resolved that when and if the United States completes construction of the new aircraft carrier known as the USS *United States,* the construction of which was discontinued on April 25, 1949, or the aircraft carrier authorized in Public Law 3, 82nd Congress, it shall be named the *Forrestal.*" (James V. Forrestal had been Navy Secretary from 1944–1947 and the nation's first secretary of defense in 1947. He served until his death from suicide in 1949.)

The *Forrestal*'s keel was laid in 1952 and it incorporated many new and provocative ideas, representing a complete departure from previous designs, with four catapults and four elevators. The Navy placed such reliance on new carriers that the department sacrificed other combatant ships to order a sister ship to the *Forrestal* to be called the USS *Saratoga.*

The *Forrestal* was launched on December 11, 1954. Its deck covers four acres and is 1,016 feet long and 252 feet wide. Displacing 59,650 tons, the huge ship with its 3,500 officers and men and its air group has a speed of over thirty knots.

Another ship of this class, the USS *Ranger,* is ten feet longer.

The USS *Independence* became the fourth carrier of the *Forrestal* class to join the fleet when it was commissioned on January 10, 1959.

The *Kitty Hawk* and the *Constellation* were designed along the same lines, but a new Terrier antiair missile capability gave them a designation as the *Kitty Hawk* class. An angled deck forty feet longer, and larger catapults and elevators, increased their operational efficiency. The USS *America,* commissioned in 1965, has an even longer deck.

Almost three years later another conventional attack carrier, the *John F. Kennedy,* was commissioned.

The world's first nuclear-powered carrier was named the *Enterprise,* to perpetuate the World War II carrier and her six predecessors. *Enterprise* displaces 85,350 tons, and her eight nuclear reactors permit her to cruise twenty times around the world without refueling.

Three more nuclear attack carriers—the *Nimitz,* the *Dwight D. Eisenhower,* and the *Carl Vinson*—have since added to the Navy's air warfare potency. The *Nimitz* was commissioned in 1975, the *Dwight D. Eisenhower* in 1978, and the *Carl Vinson* in 1982.

In 1947 the Bureau of Aeronautics was considering a heavy attack airplane for the proposed supercarrier the *United States.* Captain J.N. Murphy, head of the bureau's aircraft section, discussed the problem with his associates. For its high-altitude, long-range mission they considered it should weigh approximately 100,000 pounds.

E.H. Heinemann, chief engineer of the Douglas El Segundo Division in California, went back to the bureau with plans of his own.

Captain Walter Diehl, head of the aeronautical division of the bureau, Captain Murphy, and Heinemann sat down for a serious discussion. "I understand the mission such a plane must have," Heinemann said. "I believe a plane can be produced that will perform the mission but weigh only 70,000 pounds."

"That can't be done," Diehl said. "Ed, we've gone over this problem for months. I can assure you the plane must weigh about 100,000."

"I disagree," Heinemann replied. "I'm willing to prove the plane you need can be scaled down to acceptable weight even for service aboard converted *Essex*-class carriers."

Murphy smiled. "I agree with Walt. I don't think you can do it."

"Give us a chance to prove it," Heinemann said.

"You've got it. Put your department to work and let's see what you come up with."

Heinemann and his engineers completed preliminary designs for the A3D Skywarrior. It actually grossed 70,000 pounds, while maintaining its long-range capability for either conventional bombing or nuclear or thermonuclear missions. Diehl and Murphy were overjoyed, and readily conceded they had been in error.

The twin-jet Skywarrior joined the Fleet in March 1956, and later photo-reconnaissance and bomber–trainer versions were added to the Navy's roster of planes.

In January 1952, while Heinemann was extolling his ideas to a group of Navy officers at El Segundo about the necessity of eliminating unnecessary equipment on planes—reducing their complexity and drastically slashing their weight—he was invited to come back to the bureau in Washington by Captain H.H. "Spin" Epes, who handled military requirements for the bureau.

Heinemann brought along plans for a new lightweight interceptor. Epes shook his head. "There are no requirements for an interceptor right now, Ed." He looked at the small model Heinemann had brought with him. "Now if you could make a small attack bomber like this, we would be interested."

Heinemann thought about the design of an attack bomber he and his men had been working on. "What kind of an airplane do you need?"

Epes thought a moment. "We've been thinking about a small jet plane to replace the Skyraider. A plane that can carry a comparable bomb load, the same distance, and able to operate off all carriers. I'd say it should weigh in the neighborhood of 30,000 pounds."

Skywarrior touch-and-go operations were conducted aboard the *Constellation* by members of Heavy Attack Squadron 123. (Courtesy of U.S. Navy.)

Heinemann shook his head. "That's too heavy. We've been considering a new type that should not weigh over 12,000 pounds."

Epes protested. "A plane can't be designed for such a mission at that weight."

"Give me a few weeks and I'll be back to prove we can do it." They shook hands warmly and Heinemann returned to California.

When he returned to Washington he went first to see the head of the Bureau of Aeronautics, Admiral Apollo Soucek. "Apollo, we've designed a new plane we call the A4D Skyhawk," he said. "Let me show you some drawings." Soucek leaned over as Heinemann explained his idea for an extremely light jet bomber.

Soucek's eyes betrayed his interest as he studied the drawings, noting that the short wingspan made it unnecessary to fold the wingtips. Heinemann watched him closely while he described the airplane.

"Are you sure it will do what you say?" Soucek said at last.

"I am."

Soucek played with a pencil and sighed. "We need such a plane, but our present budget is limited. I doubt if we can afford to go ahead with a new program right now."

Heinemann's eyes were serious. "Of course you can afford such a plane," he said persuasively. "Look at it this way. You have so many dollars in the budget for new planes. All right. Divide that amount by the number you need. That will give you a unit cost."

"It wouldn't be much," Soucek said wearily. "No company could build airplanes for that amount of money."

A fully loaded A-4E prepares for takeoff. (Courtesy of McDonnell Douglas.)

"Planes have been getting too expensive," Heinemann said, his lean figure tense and his eyes serious behind his thick glasses. "It's time we all did something to reduce the cost of airplanes. Set the unit cost low, then ask the industry to bid on such a program. I think you'll be surprised at what happens."

The bidding worked out as Heinemann knew it would, and Commander John Brown became the godfather of the new A4D Skyhawk. When Douglas won the competition hands down, Brown encouraged the company to reduce the weight of everything that went into the airplane.

After the A4D entered service in October 1956, the bantam "Mighty Midget" weighed approximately 14,000 pounds. It provided a fleet "muscle ratio" that included a combat load of missiles, bombs, guns, or nuclear weapons. Ultimately its offensive load went up to 8,200 pounds on the A4E Skyhawk, although the gross weight climbed to only 24,500 pounds.

It was not only adaptable to limited war tactical missions, but the Marine Corps found it ideal for close support. Most important of all, by holding the weight down and simplifying systems, the Navy through the years saved itself three billion dollars.

The F-8 Crusader. (Courtesy of LTV Aerospace Corporation.)

The Bureau of Aeronautics issued instructions on February 16, 1955, that all Navy and Marine aircraft should be painted in new colors, with a completion date two years later. The familiar sea blue was changed to light gull gray on top and glossy white below for carrier aircraft. Water-based aircraft would be painted gray. Bare aluminum was retained for utility types and land transports, although the latter would have white tops to reflect solar heat.

New attack bombers increased steadily in performance. North American Aviation produced the A3J Vigilante for use aboard supercarriers and land bases. Flying at twice the speed of sound, with an operating range above 90,000 feet, the redesignated A5 was later employed as a reconnaissance plane as well as a bomber. A costly and limited aircraft as a bomber, the Vigilante served more usefully as a reconnaissance aircraft.

In a hard-fought competition against Douglas, Ling-Temco-Vought won a competition for an interim attack plane early in the 1960s. The A-7 was adapted from the F8U Crusader fighter that had won the Thompson Trophy on August 21, 1956, with a new national speed record of 1,015.428 mph over a fifteen-kilometer course with Commander R.W. Windsor at the controls. The A-7 proved to be an attack bomber of great versatility, giving the Fleet unheard-of attack potency.

In the fighter field few new versions were successful during the 1950s and 1960s. Douglas built 447 F4D Skyrays, an early record holder that unfortunately devel-

The delta-type wing shows clearly in this view of the Skyray.
(Courtesy of U.S. Navy.)

oped too many "bugs" during its production cycle. This bat-winged fighter had shown great early promise that was not fulfilled once it was placed in operational use.

The F5D Skylancer, a more powerful and faster fighter that could attain close to 1,000 miles an hour, also proved to have more promise than real capability. The Navy cancelled the development program after only five were built because the F-8U Crusader offered greater versatility.

Tailless airplanes like the F4D and the F5D proved themselves to be fundamentally unstable without costly devices to make them stable. Bauer had become enamored with such concepts after World War II, only to learn the hard way that such aircraft increased performance but at a prohibitive cost. There were far better ways to achieve the same results.

During the 1960s, the most successful fighter-bomber of all time was developed by the McDonnell Aircraft Corporation in St. Louis. The F-4 Phantom, a twin-engine, two-man, all-weather weapons system, was later produced in a number of versions not only for the U.S. Navy and the Marine Corps but also for the U.S. Air Force. For once the Navy and Air Force dropped their age-old rivalry and concentrated on joint procurement of a single fighter.

The Phantom was originally developed to meet a Navy requirement for a high-performance interceptor capable of all-weather operations. Although delivery of all modern air-to-ground weapons was a secondary requirement at that time, it was given careful consideration in all basic design decisions.

Two engines were considered to be mandatory to ensure crew safety and reliability, and a two-man crew was chosen to increase the probability of successful detection, attack, and kill of high-speed targets. The same factors that dictated two engines and a two-man crew have subsequently proven equally effective in providing maximum effectiveness for later reconnaissance and attack versions.

The F-4 Phantom, designed initially for Fleet defense, made its maiden flight in May of 1958. With its powerful engines it easily reached speeds over Mach 2 with a maximum altitude near 60,000 feet.

Due to its bent-up wing tips and drooping horizontal tail (both for aerodynamic stability), the F-4 has been described as "brutishly ugly in appearance." But the aircrews in the Navy and Air Force who flew the Phantom in combat—shooting down MiGs, bombing heavily defended targets, and making it home in severely crippled condition—thought it was a beautiful bird.

The F-4 became the first jet fighter to be designed and built for the Navy to go into the Air Force inventory, when it was acquired by the Tactical Air Command in 1962. A long-range inertial navigation system, air-to-ground missile capability, and flight controls in the rear cockpit were the main changes required by the Air Force for its F4C. It could carry eight tons of munitions, or other combinations of fuel tanks and armaments.

The Phantom was the star performer of all aircraft developed during the 1960s. It can fly in excess of 1,600 miles an hour, and it has flown at altitudes above 100,000 feet.

The concept of joint development and production of an aircraft for both the U.S. Navy and the Air Force has intrigued Pentagon officials for years.

When General Dynamics Corporation won the F-111 contract, Secretary of Defense Robert S. McNamara hailed the agreement to produce aircraft for both the Navy and the Air Force as one that would save taxpayers hundreds of millions of dollars.

The program failed to fulfill the exaggerated claims originally made for it. The Navy version of the P-111 was cancelled because its suitability for carrier operations was sharply limited due to its overweight problems.

The Phantom was a success for both services because it was developed initially as a fighter for the Navy. The structure of this aircraft was designed to withstand the greater stresses imposed upon carrier aircraft—more than twice the stresses placed on their land-based counterparts.

Inasmuch as the swing-wing F-111 was designed first for Air Force use from conventional land bases, the efforts to "beef it up" for carrier operations increased the weight to the point that it was operationally unsuitable for carrier use. This was a lesson the Navy had tried to teach Secretary of Defense McNamara, but the words fell on deaf ears.

U.S. naval aviation suffered severe growing pains during the 1960s, and there were many false starts on promising new programs.

Fortunately, three outstanding aircraft were available to meet a new threat to peace in the Far East that had been developing since the early 1960s. They included the A-4 Skyhawk, the F-4 Phantom, and the Grumman A-6 Intruder. The latter was the first completely new all-weather attack aircraft ever to be used in combat.

International maneuverings provided incidents that threatened world peace. In each instance, the U.S. Navy was called upon to make special efforts throughout the 1950s and 1960s to counter these threats. On different occasions the Navy evacuated refugees, particularly in the Middle East, patrolled troubled waters, and provided support for menaced nations.

The period was also marked by technological and scientific advances of such magnitude that the Navy and naval aviation underwent changes greater than any in their history. The effective exploitation of these advances enhanced the speed, firepower, versatility, and mobility of the Navy's sea and air forces. Guns were replaced by guided missiles, and the capability to deliver nuclear weapons was established, all during a period when aircraft speeds were jumping from subsonic to supersonic.

While technological advances were being made, there were extensive reorganizations within the Navy to place greater emphasis on research, while new provisions were made to utilize developments in space.

Early space achievements by the Soviet Union raised questions about America's ability to remain competitive in science, and the quality of the nation's schools and colleges came under intense scrutiny. Successful orbits by Explorer I and Vanguard provided the first of a number of achievements that indicated America was still competitive in missiles and space systems. Within months the orbit of man-made satellites became almost commonplace, and an astronaut training program was launched to put Americans in space.

Commander Alan B. Shepard became the first American to go into space on May 5, 1961, as he completed a flight reaching 116 miles high and 302 miles downrange from Cape Canaveral. His space capsule, "Freedom 7," was launched by a Redstone rocket and recovered at sea by a crew from the USS *Lake Champlain*.

In September 1962, nine pilots were selected to join the nation's astronauts. Three were Navy men, including Lieutenant Commander James A. Lovell Jr., Lieutenant Commander John W. Young, and Lieutenant Charles Conrad Jr.

A joint Army–Navy–Air Force regulation was issued on September 18, 1962, establishing a uniform system of designating aircraft similar to that previously in use by the Air Force. By it, all existing aircraft were redesignated using a letter, dash, number, and letter to indicate in that order the basic mission or type of aircraft, its place in the series of that type, and its place in the series of changes in its basic design. For example, the Crusader, formerly designated the F8U-2, now became the F-8C.

A deliberate attempt by the Soviet Union to bring the Cold War home to the United States was taken in 1962 when medium-range missiles with atomic warheads were installed in Cuba. Once they were spotted, and it became evident that they were targeted against America's cities, President John F. Kennedy denounced this action as "deliberate deception" and announced on October 22 that the U.S. Navy would enforce a quarantine on shipping to Cuba and that Soviet-bloc ships would be searched. American ships quickly formed a blockading force, spearheaded by the attack carriers *Enterprise* and *Independence* and the antisubmarine carriers *Essex* and *Randolph*. Shore-based aircraft were ordered to patrol the waters surrounding Cuba. Although these forty missiles were only a fraction of the 10,000 warheads

Aerial view of the USS *Enterprise* in 1962 and the USS *Shasta.*
(Courtesy of U.S. Navy.)

each side later aimed at one another, for a time it was feared that the world was on the brink of a nuclear war. Six days of secret negotiations ended the crisis.

On October 28 Premier Nikita Khrushchev ordered the missile sites dismantled and shipped back to the Soviet Union. In return, the United States pledged not to attack Cuba, and the crisis was resolved, although relations between the two superpowers remained at a dangerously high level because Russian troops remained in Cuba.

Part VII

Vietnam

20

First Navy Action at Sea

After World War II, fifty nations emerged as independent states. These nations had several characteristics in common: they were weak sociologically, economically, politically, and militarily. As a result they were natural objectives for an aggressor.

The United States recognized this fact in the 1950s. As part of a worldwide chain of alliances, the United States, Australia, New Zealand, the Philippines, Great Britain, Pakistan, Thailand, and France established the Southeast Asia Treaty Organization (SEATO) in 1954. In that treaty there was a simple statement that acknowledged there were weak countries in the Far East unable to protect themselves. The United States pledged to help a nation that sought assistance in an emergency.

South Vietnam, as a protocol member of SEATO, asked for help under terms of the treaty when she was invaded by the armed forces of communist North Vietnam. President Dwight D. Eisenhower sent 3,000 advisers, along with material aid. He continued this effort as long as he was president. After John F. Kennedy became president in 1961, he increased the advisory effort tenfold, and also increased material aid.

After Kennedy was assassinated November 22, 1963, President Lyndon Johnson was faced with the prospect of abandoning the nation's allies in South Vietnam or increasing the scope of the effort to a full wartime basis. His decision was made for the same reason that the United States fought World Wars I and II and Korea—to stop aggression.

Although it was not recognized as such at the time, the same problem had asserted itself in 1946 while the Japanese were moving out of the many countries they had occupied during World War II. France sent armed forces to Indochina to reestablish control over its prewar colony.

China had taken over a part of North Vietnam, and the country was in a complete state of chaos. At this time, the Viet Minh party was formed. It was a very popular national party because its aims were to drive out the French colonialists, the Japanese, and the Chinese. The French had never been popular in Vietnam, particularly in hamlet and village areas, so the Viet Minh movement quickly gained the support of a majority of the South Vietnamese people.

It took four or five years before most Vietnamese realized that the Viet Minh were inspired, planned, controlled, and directed by the communists. When they did find out, the party lost wide support because people realized that the Viet Minh were taking over the countryside not to drive the French out, but to establish communist rule.

The Viet Minh went into each hamlet, and each village, and first destroyed the government, the schools, the churches, and the market places. Then they took the young people and sent them to North Vietnam for communist indoctrination and training. They made virtual prisoners of those left in the villages, taxed them up to fifty percent of everything they made, and forced them to work two days a week to fortify their villages or haul supplies to communist forces.

After the communists under Ho Chi Minh were successful in driving the French out of Vietnam, the nation was divided in 1954 into a communist state in the north and an anticommunist republic in the south.

Ho Chi Minh sent his guerrilla forces into the Republic of Vietnam in 1959 with the aim of forcing that independent nation into uniting as one nation. Against this background, President Kennedy in early 1961 increased American aid to Saigon and dispatched more American advisers to Vietnam. Air command and ground force advisers sought to assist Vietnamese military forces to counter the infiltration of communist cadres southward and the growing insurgency within the country.

During the early 1960s, the U.S. government recognized that the North Vietnamese were actively participating in military operations, both in South Vietnam and Laos. Their interference in the affairs of Laos was considered essential to the communists because what became known as the Ho Chi Minh Trail wended its way through the Laotian panhandle into South Vietnam. Despite this knowledge, Washington officials decided that the insurgency would have to be defeated within South Vietnam and operations should not be expanded into North Vietnam. A major American objective from 1961 through 1964 was to strengthen the Republic of Vietnam and enable it to withstand the communist guerrilla effort to cripple it.

North Vietnam's strategy called for building an insurgent force in the south, then beginning widespread guerrilla operations, and finally launching an all-out offensive to destroy the Republic of Vietnam's military forces.

After December 1963, North Vietnam's leaders greatly increased their infiltration into the south, and by the autumn of 1964, they were ready to start the final campaign to overthrow the government in South Vietnam.

On the night of August 2, 1964, three North Vietnamese torpedo boats attacked the U.S. Navy's destroyer *Maddox* in the Gulf of Tonkin, at least twenty-eight miles off the coast of North Vietnam. The *Maddox* was well within international waters when it was attacked. The destroyer took evasive action and returned the fire. Her skipper, Captain John J. Herrick, sent an emergency appeal to the carrier *Ticonderoga* for air support. Aircraft were quickly dispatched and Task Force 77 went into action for the first time.

Commander James Stockdale of VF-51 responded to the *Maddox*'s appeal with four fighters. After he noted the speedboats streaking north in a ragged single file, Stockdale ordered an attack. Their first Zuni missiles failed to hit, so they wheeled back and made strafing runs, sinking one boat and damaging two others.

The next day President Johnson warned North Vietnam's government that U.S. ships "have traditionally operated freely on the high seas in accordance with the rights guaranteed by international law.... They will continue to do so and will take whatever measures are appropriate for their defense." In no uncertain terms the president warned North Vietnam to be under "no misapprehension as to the grave consequences which would inevitably result from any further unprovoked action against United States forces."

A second destroyer, the *C. Turner Joy,* was sent to the gulf, and the carrier *Constellation* was ordered to leave Hong Kong and reinforce Carrier Task Force 77, along with the *Kearsarge* with its antisubmarine warfare capability.

The two destroyers formed a task group on August 3 and entered the Gulf of Tonkin under orders to fire only in self defense.

During the night of August 4 the *Maddox*'s radar indicated that at least five high-speed contacts were thirty-six miles away. The radar blips were thought at first to be torpedo boats, and the American ships were ordered to change course and to increase their speed. Torpedo noises were reported by the *Maddox*'s sonar, and when the torpedo boats were supposedly within 6,000 yards of both ships they fired their guns and began evasive maneuvers. One report said that a torpedo wake had been sighted 300 feet to port of the *C. Turner Joy.*

Stockdale and several other pilots investigated this new report of an attack, although they found no evidence of enemy attackers.

After the first attack on August 2, Admiral U.S.S. Sharp, Commander in Chief, Pacific, recommended to the Pentagon that Task Force 77's planes be sent on air strikes against North Vietnam's torpedo boat bases.

Secretary of Defense McNamara had thought this first attack was merely the impulsive act of a local commander, but when Sharp reported the possibility of a second attack, he was convinced that it was a deliberate case of provocation against the United States by the communist leadership in Hanoi. Two hours later Sharp was given authority by the Joint Chiefs of Staff to take an immediate strike against North Vietnam at dawn on August 5.

But Sharp began to suspect the second attack had never occurred when he received several reports from Captain Herrick about whether his ships had been attacked a second time. The captain said he was beginning to suspect the two destroyers had been shooting at radar shadows. He called for a complete reevaluation of the incident.

Although the reports were not conclusive, it was finally decided that the weight of evidence, including radio intercept intelligence, was strong that a second attack had not taken place. Later, after all the evidence was in, the Navy concluded there had been no second attack.

However, an air attack was approved against four torpedo boat bases and oil storage facilities at Phuc Loi and Vinh. President Johnson told the American people that "our response for the present will be limited and fitting." It was decided, however, to delay the air strikes until the carrier *Constellation* could join the task force.

Sixty-four aircraft were launched on the morning of August 5. Ten of the *Constellation*'s planes struck torpedo boat bases at the northernmost target at Hon Gai. Farther south, twelve other planes from the *Constellation* struck PT boat bases at Loc Chao. Six fighters from the *Ticonderoga* led by Commander Robair Mohrhardt hit torpedo boat bases at Quang Khe, while twenty-six others blasted the oil storage dumps at Vinh. Eight gunboats and torpedo boats were destroyed, with twenty-nine others damaged. Smoke from the ten Vinh storage tanks rose to 14,000 feet and damage was estimated at ninety percent as a result of attacks led by Commander Stockdale of VF-51. Two aircraft from the *Constellation* were shot down by antiaircraft fire at Hon Gai. Lieutenant Everett Alvarez's A-4 Skyhawk of VA-144 was hit over the target. He ejected, bailing out over the gulf, and was captured by the North Vietnamese. He became the first American pilot to be captured by the North Vietnamese and remained the longest as a prisoner, for eight and a half years.

The immediate reaction of the North Vietnamese was to move thirty MiG-15/17 fighters from China to Hanoi's Phuc Yen airfield on August 7. In open defiance of the United States, a division of North Vietnam's regular soldiers began to deploy down the Ho Chi Minh Trail in Laos headed for South Vietnam.

Although President Johnson was now aware that there had been no second attack against American ships in the Gulf of Tonkin, he believed the first attack warranted decisive action by the United States. He sent a message to Congress on August 5 and asked for a joint resolution of support for his Southeast Asia policy. A resolution was approved by the administration and introduced by the Chairman of the Senate Relations Committee, J. William Fulbright, and the Chairman of the Foreign Affairs Committee of the House of Representatives under Thomas E. Morgan.

A joint resolution was approved that termed the attacks on the American destroyers part of a "deliberate and systematic campaign of aggression that the communist regime in North Vietnam had been waging against its neighbors and the nations joined with them." The resolution assured the president of Congress's determination to "take all necessary measures to repel any armed attack...and to prevent any further aggression" until the president determined that peace and security in the area was reasonably assured. The resolution passed the House by a vote of 416 to 0. In the Senate, it was approved by an 88 to 2 vote.

The resolution went too far in giving the president authority to take whatever steps his administration deemed necessary without a declaration of war. His later actions were taken, therefore, without a commitment by a majority of the American people, and were doomed to failure.

Now that the Vietnam conflict had escalated, with growing American involvement, the Joint Chiefs of Staff drew up contingency plans for American air operations against North Vietnam. On two occasions strikes were purely retaliatory in nature. Carrier-based planes flew into action from their station in the Tonkin Gulf in response to Viet Cong provocations in the south. On February 7, 1965, a heavy mortar attack killed seven and wounded 109 Americans near Pleiku air base. The carriers *Coral Sea, Hancock,* and *Ranger* struck North Vietnamese Army barracks and port facilities at Dong Hoi, where twelve buildings were destroyed or damaged. Three days later, Dong Hoi was hit again in response to another terrorist attack.

In late 1964, the Joint Chiefs recommended a "fast/full squeeze" sixteen-day air campaign against ninety-four targets in North Vietnam to establish American air superiority and destroy North Vietnam's ability to continue support of operations in South Vietnam. However, President Johnson and Secretary of Defense McNamara rejected the plan. They decided test bombing North Vietnam would be a supplement to and not a substitute for an effective pacification campaign within South Vietnam.

McNamara outlined the basic objectives of his plan to bomb targets in North Vietnam:

1. Reduce the flow and/or increase the cost of infiltration of men and supplies from North Vietnam to South Vietnam.
2. Make it clear to North Vietnam's leadership that as long as they continued their aggression against the South, they would have to pay a price in the North.
3. Raise the morale of the South Vietnamese people.

Smoke, dust, and mud boil skyward as A-4 Skyhawks from the *Oriskany* press their attacks on the Phuong Dinh railroad bypass bridge six miles north of Thanh Hoa, North Vietnam. (Courtesy of U.S. Navy.)

President Johnson expressed his approval, and in early 1965 he authorized the first strikes against North Vietnam. He called them a demonstration of America's determination to retaliate against military targets so that the communists would understand that they were not immune to attack.

When it became necessary to launch air strikes against North Vietnam, the first ones were flown from Seventh Fleet carriers. They were there and they were ready to fight.

Air-to-air warfare began on April 3, 1965, when a Navy strike force of four F-8Es bombed the Thanh Hoa Bridge, south of Hanoi. They were attacked by MiG-17s, and one enemy aircraft was damaged.

As the need for airpower increased faster than airfields could be built, more carriers were deployed to the Western Pacific until five were in continuous combat.

The Joint Chiefs, with only General John F. McConnell, the Air Force Chief of Staff, dissenting, directed that air operations against North Vietnam be limited in scope and not be a hard-hitting air campaign. The operations were called "Rolling Thunder."

General Hunter Harris Jr., Commander in Chief, Pacific Air Forces, proposed that the communist MiG base at Phuc Yen, north of Hanoi, be struck first, because

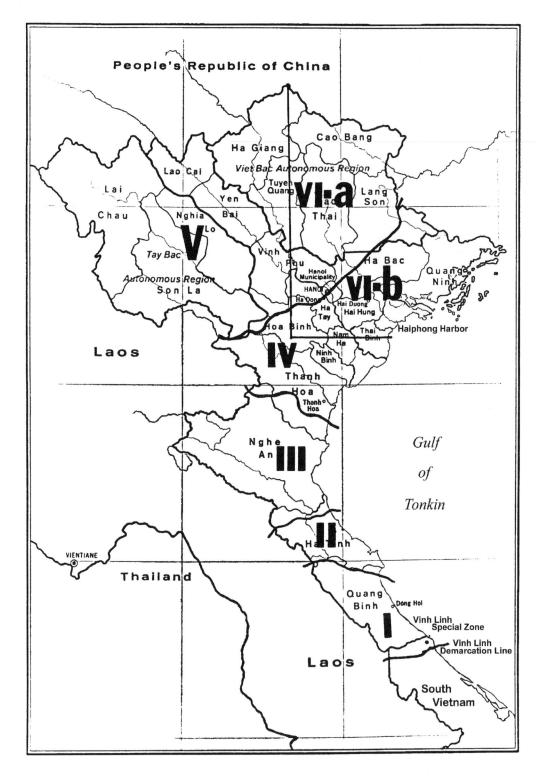

"Rolling Thunder" route package areas.

A North Vietnamese patrol boat is left burning by an aircraft from the USS *Midway*
April 28, 1965, in the Raonay River, 12 miles north of Dong Hoi.
(Courtesy of U.S. Navy.)

he considered it an immediate threat to American air operations. The attack would
be led by thirty B-52s from Guam at night, followed by an early morning attack by
tactical fighters to complete the destruction of the base. But the strike was cancelled without explanation.

Between March 2, 1965—when the first tactical air strikes were launched by
Navy and Air Force crews—until the first phase of operations ended May 11, attacks were limited to military and transportation targets in North Vietnam's southern panhandle below twenty degrees north latitude.

At this time North Vietnam's air defenses were limited to 1,426 antiaircraft
guns, twenty-two early warning radars, and four fire-control radars, so the early
Rolling Thunder attacks were flown against a defense system that was too inefficient to be effective against modern aircraft.

The early strikes, however, used the basic tactic that pilots had been taught for
the release of nuclear weapons. In other words, they flew to a target at high speed at
a low altitude and then performed a pop-up maneuver to unload their bombs and
departed at high speed. Low clouds often encountered on such missions served to
justify this tactic. Some pilots made multiple runs, and enemy automatic weapons
fire and small caliber antiaircraft guns began to take a heavy toll.

Pilots changed their tactics and dive-bombed their targets from 3,000 to 5,000 feet, which brought them above the effective altitude of most enemy guns, but it also decreased their accuracy.

With the aid of the Soviet Union and the People's Republic of China, North Vietnam's air defenses rapidly improved, particularly with the installation of 85-mm radar-controlled guns.

By April 3, several MiG-17 pilots became more aggressive, as they attacked a U.S. Navy strike force bombing a road and railroad bridges near Thanh Hoa. The following day, when American Air Force F-105s attacked the same bridge, enemy pilots downed two heavily laden F-105s and escaped at high speed before F-100s could arrive on the scene.

President Johnson three days later announced that the "slowly ascending" tempo of the Rolling Thunder operations would continue against targets outside of the effective range of the MiGs. Air Defense Command EC-121 "Big Eye" aircraft were sent to Vietnam to provide radar surveillance orbits over the Gulf of Tonkin while American strikes were in progress. They proved of enormous help by alerting American pilots to the presence of communist fighters.

Seventh Fleet carrier pilots joined the Air Force in interdiction strikes in South Vietnam on April 15 by attacking Viet Cong positions near Black Virgin Mountain. Their attacks were so successful that future "in-country" missions were assigned to the Seventh Fleet. To carry out these missions, one carrier was normally assigned to "Dixie Station" off the coast of South Vietnam. Operations continued there until August 4, 1966, when land-based aircraft were well enough established to handle most of the required air attacks in South Vietnam.

While American officials sought to get the North Vietnamese to begin peace talks, air strikes were cancelled starting May 12. When peace efforts proved fruitless, strikes were begun again on May 18, as Rolling Thunder II was authorized to hit targets north of twenty degrees latitude against bridges on the northwestern rail line between Hanoi and the Chinese border. These attacks continued through the end of 1965 under tight controls. Pilots were not permitted to enter a thirty-mile buffer zone along the Chinese border, or fly within ten miles of Hanoi and four miles of Haiphong.

MiG pilots tried to use their superior turning ability to get into a 6 o'clock position behind Air Force F-105s, but this maneuver did not work against Air Force and Navy F-4 Phantoms. These supersonic fighters were also highly maneuverable.

SAM-2 missiles were used for the first time on July 24 around Hanoi and Haiphong, and one flight of four F-4C fighters was hit, resulting in destruction of one aircraft and damage to the other three.

American reaction to the missile threat was to attack targets at an altitude of 500 to 1,500 feet in SAM-heavy areas. But aircraft losses due to other causes rose correspondingly, and strike altitudes were raised, and evasive maneuvers were adopted to avoid the SAM's lethal range.

While escorting a strike on June 17, 1965, on the barracks at Gen Phu in North Vietnam, Commander L.C. Page and Lieutenant J.E.D. Batson, flying *Midway* F-4B Phantoms of VF-21, intercepted four MiG-17s. Each Navy pilot shot down one, scoring the first American victories over MiGs in Vietnam.

That same day the *Independence,* with Air Wing 7 on board, arrived at Subic Bay in the Philippines for duty with the Seventh Fleet. Her arrival from the Atlantic Fleet around the tip of Africa added a fifth attack carrier to naval forces operating off Vietnam.

Navy strikes in August of 1964 prompted the North Vietnamese into an accelerated buildup of their MiG capability. By mid-June 1965, they had received seventy MiG-15s and MiG-17s, and in December of that year they were receiving MiG-21s. With Russian aid their early warning and height-finding radar capability increased rapidly. This new capability gave them ground-controlled intercept (GCI) coverage over all of North Vietnam and much of the Gulf of Tonkin.

The Navy played the major role in the war on the Thanh Hoa Bridge and a lesser one against the Paul Doumer Bridge, both critical elements of North Vietnam's transportation system.

Vietnam's 1,500-mile railroad system was conceived by Governor General of French Indochina Paul Doumer and was built between 1896 and 1902. By 1960 the system was a major factor in North Vietnam's movement of military supplies from China and Haiphong into Hanoi by railroad and then south to the battlefields. All supplies heading south by railroad passed over the Thanh Hoa Bridge, spanning the Song Ma River. A new bridge, completed in 1964, replaced the original French-built bridge, which the Viet Minh destroyed while France occupied the country. It was 548 feet long and 56 feet wide. The Doumer Bridge was 5,532 feet long and 38 feet wide, the longest in Vietnam.

Navy fliers used six different aircraft in their bridge operations: The A-3, A-4, A-6, A-7, F-4, and the F-8. But the A-4 Skyhawk was the Navy's workhorse. A single-engine, single-seat attack airplane, it flew 208 sorties against Thanh Hoa—more than three times that of any other Navy aircraft. The A-7 Corsair II, designed to replace the A-4, did not get into action until late 1967, and it was not used against the bridges until 1972.

Carrier planes on June 17, 1965, began a three-year effort to destroy the Thanh Hoa Bridge by attacking it with two to four aircraft sections. The bridge was hit twenty-four times by sixty-five aircraft before the end of May in 1966 with 128 tons of bombs weighing from 500 pounds to 750 pounds each.

Incredibly, the same mistake that was made at first during World War II against similar bridge structures—using bombs that were too small—was made again, with identical results, until 2,000-pound and 4,000-pound bombs were used. Despite the pounding, the bridge suffered only temporary damage, and remained usable by the North Vietnamese. But the area around the approaches to the bridge looked like a valley of the moon.

The North Vietnamese had to divert an enormous labor force to repair the bridge and position pontoon bridges in its vicinity to provide a bypass while the bridge was temporarily out of service.

The proximity of the Thanh Hoa Bridge to the gulf, approximately eleven miles away, coupled with the normal weather patterns over the northern part of North Vietnam, continued to provide very poor weather over the target for much of the year.

In some seasons of the year bad weather permitted only two to four visual missions each month. This condition worked to the enemy's advantage, as the communists repaired the bridge during these bad weather cycles.

The weather during the early months of 1966 was so poor that the Navy flew only eleven bombing sorties against the Thanh Hoa Bridge. But these strikes served to keep the railroad approaches in a constant state of disrepair.

During the thirty-seven days of the "Christmas Truce" from December 25, 1965, to January 30, 1966, all bombing of North Vietnam ceased while the president and his aides sought once again to bring the communists to the conference table. But the North Vietnamese were not interested.

It was not until September 23 that the bridge was removed from service again. On that date it was hit by twenty-two Navy attack planes that dropped fifty-seven tons of ordnance. Eighty tons of rolling stock and 1,678 tons of petroleum products were trapped in the area, and were systematically destroyed during a four-day period.

Photo reconnaissance in October indicated that the bridge was still out of service, and no strikes were scheduled. But in December the bridge was being used again and strikes were scheduled week after week but still failed to knock it out.

It had been evident since the start of hostilities in 1964 that coordination of the air interdiction effort against North Vietnam by Air Force and Navy planes was ineffective. In most cases, coordination did not exist, and each service acted on its own because of differences in operational concepts, procedures, and equipment. In the early days the Navy launched its operations from two aircraft carriers off the coast of North Vietnam in an area called "Yankee Station," while the Air Force operated from bases in Thailand. In mid-1966 the Navy added a third aircraft carrier to Yankee Station's operations.

As the number of daily sorties increased, the authorized target list expanded. As the North Vietnamese defenses grew stiffer, it became increasingly obvious that control of tactical air operations had to be established. The Commander in Chief, Pacific, had authority over all operations, but in a practical sense he delegated responsibility for day-to-day planning and conduct of air operations to the Commander of the Seventh Air Force and the Commander of Task Force 77.

When Rolling Thunder II was initiated, an Armed Reconnaissance Coordinating Committee was formed to resolve differences and coordinate operations between the Air Force and the Navy.

In November 1965, North Vietnam had been divided into six areas, called "Route Packages," and each service was assigned primary armed reconnaissance responsibility in these areas. The Air Force and the Navy were allotted three packages each. The Thanh Hoa Bridge, in Route Package IV, became the responsibility of the Navy.

Carriers operating from Yankee Station had fewer airplanes available for interdiction efforts against North Vietnam than the Air Force. The number and type of Aircraft carriers operating "on the line" varied throughout the conflict, but a typical carrier had seventy attack and support aircraft. But in 1965 eighty-nine Navy airmen were killed, captured, or missing, and 123 aircraft were lost.

At the Honolulu Conference in January 1966, Admiral Sharp insisted that U.S. Air Force and Navy aircraft be employed more aggressively against North Vietnam. If the war is to be ended soon, he said, massive air strikes on all communist bases must be approved, not only against North Vietnam but also against their bases in Laos and Cambodia. He also called for mining of Haiphong's harbor, saying it was evident that the North Vietnamese were unwilling to negotiate. The Joint Chiefs endorsed his proposals, but the Johnson Administration rejected them.

Secretary of Defense McNamara said publicly on February 5, 1966, that the United States' objectives were not to destroy or to overthrow communist China or the communist government of North Vietnam. They were limited to the destruction of the insurrection and aggression directed by North Vietnam against the political institutions of South Vietnam. This was a very, very limited political objective.

McNamara continued to insist there was only an insurrection in South Vietnam. In reality, there was a full-scale war.

Admiral Sharp and the Joint Chiefs did not agree with the secretary of defense. They pressed for a total air campaign against North Vietnam as the only alternative to a full-scale ground war, which could involve hundreds of thousands of American troops. Sharp stressed that the United States should control events and stop reacting, through the proper use of airpower. In the South, he said, the North Vietnamese maintained the initiative by fighting where and when they desired, and then retreating into the jungle to their sanctuaries in Laos and Cambodia. Sharp insisted, with the backing of the Joint Chiefs, that calling the war an insurgency was not only inaccurate but that it misjudged the situation that existed.

Establishment of the route packages did much to control Air Force–Navy attacks against North Vietnam and put them on a coordinated basis. Unfortunately, President Johnson and Secretary McNamara then placed the major targets within them off limits. There was a five-mile sanctuary around Haiphong and a ten-mile limit around Hanoi. There was also a thirty-mile-wide buffer zone along the North Vietnam–People's Republic of China border that was restricted.

The shortage of pilots that had developed in 1965 became more acute this year, along with an inadequate supply of ammunition, rockets, and bombs. The chief of naval operations announced that no naval aviator would fly more than 150 combat missions during a seven-and-a-half month deployment, or go on more than two such tours. Many shore jobs were eliminated to release pilots for Vietnam, and training increased, but the situation was not immediately alleviated because it took eighteen months to train new pilots.

The tours were not only long, but were also becoming increasingly dangerous. While flying an A-4E against petroleum facilities in North Vietnam, Commander Wynne F. Foster of VA-163 was struck by a fragment of a shell that almost severed his right arm below the shoulder. Bleeding badly, he managed to radio his condition while flying his Skyhawk with his knees as he held the stump of his shattered arm to restrict the flow of blood. He made a left-handed ejection and was picked up by a destroyer's whaleboat. He was rushed to the *Oriskany* where his arm was amputated, but he survived the ordeal.

The campaign against petroleum stocks had been started too late to be totally effective, even though the largest oil storage facility in North Vietnam had been wrecked. By the end of the year, incoming fuel was reduced to marginal levels. Sufficient reserves were available, however, buried underground in drums and in caves. McNamara's indecision had given North Vietnam a chance to disperse its stocks.

After another peace proposal predictably failed, the president authorized the third phase of Rolling Thunder, to begin on January 31, 1966. Although strikes were renewed, tight controls were maintained by the president over the operations. Strikes were restricted to transport supplies to the fighting fronts in the south and limited to 300 sorties a day.

In April, Navy and Air Force efforts against North Vietnam were further coordinated so that each service had permanent areas of tactical responsibility. This system resulted in a sharp reduction in aircraft losses and a general increase in successes against targets, as pilots became more familiar with the defenses in their assigned areas.

Yankee Station was moved farther north in the Gulf of Tonkin to reduce the time to and from targets in the central panhandle and around Haiphong.

The southward flow of men and supplies was drastically reduced during this period. Navy jets concentrated heavily on railroad lines and yards. At Phu Can, a *Hancock* flight hit a 25-car ammunition train, which exploded like a string of fireworks. Pilots reported, "Large red fireballs completely disintegrated the cars and there was heavy smoke, many fires and numerous rail cuts."

During April, May, and June, American planes encountered increasing resistance from SAMs, MiGs, and antiaircraft artillery. As a result, improved tactics, weapons, and newly developed electronic countermeasures had to be quickly put into use. Despite increasing resistance, a kill ratio of three MiGs for each carrier plane shot down was maintained.

The arrival of Air Force F-105s—nicknamed "Wild Weasels"—in Thailand in 1966, to perform flak and missile suppression over Vietnam was welcomed by crews of strike aircraft. They were equipped with radar homing and warning sets to enable F-105 crews to home in on an SA-2's radar guidance signals and to mark their location with rockets for strikes by accompanying Navy A-4s, nicknamed "Iron Hand." The F-105s also gave early warning of an impending SAM firing.

The fourth phase of Rolling Thunder began in April 1966. Now all of North Vietnam—except specific sanctuary areas, which usually included the most worthwhile targets—was open to attack. Air Force planes concentrated on seven major bulk petroleum storage areas in the Hanoi area, and the Navy concentrated on the Haiphong area.

The North Vietnamese finally ordered their air force to resist these attacks. A major confrontation occurred on April 23 when two MiGs were shot down. MiG-21s entered the air battle on April 25 and 26 for the first time, as they launched a high-altitude attack against EB-66s.

The first MiG-21 was shot down by two Sidewinder missiles on the 26th day by an F4-C flying on combat air patrol. Although Air Force and Navy crews flew a large number of missions this spring, enemy air opposition was sporadic, even when the president authorized attacks against North Vietnam's major airfields, which had been off-limits.

Attacks by A-4s from the *Hancock* on June 16, twenty-four miles west of Thanh Hoa, initiated the first carrier strikes against petroleum centers, and were the start of a systematic destruction of such facilities.

In June, carriers and Air Force wings were turned loose against the most significant targets they had yet been assigned—the extensive petroleum storage system in North Vietnam. On June 29, Navy pilots hit the Haiphong petroleum storage areas, causing three huge fireballs and leaving columns of black smoke rising to 20,000 feet at the complex. "It looked like we had wiped out the entire world's supply of oil," one pilot said. He could still see the smoke from as far as 150 miles away. Twenty storage tanks, and a pumping station constituting eighty percent of the installation, were destroyed.

During succeeding strikes, Navy aircraft swept farther inland to obliterate petroleum storages at Bac Giang, northeast of Hanoi. As a result, no more tankers were pulling into Haiphong in July because there were no storage tanks left.

On the first of July, the North Vietnamese reacted with an attack against two American search and rescue destroyers, the *Coontz* and the *Rogers,* steaming fifty-five miles southeast of Haiphong. Navy planes counterattacked and both enemy PT boats were destroyed. Ships picked up nineteen enemy survivors. These men were the first to become Navy prisoners of war.

During the summer of 1966 strike operations increased in intensity. In August, three carriers became permanently available at Yankee Station.

A North Vietnamese general now announced his nation's goals: "Our basic intention is to win militarily. We must gain military victories before thinking of diplomatic struggles. And even when we are fighting diplomatically, we must go on with our war effort; we must multiply our military victories if we want to succeed diplomatically." Although the Johnson administration failed to understand the situation, this general made it clear that North Vietnam saw no reason to negotiate because it was not suffering badly due to restrictions imposed upon American airpower by officials in Washington.

While the buildup of American troops continued, General Westmoreland initiated search and destroy operations to clear main force communist units from population centers and to drive them into areas where they could be destroyed. These operations were conducted to create favorable conditions for pacification of the countryside. Airpower was used extensively to bomb the Ho Chi Minh Trail—an expensive and ineffectual use of airpower.

At Yankee Station Task Force 77 concentrated on supply dumps and railroad lines to the south in the fall of 1966. When a morning reconnaissance pilot noticed a series of railroad cars passing through the town of Ninh Binh, Rear Admiral David C. Richardson, task force commander, advised his chief of staff, Captain James D. Ramage, to concentrate all airborne aircraft on this target.

Later it was reported there were possibly three trains of one hundred cars in the Ninh Binh area. After a reconnaissance leader wisely dropped his bombs to cut the line north of the town, the trains became immobilized. Ramage advised Richardson that he'd like to order an attack but some of the rolling stock was in the town. "The target is the trains," Richardson said. Ramage ordered strike control, "Tell the pilots to take them all." Three carriers sent their strike aircraft to bomb the trains, creating havoc on the ground for the next three days as the North Vietnamese frantically tried to unload their contents into trucks.

Admiral Richardson was disturbed by periodic orders from Washington that limited the number of aircraft to be assigned to a target because it increased the risk for the crews. He voiced vigorous objection to such orders, but McNamara's officials said that if there were too many planes some of them might miss the target and kill civilians.

After one attack near Hanoi, Richardson received a message saying, "Did you kill any civilians?" He was irate, telling his staff that he never received a message asking how many pilots he had lost. In this case, no civilians were killed, but inevitably some were on other missions.

But carrier officials knew that a higher number of strike aircraft were needed to reduce the risk. Richardson complained to Admiral Roy Johnson, Commander of the Pacific Fleet. "Why send a message out like this? Doesn't anyone consider our pilots?" Admiral Johnson replied that the message in question had come from the White House.

By September, carriers were averaging over 300 strikes a day—a level they maintained until mid-October. But during these months they inflicted heavy damage against targets in North Vietnam, knocking out thirty major bridges, more than 1,000 supply boats or barges, 450 railroad cars, and 350 trucks. In October, despite the bad weather, twenty-one bridges were dropped, and 600 supply boats, 130 railroad cars, and 100 trucks were destroyed or damaged.

During this period of high activity a fire broke out in the hangar of the *Oriskany* on October 26 while it was operating in the South China Sea off Vietnam, causing the loss of forty-four officers and men. Even greater loss was prevented by the heroic efforts of the crew against great odds.

Admiral Sharp sent another message to the Joint Chiefs on October 26 saying that self-imposed controls on the use of airpower had an adverse effect upon the reduction of North Vietnam's capability to support the insurgency in the south. He acknowledged that air attacks had disrupted the enemy's operations and prevented them from seizing significant ports of I and II Corps. He added, however, that air operations had not reduced their support to the South to a satisfactory level and that these air attacks had failed to bring the enemy to the negotiating table. He called for an air campaign that would bring the war home to the communists. In describing the year's operations, he said that of the 104 targets in the northeastern part of North Vietnam he had been permitted to hit only twenty. He claimed that the Rolling Thunder campaign had not applied sufficient pressure on the communists because of restrictions approved by McNamara and President Johnson.

Navy interdiction operations in North Vietnam had only been authorized in October, and they were limited to seventeen degrees, thirty minutes north latitude—an area of thirty miles. Although often fired upon, Navy planes suffered their first casualty on December 23 when two men were killed and three were wounded. Sharp had argued forcefully for permission to bomb targets whenever such attacks were desired. His request was turned down. The president and McNamara refused to authorize an all-out aerial offensive of North Vietnam.

After bombing halts over the Christmas and New Year's holidays, pilots were granted permission to bomb closer to Hanoi and Haiphong, although the cities remained off-limits with their valuable strategic targets.

After McNamara returned from a December visit to South Vietnam he wrote the president a memorandum. "Pacification has if anything gone backward," he said. He added that in his opinion the air war had not "either significantly affected infiltration or cracked the morale of Hanoi." He recommended a limit to ground forces in Vietnam, and consideration of a bombing halt. Either that, he said, or shift the bombing of targets from the vicinity of Hanoi and Haiphong to infiltration routes to increase the credibility of peace gestures.

The Joint Chiefs disagreed, urging that strikes be made against locks, dams, and railyards. McNamara rejected their recommendations.

The United States started 1967 with 385,300 military personnel in South Vietnam. In reviewing the previous year's accomplishments, General Westmoreland said, "Airmobile operations came of age. All maneuver battalions became skilled in the use of a helicopter for tactical transportation to achieve surprise and outmaneuver the enemy."

But almost 60,000 North Vietnamese regulars had entered South Vietnam during 1966, bringing their total strength to five divisions of 282,000 men, plus 80,000 in political cadres.

Westmoreland's request for a thirty percent increase in troops, for a total of 500,000 men, plus a reserve of three divisions in the United States, shocked the Johnson administration. Instead, McNamara authorized an increase to 469,000 by the end of June in 1968.

★ ★ ★

The American pacification plan—to assign the primary mission of supporting the countryside to the South Vietnamese Army—was not working. The failure to secure and hold outlying areas was the primary cause of its failure. It was a strategy that was bound to fail because in the months and years ahead the Americans controlled rural areas during the daytime while the Viet Cong ruled supreme at night when the Americans drew back to their fortified positions.

In early 1967 Admiral Sharp made another strong appeal to curtail North Vietnam's access to supplies from China and Russia. In his latest appeal to Pentagon officials in Washington, Sharp asked for permission to curtail the flow of men and supplies from North Vietnam into Laos and South Vietnam. He listed six basic target systems in North Vietnam for attacks by Air Force and Navy planes. They included electric power, war-supporting industries, transportation support facilities, military complexes, petroleum storage depots, and air defense systems. His appeal to the Joint Chiefs stressed that attacks against these target systems—rather than against individual targets on a nonconsistent basis—would steadily increase the pressure against North Vietnam. He reminded the Joint Chiefs that he had previously sought such permission, but that his command still faced operational restrictions.

On January 18 he sought approval to attack these targets on a package basis, and not doled out a few at a time. He cited 104 specific military targets in the northeastern sector of North Vietnam, saying only twenty had been authorized for strikes. He went even further later by designating 242 fixed targets in North Vietnam and asking permission to attack them at the rate of fifteen new targets each month.

Since Sharp had become Commander in Chief, Pacific, in 1963, he had urged the Joint Chiefs to grant him the right to attack all strategic targets in North Vietnam before the communists were able to build up their defenses. The Joint Chiefs had been supportive, but McNamara turned down each request. This was tragic, because more than eighty-five percent of the war materials that arrived in North Vietnam came by sea through Haiphong. Cargoes from Russia and her satellite countries arrived daily on a 24-hour-a-day basis. Mining Haiphong's harbor early in the war—which Sharp had long recommended—would have stopped this inflow of vital war materials. Again, McNamara vetoed the Joint Chiefs' endorsement.

When President Nixon authorized such mining in 1972, it cost less than a million dollars and not one person on either side was killed. In less than a year the North Vietnamese were forced to sign a cease-fire agreement. If the mining had been accomplished in 1965, it would have had even a greater impact on ending the war.

Sharp had also recommended that overland supply routes from China be destroyed, but this request was also denied. The Johnson administration feared that such an act would bring the People's Republic of China into the conflict. Even an elementary understanding of Far Eastern politics would have shown the falsity of such an assumption. The Vietnamese and Chinese had been mortal enemies for years. At one period China had ruled Vietnam for a thousand years. The North Vietnamese would have stoutly resisted open Chinese involvement in the war.

President Johnson stated the administration's position on the war when he gave his state of the union address on January 10, 1967. "Our adversary still believes that he can go on fighting longer than we can. I must say to you that our pressures must be sustained...until he realizes that the war he started is costing more than ever."

The administration's strategy of attrition was fundamentally wrong, because it primarily involved the destruction of North Vietnam's manpower, which was in plentiful supply—in the neighborhood of two million eligible men. If Sharp's total plan to unleash American airpower had been approved, North Vietnam's resources would have been destroyed, along with the people's will to fight.

Aerial mining of selected river areas was approved on February 23 and proved the effectiveness of the so-called destructor mines (500-pound bombs rigged to explode when a barge or sampan passed over them). Sharp's request to use heavier mines in North Vietnam's harbors was again rejected. Such large mines were not used until 1972.

The first aerial mining of rivers took place on February 28 when seven A-6A aircraft led by Commander A.H. Barrier of VA-35's "Black Panthers" placed mines in the mouths of Song Ca and Song Giang Rivers. The work was done primarily to stop coastal barges from moving supplies into immediate areas. Straight-in, low-level passes were made either at night or during periods of bad weather. These dangerous flights required a high level of skill by the crews, who all performed exceptionally well.

21

Dragon's Jaw

In the first few weeks of 1967 several strikes were flown against the Thanh Hoa or Dragon's Jaw Bridge, but it remained battered but still in service.

The Walleye glide bomb, one of the new "smart" bombs, was brought to Vietnam in March and assigned the task of destroying Dragon's Jaw. This free-fall bomb has a 1,000-pound warhead with a television camera in its nose that tracks the point of highest contrast, which becomes its aiming point. The camera relays what it sees to a scope in the cockpit through which the pilot can identify the target. The pilot then sights the target on his scope, positions a set of crosshairs over a pre-selected contrast point, identifies this point to the Walleye, and releases the bomb within its glide and guidance parameters. It has pinpoint accuracy and permits a pilot to remain out of range of ground batteries.

After familiarization flights, the commander of Task Force 77 scheduled the first Walleye mission for March 12. The pilots of Attack Squadron 212 had been given a scale model of the Thanh Hoa for use in conjunction with a movable light source (simulating the sun) to locate the best points of contrast and the time of day these conditions would occur. Army demolition experts were also on board the carriers to assist in identifying the most vulnerable spots on the bridge's structure and the sun's contrasting effect. Pilots and demolition experts agreed that 2:12 P.M. on March 12 would provide the maximum contrast for the chosen aiming point.

Pilots rode the escalators from their ready rooms to the flight deck on that date, noting with satisfaction that the sun was shining brightly. Three A-4 Skyhawks were assigned to the mission, with one Walleye each, while two F-8 Crusaders provided top cover from MiGs. The mission was planned so that each aircraft would make individual runs on the bridge from south to north in order to give each pilot as much time as he needed to locate the aiming point, identify it to the weapon, and release it. The flight was launched and joined up over the carrier prior to heading to the target. En route, the pilots completed their checks of their weapons and declared the mission a "go."

Oriskany Skyhawks strike with pinpoint accuracy during attacks on the Thanh Hoa Bridge in North Vietnam. (Courtesy of U.S. Navy.)

Over the target, each pilot dove at the bridge at 500 mph and released his Walleye. Initially antiaircraft fire was very light, but when the third pilot began his run enemy fire became intense. As this pilot searched for the aiming point he could see, in his peripheral vision, hundreds of flashes on the ground as enemy guns fired at him. After he sighted the aiming point, and it was identified to the weapon, he released the Walleye. Then all three Skyhawks streaked toward the Gulf of Tonkin.

Their planes recorded the attack on film, and when viewed later on the carrier it was noted with satisfaction that all three weapons hit within five feet of one another on the designated aiming point. But the bridge remained proudly erect.

Due to bad weather, it was another five weeks before the Dragon's Jaw was attacked again, although Walleye attacks had knocked out other bridges. Navy pilots also dropped sixty-eight more Walleye glide bombs against barracks, power plants, and bridges, scoring sixty-five hits. As for the Dragon's Jaw, it had become a symbol of North Vietnamese defiance. Further attacks began in late April after the weather improved.

In one of the largest raids during the spring of 1967, Haiphong was left without lights on the night of April 20 when strike and fighter jets from two carriers attacked a 10,000-kilowatt power plant a mile northwest of the city and a 7,000-kilowatt facility two miles east of it. Together the two plants produced nearly fourteen percent of North Vietnam's power capacity. Pilots encountered antiaircraft fire that became extremely heavy in these strikes, the closest yet made to the port city.

The first Navy strikes of the war against MiG bases were launched on April 24 with a series of attacks on Kep airfield, thirty-seven miles northeast of Hanoi. A-6 Intruder aircraft from *Kitty Hawk* hit the runway, control tower, and maintenance and support buildings, and their fighter escort downed two MiG-17s. The airfield had twenty MiGs, out of an estimated total of 120 in North Vietnam.

On April 25 and 26, in the largest combined strike of the month, Navy jets from three carriers faced heavy SAM, MiG, and antiaircraft fire to blast a Haiphong cement plant and nearby ammunition and petroleum storage centers. The cement plant was a major manufacturer of construction products used by North Vietnam to build military installations. After the raids, one pilot said kiddingly, "If they had cement made, they could have shoveled what was left into a bag."

In May, with good weather and four carriers on station, more than 3,100 missions were launched. This effort was directed mainly against electrical power sources and MiG fields, with repeated heavy strikes against the thermal power plants in North Vietnam and also its airfield. At Kep, on the first of the month, eight MiGs were destroyed on the ground, and two were shot down in the air.

Admiral Sharp was informed in mid-June 1967 that McNamara and other Washington officials would be in Saigon in early July. General Wheeler, chairman of the Joint Chiefs of Staff, told Sharp that the meeting would determine the future of the Rolling Thunder bombing campaign because McNamara was opposed to bombing above the 20th parallel.

Sharp went to Saigon to remind Westmoreland and his staff that he expected their full support for the bombing campaign. After reviewing proposals they had prepared for McNamara's visit, Sharp was critical and insisted that they redo them. He warned Westmoreland, "Westy, I don't want any screwed up things from you. I want you to support it, too."

Sharp was so concerned about the McNamara conference that he arrived on July 5, two days prior to its start, to assure himself that his subordinates would support his bombing campaign.

McNamara arrived with the Joint Chiefs and representatives from the State and Defense Departments. When Sharp addressed the conference, he said he had been recommending an air campaign centered around attacks against vital target systems in the North with sufficient strength to convince the communists that their aggression was not worthwhile. He admitted that bombing strikes had not been as effective as they should have been due to bad weather, but added that by April the weather had improved over North Vietnam, and now that additional target systems had been approved the bombing campaign was putting greater strain on North Vietnam's government than ever before.

Sharp said he objected to the administration's imposition of restrictions May 23 against targets within ten miles of Hanoi after his aircraft had successfully attacked a vital thermal power plant three days earlier. This action, he said, forced his command to remove the pressure against the communists just when they were being hurt.

He reminded McNamara and the other conferees that the sector within ten miles of Hanoi was North Vietnam's most important support area and was militarily vital to them. "Our continued failure to strike decisively in that area gives them a virtual sanctuary from which to operate." He cited North Vietnam's loss of thirty MiGs in the previous five weeks of air warfare as the reason its air force refused to contest further American attacks. He stressed that the trend in the air war was in their favor, and that the momentum of air attacks must be maintained. Imposition of prohibited and restricted zones around cities in North Vietnam, Sharp said, continued to obstruct his air operations. He reminded conferees that fifty-nine percent of North Vietnam's industrial targets were still off-limits. "To have the greatest impact on the communists," he said, "all target systems must be attacked and re-attacked if, like transportation targets, they are repaired."

Westmoreland and his top officers supported Sharp's position, although Westmoreland requested more ground troops.

Sharp summarized his remarks in closing, saying that his command had flown about 18,000 sorties in the North and another 185,000 in the South, including Laos, in the past eighteen months. He admitted that attacks against bridges would have to be continued once they were repaired, but he said that eighty-six percent of North Vietnam's power capacity had been destroyed, thirty to fifty percent of its war-supporting industries had been disrupted, while forty percent of the major military installations in the North—and all fixed installations in the South—were either destroyed or disrupted.

Sharp told them that a reliable intelligence source in Hanoi had estimated that thirty percent of the war materiel supplied to North Vietnam had been destroyed en route to Hanoi by air strikes. This informant claimed, he said, that "The effectiveness of American air action and the seemingly infinite capacity of the Americans to escalate to the degree necessary to overcome any increased support rendered, are the principal causes of the disillusionment among representatives of the Soviet and bloc countries in Hanoi, and their conviction that the North Vietnamese cannot win. They also feel that the problem must be negotiated if the complete destruction of North Vietnam is to be avoided."

Sharp stressed that Rolling Thunder did not begin to reach the level of intensity in the northeast quadrant in April that he considered necessary to do the job until late. Since late May, he said, maximum effectiveness was not being achieved because of restrictions and the lack of authority to hit the more vital targets. Sharp emphasized that the restrictions came at a time when two factors were in their favor: first, good weather months, and second, important and perhaps decisive changes in North Vietnam's capability to defend itself. He stated that a momentum had been established that was now endangered. "To retrench even further and limit our attacks to the south of the 20th parallel will have adverse, and I believe, disastrous effects. War-supporting industries in the North would be brought back to maximum output. Morale of our own forces would allow North Vietnam to operate out of a virtual sanctuary with complete freedom to move supplies to the South without damage.

"We are at an important point in this conflict and an incisive air campaign, including sustained attacks in the northeast quadrant against all target systems, would assure interrelated effects against the enemy's military, political, economic, and psychological posture. There is ample evidence that the enemy is hurting. I consider it essential that we continue this effective, successful air campaign."

His plan included recommendations that he had made since he took command in the theater. He proposed closing Haiphong Harbor to deep-water shipping by bombing and/or mining, destruction of six basic target systems, and a reduction in Hanoi's restricted area to ten nautical miles.

McNamara pointedly ignored Sharp at the conclusion of the conference, as he said, "General Westmoreland, that was a fine presentation."

Sharp told Wheeler the next morning, "Goddamn it. I'm getting sick and tired of that son-of-a-bitch making a comment like that to Westy when I'm the senior officer present and I have made a presentation and he didn't say a word to me."

F-4 Phantoms. (Courtesy of McDonnell Douglas.)

Wheeler replied, "Oley, he was furious with you. He came out here to stop the bombing above twenty degrees and you made it impossible for him to go back and say that everyone agreed."

Sharp realized Wheeler was right. McNamara often wrote his report on the way out to the Far East and tried to use these presentations to justify his decisions. This time he could not do that, and Rolling Thunder—even in its emasculated form—survived. McNamara still refused to approve all ninety-four targets approved by the Joint Chiefs, but he did not seek an end to the bombing when he talked to President Johnson. He did refuse authorization to close the ports, and targets within the prohibited zones in Hanoi and Haiphong remained off-limits.

The air war changed drastically in August when MiG-21s introduced a tactic that proved extremely difficult to counter, and evidently with their best pilots. They took off in pairs from either Phuc Yen or Gia Lam and flew at low level, keeping within radar ground clutter until they were abeam of an inbound American force. With such a tactic, the Phantom's radar could not detect the enemy presence. Once behind an American formation the MiG-21 pilots used their afterburners to climb to a high perch above the American force. Then, with the aid of ground vectoring, they launched their "Atoll" missiles (infrared seekers like Sidewinders) and zoomed back to altitude or passed through the American formation. After one firing pass, the MiGs separated, landing at an airfield in North Vietnam or China, where they were immune to counterattack.

The initial success of these hit-and-run tactics was due in part to the fact that the F-4s were used either as strike or combat air patrol aircraft. At first, the MiG-21s operated with near impunity. But by the end of August F-4s were positioned to guard the rear of strike forces against the MiGs.

A symbol of triumph. Lieutenant William T. Patton of VA-176 stands on the wing of
his A1-H Skyraider aboard the USS *Saratoga* by the painting of a MiG-21
on the fuselage in recognition of his shooting down the communist jet
with his propeller aircraft. (Courtesy of U.S. Navy.)

Commander Richard M. Bellinger, gesturing at right, returns aboard USS *Oriskany* after shooting down a MiG over North Vietnam. (Courtesy of U.S. Navy.)

In the summer of 1967, President Johnson finally decided to attack some targets within and near Hanoi. Carriers hurriedly launched an unprecedented barrage of major attacks against North Vietnam's vital northeast industrial sector.

The Air Force was assigned the task of destroying the Paul Doumer Bridge. It had never been attacked, because it lay within an area of Hanoi that the United States had chosen to leave untouched.

During August, targets were released within the buffer zone sanctuary along the border between the People's Republic of China and North Vietnam. Strikes were made against the Port Wallut naval base, the Long Son railroad bridges, and the Na Phuoc railroad yard. In that month, the Navy lost sixteen aircraft, its highest toll for any month of the war.

During the first six months of 1967, American aircrews scored fifty-four confirmed MiG kills at a cost of eleven aircraft, and the North Vietnamese lost another nine aircraft on the ground.

The flight deck of the *Forrestal* was swept by fire as aircraft were being readied for launching on July 29. Flames engulfed the fantail and spread below decks, touching off bombs and ammunition. It was some time before the fires were brought under control, and damage to aircraft and the ship was severe. The final casualty count was 132 dead, two missing and presumed dead, and sixty-two injured.

The Preparedness Subcommittee of the Senate Armed Services Committee held extensive meetings in August about the progress of the war now that there were about 470,000

troops in South Vietnam, with the war costing $2 billion a month and no end in sight.

Senator John C. Stennis informed McNamara that he would be expected to justify the war's direction, along with the nation's top military leaders. The committee was sympathetic to the plight of the nation's military leaders, and Stennis and most members viewed bombing restraints as an irrational shackling of a major instrument that could help terminate the conflict on a satisfactory basis.

During Admiral Sharp's appearance, he presented much the same line he had offered to McNamara in the Saigon conference. He cited progress in the air war, although he said that the earlier restrictions imposed by McNamara were still in effect. He ended his comments by saying, "The best way to persuade the ruling elements in North Vietnam to stop the aggression is to make the consequence of not stopping readily and ever more painfully apparent."

McNamara appeared on August 25. He defended the restrictions on the bombing campaign, saying they were carefully tailored to limited objectives in Southeast Asia, primarily to strike areas of infiltration. "The bombing of North Vietnam has always been considered a supplement to, and not a substitute for, an effective counterinsurgency land and air campaign in South Vietnam. The bombing campaign has been aimed at selected targets of military significance, primarily the routes of infiltration." He asserted that when the bombing was viewed in this light, it had been successful.

The committee disagreed with McNamara. It said the administration had failed to defer to the authority of professional military judgments instead of McNamara's civilian analysts. The report advocated an escalated pressure against North Vietnam. These hearings demonstrated that there was a rift within the administration, and President Johnson was forced to overrule some of McNamara's defense strategies.

The fires aboard the *Oriskany* in 1966 and the *Forrestal* this July were investigated in August by Admiral James S. Russell, former vice chief of naval operations, who was recalled to active duty to review carrier operations and enhance their safety.

Shortly after he began his investigation, he realized that both fires could have been prevented. An inexperienced seaman on the *Oriskany* had mishandled a Mark 24 flare by dropping it and thus activating its ignition cord. (The flare contained thermite, which needs no external oxygen.) Instead of taking it to the adjacent hangar deck and throwing it into the sea, he put it in a locker with other Mark 24 flares and dogged down the door. When the flares exploded under pressure the flames went through the ventilation system and killed a number of pilots asleep in their bunks. The premature firing of a five-inch rocket on the *Forrestal* in July had been caused by storing rockets on the outside of the carrier's island where there was no sprinkler system to keep bombs and rockets cool, or flood them if there was a danger that they might explode. The pace of operations had been so heavy that this type of storage had been adopted.

Russell and his committee made ninety-six recommendations to enhance safety. Through his efforts the Navy adopted a sprinkler system for flight decks, and its control switches were placed at the air officer's station in Primary Fly Control and, as a backup, on the navigating bridge.

A new target list, Rolling Thunder 57, contained six targets within a ten-mile radius of downtown Hanoi, and the Doumer Bridge had a high priority. The new list had been released after Senator Stennis had announced he was holding meetings about the war. Thus the stage was set for a major effort against the North Vietnamese transportation system, and the Doumer Bridge became a prime target.

During the planning stage it was decided that F-105s of the 355th Tactical Fighter Wing would lead three wings on the strike. Each Air Force fighter-bomber would carry 3,000-pound bombs.

On August 11 the strike force came off the east end of Thud Ridge and the pilots noted that they could clearly see the bridge. It looked like a black snake spanning the brownish waters of the Red River.

Time was short as the last turn was made to the south by each flight of four aircraft and the climb to 13,000 feet was begun for the bombing run. The pilots called this climb the "pop" for the roll-in. It was a maneuver wherein a flight could evade flak, climb to its bombing altitude, and position all four aircraft into echelon formation. At precisely the correct point over the ground calculated to produce a forty-five-degree dive angle for the final run, aircraft one and two, followed by three and four, commenced their roll-in and dive-bombing run. During the run the aircraft were pointed at the ground, and aimed at the target with engines on afterburner. Each wingman flew formation just off the wing of his leader.

During the seven-second bombing run the flak was very thick, and a number of SAMs headed their way. Meanwhile, fighter patrols roamed above them to protect their backs while Wild Weasels pummeled flak and missile sites.

As the lead flight maneuvered to the east at supersonic speed, the leader of the second flight saw a span drop into the Red River. Only two F-105s were hit, and they all managed to return to their bases. At the end of the strike one railroad span was destroyed, along with two highway spans on the northeast side of the bridge. The superstructure was damaged and the highway portion on the north side of the bridge, where it crossed the island in the river, was cut. The movement of an average of twenty-six trains per day ceased for the time being.

On August 14, a 1,150-foot highway bridge across the Red River, seven miles north of the center of the city, was blasted by A-6 Intruders and A-4 Skyhawks. Flying from the *Constellation,* they were successful in dropping two spans into the river.

In succeeding months, strikes were made against bridges, railroad lines, and warehouses near Haiphong to isolate the port, although its facilities were not directly struck.

In October, President Johnson finally authorized the Air Force to hit Phuc Yen, the major North Vietnamese base that had incredibly become a privileged sanctuary. Before the year ended, all jet airfields in North Vietnam, except for Hanoi's Gia Lam International Airport, were attacked. As a result, all but twenty MiGs were driven into China, although hit-and-run attacks by MiG-21s continued to take a heavy toll because they could not be pursued.

Navy pilots spent the last two months of 1967 maintaining pressure against MiG bases, thermal power plants, and lines of communication throughout North Vietnam.

Forty-four Navy Skyhawks and Air Force fighters made a major attack on October 28 against the Dragon's Jaw Bridge. For three and a half hours they dropped three tons of bombs every four and a half minutes. The bridge lived up to its reputation. It suffered only superficial damage to its superstructure, although its girders were twisted and bent. The southern approaches to the bridge were severely damaged and its railroad tracks lay astride their beds twisted out of shape. The bridge was easily repairable, and this was done as bad weather prevented further attacks in the immediate future. Throughout the remainder of the year the bridge was knocked out every time it was repaired. Railroad traffic again was temporarily halted.

But the once-massive structure of the Dragon's Jaw—a central railroad flanked by concrete highways—was a charred, dented, and twisted maze of steel girders. The central railroad bed had become a patchwork of flimsy boards upon which rested the twisted and bent rails. Both approaches to the bridge were so cratered that the movement of vehicular traffic became impossible.

Two prerecorded tapes were seized from the Viet Cong on the afternoon of January 29, 1968, at Qui Nhon and transmitted to II Corps headquarters. Their contents were revealing, indicating a massive communist offensive. General Cao Van Vien, chairman of South Vietnam's Joint General Staff, immediately warned all corps commanders of imminent enemy attacks and to take appropriate defensive measures. His words were not taken seriously, and some troops slipped off to join their families for the Tet holiday.

As the South Vietnamese began celebrating Tet at midnight on January 30 the first communist attacks began at Nha Trang. During the night all five provincial capitals came under heavy attack.

The communist attacks were not fully coordinated—probably to deceive the South Vietnamese—as twenty-eight out of forty-eight cities and provincial capitals were targeted.

In Washington, the Joint Chiefs urged President Johnson to authorize air strikes against Hanoi and Haiphong. They also pressured him to call up the reserves, but Mr. McNamara approved only a 10,500-man increase for Vietnam and refused to call up the reserves.

Another Rolling Thunder phase had begun on January 3, but it was severely limited due to adverse weather around Hanoi and Haiphong. Other than all-weather A-6 flights, most aircraft were diverted to the besieged Marines at Khe Sanh below the demilitarized zone.

The military situation in the Far East had worsened on January 23 when the Navy ship *Pueblo* was captured by North Korean patrol boats. The carrier *Enterprise* was in the East China Sea. It was ordered to reverse course and headed for the Sea of Japan to establish a formidable American presence in the area.

Commander Jerrold Zacharais led his VA-75 squadron from the *Kitty Hawk* on the night of February 21, 1968, toward a vital port facility on the southeast corner of Hanoi along the Red River and south of Gia Lam airfield. The Navy's A-6 Intruders were used on this strike because the northeast monsoon had seriously curtailed daylight operations.

They approached Hanoi from the mountains in the northeast using their terrain-avoidance radar while flying at 500 feet above the peaks. After clearing the mountains, they lowered to 200 feet to avoid missile radar sites. Then, just ahead, Zacharais spotted a SAM launched ten miles away directly at them. He ordered a descent to 100 feet but overshot the mark and instead was down to forty feet. He

Skyhawks head for North Vietnam from the carrier *Enterprise* on Yankee Station in this artist's conception by R.G. Smith. (Courtesy of McDonnell Douglas.)

quickly flew back to 100 feet. Now a second SAM was headed their way, off to the right. He made a degree course change to the right to place the missile on their left side where he could keep it in view at all times. Five seconds later the missile made a course change to compensate for his heading change. Zacharais waited a moment and then broke sharply up and to the right. At 500 feet, and in a seventy-degree bank, the missile exploded right underneath where he had been. His Intruder rocked violently as he descended to 400 feet. He headed for the target, noting that missile firings had ceased. His radar officer now informed him he had picked up the target, just as antiaircraft guns all seemed to fire at once. He brought his plane to 780 feet and dropped their twenty-four bombs. Hanoi was blacked out, but the gun flashes on the ground showed him the river and he knew they were on target. Zacharais broke left and went lower as he crossed Hanoi at 530 knots. He returned by a different route to avoid the heavy antiaircraft sites he had spotted during the flight in, and landed without further incident on the *Kitty Hawk*.

North Vietnam suffered a resounding military defeat after its 1968 Tet offensive failed when it lost every battle, but they had a psychological victory in the United States when this fact was not understood. Basically, the South Vietnamese armed forces did most of the ground fighting, and most units fought well—particularly those that were well led. Not one Army unit broke under the intense pressure,

The USS *Kitty Hawk* ready for action in the Gulf of Tonkin. (Courtesy of U.S. Navy.)

or defected to the North Vietnamese. In the first phase of the attacks, which ended on April 3, the communists lost 32,000 killed and 5,800 captured. By the end of the second phase in May, another 5,000 had died. The communists failed to hold any city except Hue, and that was only for twenty-five days. The general uprising that they had anticipated failed to occur. It was a military defeat of the first magnitude because the people of South Vietnam had refused to join the uprising.

Airpower played a key role, due in large part to allied air superiority. The United States Seventh Air Force had a large force of 650 fighter-bombers, while the Vietnamese Air Force had a large force of attack planes and helicopters. United States Army aviation had an impressive fleet of gunships and airlift transports, while Marine aviation had been used almost exclusively for support of their two divisions in the northern provinces. In addition, planes of Task Force 77 had played a prominent role.

After the Tet offensive failed, both sides changed their strategy. Due to their heavy losses, the North Vietnamese were forced to retire to their sanctuaries. General Creighton W. Abrams succeeded Westmoreland in June 1968, and gave the South Vietnamese a broader role in the war, as all forces engaged in small-unit actions to defeat the enemy's remaining forces. Meanwhile, the bombing halt continued above the 20th parallel, and the North Vietnamese started to repair their railroad network.

The intensity of air operations in the first months of 1968 after the communist offensive was launched took its toll in casualties. During 1967 and early 1968 almost half the combat missions over North Vietnam were flown from the decks of Navy carriers. Only five MiGs were shot down with a loss of eighteen American aircraft in February, compared to a loss of twenty-four MiGs in August of 1967 and the loss of only six American planes. The situation had almost reversed itself.

President Johnson announced on March 31, 1968, that he was suspending air operations against North Vietnam the following day. He said he had received assurances from the North Vietnamese that they would respect the demilitarized zone between North and South Vietnam, and would begin peace talks in good faith. The North Vietnamese further agreed to permit the flight of unarmed reconnaissance aircraft over North Vietnam, with the understanding that armed escort fighters would return the fire if they were attacked.

Strike aircraft for the most part were diverted to the besieged Marines at the Khe Sanh base below the demilitarized zone. The Navy's bombing campaign was concentrated on the northern part of the southern panhandle, mostly between Vinh and Dong Hoi. Marine Squadron VMO-2 arrived to aid operations in the south on July 6 with their OV-10A Broncos. They were used as forward air controllers, for visual reconnaissance, and as helicopter escorts.

Although the North Vietnamese accepted the offer of peace talks, they did so only to strengthen their defenses in the North, and to build up their offensive capabilities in the South. Thus began a four-year period that was one of the most frustrating and divisive in American history. During this "bomb-free" period, North Vietnam's transportation system was restored to full capacity and the input to the Ho Chi Minh Trail reached new heights. The bridges that had been destroyed or damaged prior to March 31, 1968, were repaired and traffic again was heavy.

A fire on the *Enterprise* off Honolulu on January 14, 1969, resulted from detonation of a Zuni rocket's warhead overheated by exhaust from an aircraft-starting unit. It took twenty-seven lives, injured 344 men, and destroyed fifteen aircraft.

When Richard M. Nixon became president on January 20 he had high hopes that peace talks in Paris would lead to a supervised cease-fire, and guarantee political self-determination for the people of South Vietnam. He directed that the Joint Chiefs of Staff expedite the military training and equipping of South Vietnam's armed forces to assume responsibility for the defense of their country and to permit the phased withdrawal of all American forces. He called his policy "Vietnamization." The North Vietnamese had other plans for the future.

On February 22 the North Vietnamese launched the fourth phase of their general offensive. It had started a year earlier during the 1968 Tet holiday. This year they struck at a hundred places throughout the country, against Saigon and provincial capitals, military installations, and outposts. These attacks also failed to improve their situation in South Vietnam, and again their losses were heavy.

The communists' unsuccessful attacks during four phases of their general offensive brought them again to the peace table to continue alternate fight-and-talk tactics.

North Vietnam's General Giap now revealed that 500,000 of his men had been killed while fighting the U.S. and South Vietnamese armed forces. He vowed that his country would "fight on as long as necessary—ten, fifteen, twenty, fifty years."

The danger of the war spreading beyond Indochina was enhanced on April 14 when North Korea's armed forces shot down an unarmed EC-121 on reconnaissance patrol over the Sea of Japan. Thirty-one of its crew were killed. Task Force 77 was activated to protect further such flights over international waters.

One of Admiral Russell's committee recommendations to prevent uncontrolled fires on carriers was installed when the *Franklin D. Roosevelt* put to sea on May 26 after an eleven-month overhaul. A deck-edge spraying system utilizing the new seawater-compatible firefighting chemical "light water" was installed. It marked a major development in fire prevention on board the Navy's carriers.

Colonel Steve Furimsky, commander of Marine Air Group 11 at Da Nang, led a flight of A-6 Intruders in July 1969, while seeking targets along the 17th parallel—particularly along the network of roads where Laos formed one of South Vietnam's borders and North Vietnam the other. He was running low on fuel when he heard from the ground controller that a supply dump had been located inside a hill in a region controlled by a division of North Vietnamese regulars and opposed only by a regiment of American troops. An observer on the ground relayed this information to the planes above them and asked their pilots to drop bombs one at a time to dig into the hill and explode the munitions stashed in caves inside it.

Furimsky told the ground observer that he had 28 500-pound bombs but that they all had instantaneous fuses that would explode them on contact. Although he knew that bombs with delayed fuses would do a more efficient job of digging into the hill, he said he would start the digging process with his bombs. He warned that he needed to refuel before he could start the operation. He promised to get more planes with proper fusing on their way as soon as possible.

The ground controller, knowing the need was urgent, called, "I need you, but go up and refuel and come back. I'll put you through right away." Furimsky agreed and headed for the aerial tanker. When he hooked up, he found it difficult to fly at the C-130's refueling speed of 220 knots because he was so heavy with his full load of bombs. He fell off the drogue (the funnel at the end of the fuel hose) because the fuel added to his problems by increasing his gross weight.

He asked the pilot of the refueling plane to go into a slight dive to increase his speed to 230 knots. The pilot in the tanker agreed, with one provision. They were flying at 20,000 feet, and he warned Furimsky, "I won't go below 10,000." Furimsky readily agreed and they hooked up again and began their downslope ride while the gasoline flowed into the A-6's tanks.

When he fell off a second time his tanks were almost full, so Furimsky pulled away and thanked the cooperative tanker pilot. He reported back to the area he had earlier had to leave, and the controller told him he was clear to drop his bombs one at a time on the hillside.

After the eighth drop there was a tremendous flash on the ground, and the hill seemed to explode. The various planes that had bombed the hill had finally reached the munitions. There was a lot of red smoke, indicating the weapons that detonated had some kind of chemical in them. Furimsky got a call from the ground, "You got a flasher that time." Later, after more planes returned to attack the dump, a large part of the hill was blown away and the supply dump was obliterated.

During the summer and fall of 1969 combat operations were increasingly turned over to South Vietnam's armed forces as American troops began their withdrawal under President Nixon's new plan to "Vietnamize" the war.

American troop strength had peaked in April at 541,500, but the number of military men and women dropped to 505,500 by mid-October. Now there were only scattered enemy attacks as the communists tried to recover from their heavy losses. These attacks were made largely against South Vietnam's armed forces.

Although American ground action was sharply reduced, the skies over Vietnam were still deadly after 1970 bowed in. As Lieutenant Lyle G. Bien, F-4J radar intercept officer for Lieutenant Pat Noonan, took off from the *Ranger* in the Gulf of Tonkin on February 10 for another interdiction of the Ho Chi Minh Trail in Laos, they both felt the same sense of frustration that all crews experienced. Since President Johnson had initiated a bombing halt to coax the North Vietnamese to the bargaining table, they had been relegated to these senseless attacks, knowing that the bombing restrictions merely helped the North Vietnamese to rearm.

Most of their bombing sorties had been assigned to cut the supply lifeline to the North Vietnamese and Viet Cong forces fighting in the south. He knew—they all knew—that despite their best efforts they were merely expending a lot of bombs while achieving few positive results. American airmen had struck repeatedly at mountain passes such as Mu Gia where a multitude of trails funneled into a single artery. Although these passes—like Yankee Station in the Gulf of Tonkin—were well north of the demilitarized zone, on previous missions the restrictions imposed by former President Johnson had prevented them from flying into North Vietnam, so they had to make a more circuitous route south of the demilitarized zone, then west to the Laotian border, and north to where the Ho Chi Minh Trail exited North Vietnam.

Admiral Sharp had argued for years that his planes should at least be allowed to fly a more direct path from Yankee Station over southern North Vietnam in the vicinity of Vinh, and directly to the passes that his crews were ordered to strike.

Navy air crews were convinced that the prohibition against flying over North Vietnam was ordered not to avoid provoking the North Vietnamese, but to prevent aircraft from dropping bombs on the hundreds of excellent targets along the way.

On the 10th of February 1970, they were finally given permission by the Nixon administration to fly a direct route from their carrier to Mu Gia and Ben Harai passes on the North Vietnam–Laos border. But they were given explicit directions before their departure not to bomb any targets in North Vietnam, except to return fire in self-defense.

Bien and Noonan, attached to the Black Knights of VF-154, were part of a formation of eight A-7s, six A-6s, five F-4s, and one RA5C. After takeoff they rendezvoused quickly and departed at 3:30 P.M. for a late afternoon strike. There was a thin layer of clouds just west of Vinh as they went "feet dry" (reached land) and fairly heavy antiaircraft fire was visible. As they flew westward the clouds thickened and visually directed radar diminished.

Bien listened anxiously for word that an SA-2 missile had been launched, but indications were confined to warnings. Noonan, flying wing to his section leader, proceeded to Mu Gia, where they dropped their bombs. Other sections, meanwhile, dropped at Ben Harai. The bombings were completed without incident and the flight rejoined fifty miles east of the targets to return to their carrier in a loose formation on essentially the same route.

Near Vinh, Bien noticed a flight of Air Force F-4s on combat air patrol to protect them from MiGs near the coast. As they approached at 18,000 feet, Bien noticed a single F-4 five miles on their right turning in their direction. Their flight of four F-4s trailed the A-6s and A-7s, while their fifth F-4 was 200 miles behind them on a photo run to protect the RA5C. With the Air Force F-4s appearing to join them, although there were no radio calls, the Navy flight turned in his direction to aid in his rendezvous. Bien noted that the cloud cover over Vinh was a solid layer up to 10,000 feet. As the Air Force fighter came closer he noted that it was a USAF F-4E.

Bien glanced to his left to check on the position of their lead plane and then looked again at the F-4. He stared in horror as the Air Force plane was hit by a North Vietnamese missile. There was almost total and instantaneous disintegration. This was the first time he had seen such a hit, and he was badly shaken. He watched carefully for parachutes but there were none. He learned later that both crewmen were killed. Another Air Force plane assumed search and rescue duties, so the Navy formation proceeded to the *Ranger*.

Late that night many crew members were interrogated individually about the strike in Laos. At first Bien thought these continuing inquiries were to gain more specific information about the loss of the Air Force F-4E. It was not until they were in the Philippines three weeks later that his commanding officer told Bien, with a hint of a smile, that all the extra debriefing was an attempt to put a serious face on some extracurricular activity off-target. Apparently, he said, four of the A-7s only dropped one bomb each on the pass, and saved their other five bombs for two large trains they had spotted thirty miles east of the target. A-7 pilots claimed that due to the worsening weather and approaching darkness they could not be sure which target they had hit. Bien's commander said that a photo reconnaissance plane the next morning confirmed the total destruction of both heavy-laden southbound trains due to derailment and fires.

The photographs left the ship the following day, and no further mention was made of this off-target strike. Perhaps not coincidentally, they were not permitted to strike targets in Laos via North Vietnam again until the bombing halt was lifted two years later.

The Naval Air Systems Command and Great Britain had executed a memorandum of agreement on October 22, 1969, whereby the Hawker-Siddely Harrier, a vertical takeoff and landing aircraft, would be purchased. This year twelve aircraft were procured, with initial delivery scheduled for January 1971. The Harrier was designated the AV-8A and was procured for use by the Marine Corps after its pilots had flown it earlier and strongly recommended it.

As a result of a crisis in the Middle East caused by Palestinian commando attempts to unseat the monarchy in Jordan, the *John F. Kennedy* joined the *Independence* and the *Saratoga* in the Mediterranean on September 25, 1970, followed by seven other ships to strengthen the Sixth Fleet to fifty-five ships. This standby force was created in case military protection was necessary for the evacuation of Americans, and as a counterbalance to the Soviet Union's Mediterranean Fleet.

The attack carrier *John F. Kennedy* patrols the sea-lanes in the Atlantic during the Vietnam War. (Courtesy of U.S. Navy.)

In response to attacks on unarmed American reconnaissance aircraft, 200 American planes conducted protective reaction strikes against North Vietnam's missile and antiaircraft sites south of the 19th parallel on November 21–22. Marine aircraft participated, in addition to Navy planes from the *Hancock,* the *Ranger,* and the *Oriskany.*

There was growing concern within the Department of the Navy that its worldwide commitments were spreading its available ships much too thin, but there was nothing that could be done about it.

22

Bridge Busters

Throughout 1971 there was only light ground action in South Vietnam, with limited American contacts with the North Vietnamese. American troop withdrawals continued and the communists bided their time until most Americans had departed for home. Even air operations from Yankee Station remained at a reduced level, with attacks concentrated on interdiction targets—particularly truck convoys. This state of affairs continued through the first three months of 1972. During January only eight Navy tactical air sorties were flown.

While Commander Robert Arnold of VA-97 and his wingman were over the demilitarized zone during the fall of 1971, he called the forward air controllers: "We've got some 20 mike mike left (20-mm ammunition.) Is there anything you want us to hit?" He had fuel for another forty-five minutes before he had to return to the *Enterprise*. The forward controller told him there was nothing worth hitting.

Arnold next tried to contact the airborne control plane Hillsborough. After a delay in responding, the pilot asked. "What's your ordnance?"

"I'm out of bombs but we still have our twenty mike mike."

"We've got someone who can use you. Go south of Da Nang and contact the Birdseed forward air controller."

Arnold turned south, and put in a call.

"Where are you?" the controller shouted. "War Ace, where are you?"

"We are on our way. I think we are sixty miles from you."

"We really need you, War Ace."

"Okay. All I've got left are twenty mike mike."

"Hurry up! We need you down here."

Listening to the radio, Arnold heard, "Charlie, how are you doing down there?"

"When am I getting out of here?" Arnold realized that the controller was in contact with a man on the ground who was in an observation post. It was located on a ridgeline and the Viet Cong evidently were getting ready to charge up the hill to capture him.

A heavily loaded A-7E Corsair prepares for launching to take part in an air strike over South Vietnam April 25, 1972. Assigned to the *Constellation*'s VA-146, the Corsair was an integral part of the United States Navy's attack squadrons. (Courtesy of U.S. Navy.)

The controller responded to Charlie, "We've got the choppers on the way, Charlie. Hang in there." The emotion in the controller's voice betrayed the urgency of the situation.

"War Ace, what's your position now?"

"I'm thirty-five miles from your position," Arnold responded. He glanced at the countryside, noting there was just a green expanse. "I think I've got you spotted." Arnold said. Rock your wings."

The controller did so.

"Okay, I see you."

"I'm putting smoke in," the controller said.

Arnold spotted the smoke as they came in from the north.

The controller called, "I want you to make your runs from the south to the north in a left-hand pattern. Hit my smoke, and hit west of my smoke. Don't hit east of my smoke. We've got friendlies on the east side."

Arnold acknowledged, and signaled to his wingman to follow him. He heard the controller call. "How you doing, Charlie?"

"When in hell am I getting out of here?" The man's voice shook and he kept repeating the words.

"We've got the Navy coming in, Charlie. Hang in there." Then to Arnold, "Hurry up, War Ace. We need you down here now."

"I've got your smoke and I'm ready to roll in," Arnold said.

He and his wingman made six runs shooting at the spot marked by the smoke, flying just above the treetops. He couldn't see they were doing any damage, but then the controller yelled, "You got 'em! You got 'em! They're going the other way." He called Charlie again. "How you doing?"

"When in hell am I getting out of here?"

"The choppers are on their way."

Arnold and his wingman by now had shot practically all their ammunition and they were late in returning to their ship. The last word he heard from the forward air controller as they turned away was, "Hey, Navy, you saved our lives down here today."

For Arnold, who had flown 336 combat missions in Southeast Asia, these words made up for all that he had undergone. These kinds of action continued into 1972, but few were as fulfilling as Arnold's mission helping to save a forward observer.

Lieutenants Randall Cunningham and William Driscoll shot down a MiG-21 on January 19, 1972. This was the first enemy aircraft destroyed since March 28, 1970. At that time Lieutenants Jerome Beaulier and Steven Berkley had shot down another MiG-21 in their F-4 of the *Constellation*'s VF-142. Cunningham and Driscoll—from the same carrier but assigned to VF-96—had been sent out in response to the firing of North Vietnamese antiaircraft guns and SAMs at an unarmed RA5C reconnaissance plane. This was the Navy's thirty-third MiG shot down in the Vietnam War. The first was on June 17, 1965, by Commanders Louis Page and John Smith in an F-4 of VF-21 from the *Midway*.

During the four-year bombing halt that began in 1968, American fatalities had increased by more than 25,000 and stood at 45,679. But during these same four years President Nixon's "Vietnamization" policy had reduced American forces in South Vietnam by the spring of 1972 from more than 540,000 men to approximately 70,000.

On March 30, 1972—almost four years to the day after President Johnson's initiative for peace—the North Vietnamese attacked across the demilitarized zone into Quang Tri Province in their biggest invasion of the war. American tactical airpower in Thailand was used to stall the offensive while units that had gone home were redeployed to Southeast Asia. Navy carriers were returned to the South China Sea. By the end of July, the largest air armada of recent years had been assembled. Once again the systematic aerial bombardment of North Vietnam was under way and the transportation system and the bridges were key targets.

After the North Vietnamese invasion it soon became obvious that the communists had no desire to accept any settlement other than one dictated by military victory. On April 6 American aircraft were once again sent north of the demilitarized zone to carry out a coordinated interdiction campaign against North Vietnam's logistics network. Two of the targets were the Thanh Hoa and the Paul Doumer bridges. Since the bombing halt in 1968 they had been repaired, and the railroad lines crossing the bridges were being fully utilized.

A new family of "smart bombs" was now available, consisting of Electro-Optical Guided Bombs and Laser Guided Bombs in the 2,000-pound and 3,000-pound classes. The first was a contrast weapon, similar to the Walleye used by the Navy in 1967. The optically guided system was a 2,000-pound bomb that displayed a picture of

R.G. Smith's painting of a Skywarrior refueling a Skyhawk depicts a scene
repeated thousands of times. (Courtesy of McDonnell Douglas.)

what it was viewing on a scope in the attack aircraft. The pilot pointed his aircraft
and weapon at the target to allow his Weapons System Operator (WSO) in the rear
seat of his F-4 to find the target on the scope, refine the contrast aiming point and
designate the target to the weapon. Once this was accomplished the pilot released
the bomb and raced away while the bomb guided itself toward the designated aiming point. Target weather and cloud cover posed problems, but if the weapon could
see the target when it was released it would usually impact the aiming point.

The laser-guided bomb used a laser sensor in the nose of a 2,000-pound or 3,000-
pound bomb to guide itself toward a target illuminated with low-powered laser energy. This was provided by a pod beneath the fighter. The pod contained an optical
viewing system with laser-emitting capability. Both were operated in the rear seat
by the WSO. The pilot merely pointed his aircraft toward the target while his WSO
optically located the precise target's aiming point and illuminated it with his laser
equipment. The pilot then released his bombs and headed away at high speed while
the bomb guided itself to the target. This system had a distinct advantage over the
optically guided system, because more than one aircraft could drop bombs at the
same target, with all weapons using the same illumination point to guide on.

Both systems provided less crew exposure to ground fire and greater accuracy
than earlier weapons. But the laser-guided bomb had to be continuously illuminated to be effective. If clouds obstructed the view of the illuminating pod, the bomb
became unguided and missed its target.

The Air Force's 8th Tactical Fighter Wing was given the new bombs. The wing, commanded by Colonel Carl S. Miller, earned the nickname of "bridge busters" because they destroyed 106 bridges, including the Paul Doumer and the Dragon's Jaw or Thanh Hoa Bridge, with their new guided bombs from April 6 to June 30, 1972.

In Paris, meanwhile, North Vietnam's negotiators refused to agree to substantive negotiations. Their military leaders had used the four-year bombing halt to develop their military potential. Now North Vietnam's air force extended radar control down its panhandle and expanded its fighter inventory to 93 MiG-21s, 33 MiG-19s, and 120 MiG-15s/17s, for a total of 246 aircraft. Both missile and antiaircraft sites were now established in the South, and increasingly they fired at American reconnaissance aircraft and extended their operations in Laos. Between November 1, 1971, and January 1972, there were fifty-seven MiG incursions into the panhandle of Laos, where American airmen continued their attacks against North Vietnam's men and supplies.

Now the North Vietnamese were engaged in an all-out attack on South Vietnam. They moved first into Quang Tri Province, then followed up this invasion with attacks from Laos and Cambodia into Kontum and Binh Long provinces in South Vietnam. General Giap confidently believed that these division-level assaults with heavy armor would overwhelm South Vietnam's forces now that most American ground troops had withdrawn, permitting North Vietnam to install a coalition government in South Vietnam.

American airmen were authorized to resume attacks, nicknamed "Freedom Train," as far north as twenty degrees latitude. This operation was expanded on May 8, when President Nixon approved the aerial mining of North Vietnam's ports and a resumption of air and naval strikes against military targets in North Vietnam, with the exception of those still listed as off-limits. The president announced that the United States would halt all offensive operations only after the communists agreed to release all American prisoners of war and to accept an internationally supervised cease-fire. The Joint Chiefs of Staff gave the Seventh Air Force responsibility for attacking prevalidated targets in Route Packages 5 and 6A (See map in Chapter 21), where the communists had concentrated their heaviest defenses.

"**O**peration Pocket Money," the mining campaign against principal North Vietnamese ports, was launched on May 9, 1972, and a total of thirty-six MK-2 mines were laid in the outer approaches to Haiphong Harbor. They were set for seventy-two-hour arming delays to permit the departure of merchant ships or to change the destination of inbound ships. This procedure was personally announced by President Nixon in Washington. These mines played a significant role in bringing about an end to the American involvement, by hampering North Vietnam's ability to receive vitally needed war weapons.

During the five and a half months of "Linebacker I," the Navy contributed more than sixty percent of the total sorties to the north. Tactical air operations were most intense during July through September, when a quarter of the 12,865 sorties were flown.

Nine ships at Haiphong took advantage of the grace period to leave the port after it was mined, but twenty-seven others remained. Soviet bloc ships chose to divert other ships from Haiphong and thereby avoid a direct confrontation with the United States over its mines.

On May 18, the scope of the air war changed when the Uong Bi electric power plant near Haiphong was bombed. This marked the beginning of strikes against targets previously off-limits, and also included shipyards and the Haiphong cement plant. Navy pilots shot down sixteen enemy MiGs during May, while the Navy lost six planes.

With the *Saratoga* joining the other five carriers on Yankee Station during May, carrier strength reached the highest number attained since the war began.

By the end of the month, the South Vietnamese army was still holding and regrouping, but it was obvious the forward thrust of the North Vietnamese army had been halted.

Commander Sam C. Flynn, executive officer of VF-31, and Lieutenant William H. John, his radar intercept officer, took off from the *Saratoga* on June 21 in their F-4 Phantom to help protect a strike force of A-6 Intruders and A-7 Corsair IIs whose mission was to bomb targets between Haiphong and Hanoi.

Forming a Combat Air Patrol to protect the strike aircraft, with Flynn as their leader, they positioned themselves between the strike force and the airfield from which MiGs were likely to come.

After the control ship, the USS *Long Beach,* reported that MiGs were airborne, it wasn't long before North Vietnamese ground crews were firing SAMs at the Navy planes. The *Long Beach* vectored them north to attack the MiGs fifteen to twenty miles away.

When the two F-4s spotted the MiGs at a higher altitude, they attacked immediately, and in an old-fashioned dogfight reminiscent of World War II, the MiGs rolled in from above, catching the Phantoms at a disadvantage. Flynn moved in behind the lead MiG's wingman, following him relentlessly through a series of loops, turns, and high-speed maneuvers. After John locked on the MiG, they attempted to fire a Sparrow missile but it malfunctioned.

Meanwhile, the lead MiG had maneuvered behind Flynn's wingman and was about to fire an "Atoll" missile.

John, once he had the advantage over the North Vietnamese wingman's plane, switched his missile system to the Sidewinder mode. John momentarily took his eyes off his radar display and looked at the MiG and their wingman's plane ahead of them. The MiG was at the wingman's seven o'clock position and firing. By radio John warned the wingman that he had a MiG on his tail that was firing at him. He called Flynn and recommended they switch off the MiG they were chasing, reminding him their wingman was in trouble. Flynn broke off immediately to protect his wingman, who had already evaded two missiles before Flynn's Phantom could rush to his rescue.

Flynn called his wingman, "Keep pulling on the stick and keep it in afterburner!"

Now Flynn was in a good position, but his wingman was too close for him to fire at the MiG. He was fearful his missile might just as easily lock on his wingman's plane as on the MiG. Flynn fired a Sidewinder off to the right, hoping the MiG pilot would turn his attention to him.

Flynn knew the MiG could outturn the Phantom, and that he should not remain in the Sidewinder's envelope. He was about forty-five degrees off the MiG's tail and

An A-6A Intruder of VA 75. (Courtesy of U.S. Navy.)

would soon draw away from his own missile's envelope. Then the MiG fired a third missile at his wingman. At that instant the wingman pulled back hard, causing the MiG to wallow and permitting Flynn's F-4 to fall in right behind it.

Once Flynn was in position he fired two more Sidewinders. The first missed but the second went right up the MiG's tailpipe and blew off part of his empennage (tail assembly). The North Vietnamese plane went into a flat spin. Its pilot tried to recover a couple of times but he was forced to eject at 1,000 feet. The entire fight lasted only ninety seconds, and both Navy F-4s survived the encounter.

The quality of MiG-21 pilots had improved to the point where they became deadly adversaries. They would take off, cruise at low altitude, pick up a lot of speed and energy, move in on the tail of an American plane, fire their missiles, and then streak home. Although American fighters in the first half of 1972 enjoyed a kill ratio of two to one, by summer their losses were even.

Two Navy fliers, Lieutenants Randy Cunningham and William Driscoll, destroyed three MiGs on May 10, bringing their score to five, and making them the first aces of the war.

After the Air Force and Navy developed a new command and control capability, and refined their tactics, the kill ratio changed to four to one from August 1 to October 15.

★ ★ ★

While negotiations in Paris dragged on—with the North Vietnamese raising technical objections to propositions previously approved—a decision was overdue by American negotiators about how best to handle the situation. Warnings that American air strikes would resume in full force against North Vietnam were not taken seriously by the communists. The northeast monsoon season was approaching, and communist leaders believed it would seriously hamper American tactical fighter attacks.

Dr. Henry Kissinger, Nixon's assistant for National Security Affairs, was confident by mid-October that peace arrangements would shortly be accepted by the communists. The president on October 22 called for an end to Linebacker I's bombing campaign as a gesture of goodwill to get the peace talks moving again, although close air support operations were continued in South Vietnam on a lessened basis.

Then the government in Saigon insisted that they would not agree to an armistice that left tens of thousands of North Vietnamese soldiers in their country. Dr. Kissinger told the press that it was decided to bring home—really to both Vietnamese parties—"that the continuation of the war had its price."

When the North Vietnamese stiffened their demands for a cease-fire, the president ordered that Linebacker II be launched on December 18. This intensive Air Force and Navy day-and-night campaign for the first time in the war attacked targets that had previously been immune to attack. Electric power plants, broadcast stations, railways and railyards, port and storage facilities, and airfields around Hanoi and Haiphong were attacked through December 29, with the exception of Christmas Day. For the first time the Strategic Air Command used its B-52s to strike strategic targets in the heavily defended Hanoi and Haiphong areas.

Linebacker II's intensity disrupted North Vietnam's air defenses, and this time there was no respite until the job was done properly and her most vital industries were either destroyed or seriously damaged. SAM direction radars were jammed successfully, but the North Vietnamese fired nearly a thousand SA-2s at the big bombers, knocking down fifteen of them.

President Nixon terminated Linebacker II on December 29, 1972. He announced that Dr. Kissinger would resume negotiations with the North Vietnamese in Paris on January 8, 1973.

The effect of Linebacker II hastened the conclusion of peace negotiations. Dr. Kissinger put it succinctly when he said, "There was a deadlock in the middle of December...there was a rapid movement when negotiations resumed...on 8 January."

On January 23, Kissinger and North Vietnam's Le Duc Tho initialed the agreement, which provided for a supervised cease-fire, return of all American prisoners of war, and political self-determination for the people of South Vietnam.

Admiral Thomas H. Moorer, chairman of the Joint Chiefs of Staff, described Linebacker II as the "catalyst for the negotiations.... Airpower, given its day in court after almost a decade of frustration, confirmed its effectiveness as an instrument of national power...in just nine and a half flying days."

After the cease-fire was announced on January 24, 1973, the carriers *Oriskany, America, Enterprise,* and *Ranger* cancelled all combat sorties into North and South Vietnam. During the war the Navy had lost 529 fixed-wing aircraft and 270 helicopters to enemy action. The Marine Corps lost 193 fixed-wing aircraft and 270 helicopters. Navy and Marine pilots destroyed fifty-six MiGs.

Skyhawks head away from the carrier *America* off Vietnam. Painting by R.G. Smith. (Courtesy of McDonnell Douglas.)

"Operation Homecoming," the repatriation of American prisoners of war, began on January 27 and continued until April 1 as 591 prisoners were released. Of this number, 145 were Navy personnel, all but one of whom were Navy aviators.

Aircraft from the *Enterprise* and the *Ranger* flew eighty-one combat sorties on the first day of the Vietnamese cease-fire against lines of communication in Laos. The corridor for overflights was between Hue and Da Nang in South Vietnam. These combat sorties were flown in support of the Laotian government—which had requested this assistance—and it had no relationship to the cease-fire in Vietnam. For these strikes, the carriers moved Yankee Station southward to operate off South Vietnam.

Airborne minesweeping began on February 27 off Haiphong Harbor. This was a "first" in mine warfare as airborne minesweeping was conducted by CH-53 Sea Stallions from HM-12. All operations were halted and the minesweeping task force moved to sea after two sweeps while President Nixon demanded an explanation for North Vietnam's action in stopping the release of war prisoners. He called for an explanation "on a most urgent basis" before he would order resumption of the clearing operations.

When the North Vietnamese resumed their release of prisoners the withdrawal of American troops from South Vietnam was also resumed, and the minesweeping force returned to its position off Haiphong. The harbor was quickly cleared and reopened, after being closed for ten months since May 1972. The *America* was now ordered to leave the Far East as part of an initial move to reduce the number of carriers serving in Southeast Asia from six to three by mid-June 1973.

The remaining U.S. combat forces left South Vietnam on March 29 and the Military Assistance Command, Vietnam, was disbanded, officially ending American military involvement in South Vietnam.

The last phase of "Operation Homecoming" was concluded when the final group of 148 American prisoners of war was released by the North Vietnamese.

North Vietnam's air battles were unique in American history because U.S. Air Force, Navy, and Marine pilots operated under stringent rules of engagement. With few exceptions, MiG bases could not be struck. The rules generally forbade bombing or strafing of military and industrial targets in and around the enemy's heartland, encompassing the capital of Hanoi and the port city of Haiphong. These restrictions gave the North Vietnamese a substantial military advantage. Free from American attack, and helped by its Soviet and Chinese allies, the enemy was able to construct one of the most formidable antiaircraft defenses the world has ever seen. It included MiG bases, surface-to-air missile (SAM) batteries, heavy concentrations of antiaircraft artillery units, and an array of early warning radar systems. These elements sought to defeat the American bombing campaign against North Vietnam's lines of communication and its military and industrial bases. The primary mission of American fighter pilots was to prevent the North Vietnamese MiGs from interfering with American bombing operations.

The war in Southeast Asia did not provide the opportunity to amass the high victory scores that were common in World War II and, to a lesser extent, in Korea. One reason was that enemy pilots refused to engage American aircraft whenever they were at a disadvantage. Their aircraft were furnished by the Soviet Union and the People's Republic of China, but in such limited quantities that the United States early in the conflict attained air superiority. Instead, the North Vietnamese relied heavily upon SAM missiles, conventional antiaircraft guns, and small-arms fire. Another important factor, which limited the number of aerial victories, was the four-year standdown of American air operations over North Vietnam. With few exceptions, it began in 1968 and lasted until the spring of 1972, when North Vietnam launched a massive ground assault on South Vietnam. Many airmen completed their one-year tours without once engaging North Vietnamese pilots in the air.

Part VIII

The Silent War

23

"Break Off MAD Cloverleaf"

While the war in Southeast Asia was being fought, another kind of conflict was under way against the Soviet Union's submarine fleet. A silent war, without the firing of weapons, it was conducted with deadly seriousness. From day to day no one knew whether it would erupt into World War III.

Lieutenant Richard A. Shultz settled his six-foot three-inch frame more comfortably in his chair. He had awakened at 3:30 A.M., and despite several cups of coffee and a big breakfast, he was only now slowly coming to life. He looked up with only casual interest as Duty Officer Lieutenant Stephen Larrabee walked out on the platform from the teletype room and began to shuffle papers at the podium.

Shultz pushed up his glasses and rubbed his tired eyes. He had already flown almost ninety hours during the month, but it was not the past month's flying that had drawn deep lines on his young face. He was fatigued by the years of flying antisubmarine warfare patrols month after month during the late 1960s from their base at Barbers Point on the southeast tip of Oahu in the Hawaiian Islands.

Waiting for the crew members on the flight line was their P-3 Orion, the most sophisticated submarine hunter the world had ever known. Earlier antisubmarine efforts had been crude by comparison. As a result, two world wars had been almost lost shortly after they began because people and nations of the free world had underestimated the dire menace of submarine wolfpacks and the need to keep them exposed and controlled in a virtually limitless battleground of the 140 million square miles of ocean on the earth's surface.

The duty officer cleared his throat. "Message 040005 from Commander Fleet Air Wing Pacific authorizes patrol of area X-ray Yankee. Time on target is 1000 Zulu. Duration: eight hours." The crew members made notes, while looking at their maps, and mentally calculated the time it would take to get on station.

Shultz and the others were now alert. The words of the briefing officer were not only vital to the success of the day's mission, but important to their survival. One never knew when a quiet peacetime flight could turn into the real thing.

"Here is the communications brief." Larrabee said. While the duty officer gave details of the day's code—using a highly secret book that, if lost, caused all such

The P-3C Orion. Note the sonobuoy tubes. (Courtesy of Lockheed.)

books throughout the United States Navy to be recoded—Shultz glanced at the serious face of his pilot, with whom he had flown for fourteen months.

Lieutenant Dewey Barnes, patrol plane commander, had the usual pipe stuck in his mouth. A career naval officer, he had served on destroyers as an enlisted man before he got his wings. Born in Helena, Montana, and the father of two children, Barnes was solidly built, with brown hair and eyes. Shultz felt eternally grateful that they had such an outstanding pilot at the controls of their P-3 Orion.

Shultz's mind snapped to attention as Larrabee said, "Unidentified subsurface target reported at 24 degrees, 32 minutes north and 142 degrees, 54 minutes west." Shultz glanced quickly at his map. The target was east of Barbers Point toward the U.S. mainland.

"Last known position heading 070 degrees true," Larrabee said, "SOA six knots."

Shultz studied the map intently. Inasmuch as the "speed of advance" of the unidentified sub was only six knots, he knew it was making evasive maneuvers as it proceeded eastward.

"Here's your patrol area," Larrabee said, pointing to a section due east of Hawaii on the large map of the Pacific at the head of the war room. Shultz judged the area to be about 150,000 square miles.

Plexiglass covered the map, on which symbols of ships and submarines were mounted, showing their last known positions. While the briefing was in progress, antisubmarine patrol planes were covering their own areas on a round-the-clock basis all over the world. Meanwhile, ships of the Pacific Fleet had their own protective screen of destroyers and antisubmarine carriers to protect them at all times.

While war raged in Vietnam, throughout the Pacific and Atlantic Oceans lonely airborne vigils were kept by hundreds of antisubmarine warfare (ASW) aircraft from many nations. Actually, this little-known war was an arena in which American forces daily faced Soviet sea forces frankly in a cat-and-mouse game of deadly intensity.

Few Americans comprehended how large and formidable the Soviet submarine threat was at the time. For example, the Soviet Union in 1973 was capable of launching more than 400 nuclear missiles at targets on the U.S. mainland—from the Atlantic and the Pacific oceans.

Antisubmarine missions were designed by the U.S. Navy for one reason only in peacetime—to deny the Soviets undetected access to its waters. Missions such as this one in the late 1960s were to detect all unidentified "objects" in the waters around the periphery of the United States.

At first thought, the task seemed insurmountable. Nuclear submarines, in particular, can travel faster underwater than on the surface because they are primarily designed for underwater passage. With their revolutionary hull designs, propulsion systems, and nuclear-tipped missiles, they had become the greatest single threat faced by Western civilization. Land-based antisubmarine aircraft bore the brunt of the task of searching for stealthy intruders far out to sea.

Shultz and his fellow crew members had long since taken the threat seriously—otherwise they would not be sitting in this austere war room awaiting another bone-tiring, eyeball-searing mission over the Pacific.

As Larrabee continued to explain their patrol area, Lieutenant Tom Roskens, Crew 8's navigator, made rapid notes on a pad. A former guard on a Nebraska college football team, everything about him seemed large, including his big but efficient hands. For now all Roskens needed were the coordinates of the search area—roughly a huge square in which the aircraft would fly back and forth on prescribed courses that he would soon lay out. High above listening devices beneath the sea, Roskens would issue the instructions for each course as they cruised several miles above the water seeking suspicious sounds from the ocean far below.

Shultz glanced at the insignia of VP-22, and although he had seen it countless times, he felt a stirring of pride. He knew, even though the majority of Americans cared little or didn't even know about their existence, that they were performing missions of the utmost importance for the preservation of freedom. He and thousands like him all over the world knew in their hearts that someday the work they were doing might be called the conflict that was fought instead of World War III.

One of the accomplishments Shultz was particularly proud to have participated in was the successful detection, tracking, and simulated torpedo kill of an exercise submarine using only electronic equipment. In other words, the submarine was never seen during the exercise. Only one or two crews of the squadron's twelve members could boast of such an accomplishment. He knew, as he half-listened to the radio briefing, that their efforts had been unusually successful because they had worked so long as a team.

Shultz's thoughts were interrupted by Larrabee. "Op Con has given us these final details about sightings in the area." Operations Control was responsible for gathering around-the-clock information from a variety of sources so that movements of ships in its area could be determined. Larrabee discussed these secret details at great length. When the briefing was over, they filed out without comment and headed for the flight line.

Shultz, from force of habit, carefully checked the survival equipment in his "Mae West." He knew each item could mean life or death in the event of a crash at sea. The shark repellant, the individual flares, and the strobe light were all in their proper places. He fingered the police whistle, chuckling even now at the memory of his amazement when he found it in his kit during his first assignment. Thankfully, he had never had to use it, but he knew from others how valuable it could be. It was uncanny how far the sound could be heard over the ocean.

He glanced at his personal survival kit to make sure it was strapped securely to his flight harness, for it contained life-sustaining saltwater conversion tablets and medicinal equipment.

At the aircraft Shultz saw their copilot, Lieutenant Jerry Turell, boarding their P-3. He had not attended the briefing, but instead had been in Operations to get the latest weather and to file their flight plan.

The stubby gray and white Orion was not a pretty airplane, but it looked formidable squatting on the ramp. Its four turboprop engines permitted it to loiter over a wide area at speeds as low as 180 miles an hour, or dash during a combat run at more than 450 miles an hour.

Shultz had always admired the Orion's rugged lines despite its many bulges for radar and sensing equipment and the odd-looking "stinger" protruding from its tail that helped to locate submerged submarines by detecting deviations in the earth's magnetic field.

Shultz walked through the cabin to the tactical compartment, where, as tactical coordinator, he was the key man of a well-coordinated team. He sat between Mike Miller, whose operations were so secret that no name could be given to the work he did, and navigator Lieutenant Tom Roskens.

He noticed with a wry grin that Mike's pipe was already in his mouth, ready to light as soon as they were airborne. They always accused him of smoking tobacco that smelled as bad as the paper in front of him, which was scorched as an electric stylus made tracings on it.

As Shultz settled back in the high, padded seat, he glanced professionally at the maze of switches, control buttons, and instruments clustered around the master monitor display tube. All were as familiar as the back of his hand. Here, in front of him during the flight, all tactical and navigational data from the other systems were processed and integrated at his station, then displayed on the ten-inch cathode ray tube in front of him. He would know at all times the aircraft's complete tactical situation and flight path.

Glancing down to his right, beyond the navigator's station, he nodded to "Julie" operator Willie Williamson. This quiet, blond man's job became really important during the final minutes of submarine interception because his complicated equipment pinpointed underwater targets.

At the right of the tactical compartment, radar/MAD operator Charles Lamb checked his equipment, particularly the Magnetic Anomaly Detector (MAD), which located underwater vessels by electronically reporting changes in the earth's magnetic field caused by the metal hull of a submarine. A heavyset man, Lamb always wore a jacket—even on the hottest days when the others were in shirtsleeves—and he seemed to wear a perpetual grin.

While flight engineer Robbie Robertson took his seat behind the pilot, radio operator George Tichnor adjusted his equipment just aft of the pilot's compartment, and ordinanceman Dean Ramsay and assistant flight engineer Doug Gillette seated themselves at observer stations fore and aft on the right side of the aircraft.

In the pilot's compartment Dewey Barnes signaled for the start of number two engine. The big paddle-like prop turned slowly at first, then with a muffled roar from the engine, it became a blur.

Barnes received permission to taxi from the control tower and headed for the end of the runway. There he pushed all four throttles forward for one final engine check. After clearance to take off, he headed the Orion down the long runway. He could feel the tremendous surge of power as the aircraft gained speed while the engines revved up to their full power of nearly 20,000 equivalent shaft horsepower. A glance at the instrument panel gave him instant reassurance that all was well. Crossing the shoreline he marveled again at the spectacular beauty of the coral reef as the bright colors showed through the crystal clear water.

Carefully, following the prescribed route and altitude to clear the tightly controlled defense area, Barnes swung the aircraft around Diamond Head with only a passing look at the familiar landmark.

Each crew member checked his equipment again as the Orion headed for an altitude of 20,000 feet. At the tactical coordinator's (TACCO) station, Shultz kept himself busy checking his own equipment. The boredom was most acute en route to the search area, and he liked to keep busy. At two minutes to ten, they reached their control area.

"TACCO to ordnance. On station. Prepare to drop sonobuoy channel 10."

"Roger. Channel 10," Ramsey said.

He stood in the rear of the aircraft with a sonobuoy (sonar-powered buoy) poised over a pressurized "chute" waiting for the signal to fire. Initially, sonobuoys were dropped in a controlled approach so their sensitive listening devices could transmit underwater sounds back to the aircraft.

"TACCO to ordnance. Drop Channel 10—Now!"

Ramsey fired his sonobuoy down the single chute, which was used only when they were pressurized. At lower altitudes, where pressurization was not a problem, up to nine sonobuoys could be dropped quickly through multiple tubes.

"Channel 10—Away!"

Then other sonobuoys were dropped miles apart so that tens of thousands of square miles could come under sonar surveillance. As tactical coordinator, Shultz was responsible for establishing a pattern based on the very latest intelligence and oceanographic reports.

The expendable sonobuoys were laid to search for a snorkeling long-range submarine. This factor in itself posed a problem because of temperature variations of the water. Normally there is a predicted drop in temperature for each foot of ocean depth. Sometimes, however, a variation in temperature occurs between 25 to 200 feet that reduces the efficiency of sound transmission. It is most important to determine where this "thermocline" is because it acts like a brick wall for accurate transmission of underwater sounds.

Pilot Dewey Barnes looked out over the vast stretch of ocean. There wasn't a ship in sight, and only the breaking of light swells disturbed the surface for miles around them. His job now was a quiet one of watching his instruments while the tactical compartment sought shadowy objects deep in the ocean. Occasionally he had seen submarines traveling just below the surface with only their snorkels creating feather-like wakes to identify them. Often, however, the "feather" was impossible to spot. Today there was nothing to disturb the vast emptiness of ocean clear to the horizon; even the sonobuoys had nothing unusual to report.

Back and forth they flew on precisely defined courses, with Tom Roskens giving Barnes new headings to make ninety-degree turns at each end of their patrol area.

George Tichnor monitored his radios to pick up any late word from "Op Con" in Oahu. Otherwise he obeyed radio silence so submarines in the area would have no inkling they were waiting far above, ready to pounce if necessary.

After several hours of search, starting within the 60,000-square-mile area where the unidentified submarine had been reported, the Orion tightened its search to a five-square mile area.

Although no positive contact was made, a few suspicious sounds were coming from buoy 14, so there was a visible rise of emotion in the tactical compartment as they prepared to drop four more sonobuoys around the suspicious 14.

"TACCO to navigator. Give pilot heading." The plane turned gracefully to a more northerly heading.

"Navigator to TACCO. On top. First drop position."

Ramsey released sonobuoy 16 and reported.

"Pilot to navigator. Heading to second drop."

Mike Miller dialed Channel 16 and reported to Shultz. "Channel 16 is up."

Shultz nodded, knowing that the first sonobuoy was working properly and was being monitored. It took twenty minutes of precise flying to lay the new field.

Shortly thereafter, Miller's body tensed. One or more buoys were signaling. He swiftly checked the other buoys. Now he was sure. There were readings coming from 16, 14, 12, and 13. He quickly alerted Shultz.

Barnes called, "Maneuver flaps." After a pause, "Gear."

The 135,000-pound Orion shuddered as flaps and landing gear were extended to slow the aircraft for rapid descent. The pilot's altitude meter swung rapidly as the Orion lost altitude at the rate of 5,000 feet per minute.

There was a new alertness throughout the aircraft—not apprehension, just a controlled excitement.

Barnes leveled off from their sickening plunge only 500 feet above the ocean. Now they had to pinpoint the location of the submarine by dropping more buoys around its last known position. In this manner, through triangulation, the sub's position would become better known.

The years of practice were now apparent as they worked swiftly and surely while Barnes flew a course through each group of buoys, trying to cross the predicted position of the submarine beneath the surface.

The submarine still had no idea it was being so carefully watched as the P-3 spotted it by instrumentation as it moved just beneath the surface. For the tactical compartment it was like shooting a series of three-star navigation fixes: where the lines crossed was the submarine's position in the ocean.

It was time for a final move to locate the submarine precisely. Dean Ramsey dropped buoys with explosive charges around the last known position of the submarine. These quickly generated their own sound sources, which reflected off the metal hull of the submarine and gave new lines of position that accurately determined the sub's position.

"Julie" operator Willie Williamson could clearly hear the odd "chew" sound as each sonobuoy's charge exploded, and then the echoing "chew" as the sound bounced off the submarine.

Faces were tense but hands were sure as they narrowed the range to 2,500 yards. Mike Miller listened to the "pinging" of other buoys' sonars every ten seconds. He knew they were very close to the submarine as Barnes swung the Orion in three tight cloverleafs, using the position of the submarine as the center.

With the submarine identified as friendly, the Orion pulls away.
(Courtesy of Lockheed.)

"TACCO to ordnance. Load another pinger."

Shultz's blue eyes were red-rimmed with excitement. He knew how the submarine crew must feel, because he had spent four and a half years in submarines before he had joined Naval Aviation.

They closed in for the "kill" and the crew was in a fever of excitement. Barnes swung the P-3 in ever tighter cloverleafs until they were down to 500 yards from the predicted position of the submarine. It was evident that it was twisting frantically at high speed to shake them off. Once the explosions were made by the sonobuoys' charges in the water around it, the submarine's crew knew they had been pinpointed.

Barnes headed the Orion right across where the target should be.

"Madman! Madman!"

Lamb's voice had a sharp edge to it—not of fear, but of excitement—as his instruments detected the submarine directly beneath the aircraft.

A smoke marker was dropped, and Barnes flew another cloverleaf.

"Madman! Madman!"

They had crossed over the submarine, so another smoke marker was dropped. Barnes swung the aircraft over on one wing and again they were directly over the submarine. He glanced at the selector switch for depth bombs and opened the bomb bay doors, heading in for the final run.

"TACCO to pilot. Friendly sub. Discontinue contact. Break off MAD cloverleaf."

They sat back and grinned at one another, knowing with satisfaction that the entire operation had been faultless. There were no bombs, torpedoes, depth bombs, or special nuclear weapons aboard, because the United States was still at peace with the Soviet Union. During wartime the submarine would certainly have been destroyed. Every move they had made was an exact duplicate of those necessary in wartime, except for the release of submarine-destroying weapons.

Part IX

Middle East Crisis

24

Carriers Respond to Hot Spots

The nation had no sooner ended its long involvement in Vietnam than it faced a growing crisis in the Middle East. Fighting broke out between Lebanese army units and Palestinian guerrillas in Lebanon. Martial law was proclaimed on May 3, 1973, and a cease-fire agreement between Lebanese and Palestinian negotiators stabilized the situation six days later. Among U.S. forces in the Mediterranean, the carriers *Kennedy* and *Forrestal* were alerted for any possible contingency.

On May 25 Skylab II, carrying a three-man, all-Navy crew of Captain Charles Conrad, Commander Joseph Kerwin, and Commander Paul Weitz rendezvoused with the earth-orbiting Skylab I workshop. Among the crew's first tasks was repair of Skylab's meteoric shield and solar array system. It had been damaged during the launch. The crew boarded the workshop, made repairs, conducted medical experiments, and studied solar astronomy and earth resources for twenty-eight days before returning to earth on June 22.

After intensive bombing for more than six months, the United States ended its combat in Cambodia on August 15, 1973. Congress had voted to end the funding for the operation on June 30. Aircraft from the carriers *Ranger* and *Oriskany* had flown combat sorties during February. After March they conducted carrier air patrols and intelligence surveillance.

After October 6, when the fourth and largest Arab–Israeli War broke out, the situation worsened in the Middle East, so the United States government on October 27

ordered a worldwide "precautionary alert" of its military forces. For a time it was feared that the Soviet Union might interfere with its armed forces. The following day three American aircraft carriers were positioned off Crete along with two amphibious assault carriers. But the war ended without American involvement when a cease-fire was signed between Egypt and Israel on November 11 under American auspices.

Two days later the Defense Department announced that a naval task force, centered around the *Hancock* and her escort ships, had been ordered to the Indian Ocean. This was prompted by the Middle East war and the consequent oil embargo initiated by Arab nations. It was the start of a series of deployments by four task groups to the Indian Ocean in 1974 focusing on such areas as the entrance to the Persian Gulf and the Red Sea.

On January 18, 1974, the Secretary of the Navy officially named the Navy's fourth nuclear-powered carrier the USS *Carl Vinson*. It was chosen in honor of Vinson's contribution to national defense during his fifty years in the House of Representatives.

On February 2, Lieutenant Barbara Ann Allen became the Navy's first female aviator when she received her gold wings in a ceremony at Corpus Christi. This was a major milestone for the Navy, whose bureaucrats had long resisted such a move.

The carriers *Midway, Coral Sea, Hancock, Enterprise,* and *Okinawa* were ordered to Vietnamese waters on April 19, 1975, for possible evacuation of Americans from South Vietnam. The North Vietnamese had ignored the 1973 cease-fire agreement and invaded South Vietnam. This was a violation of the 1973 Paris Peace Accords, and it soon became evident that the massive invasion would take over all of South Vietnam.

In mid-April, Lieutenant R.A. "Tony" Macdonald, assigned to Attack Squadron 93 on board the *Midway,* joined some of his squadron's pilots in flying their A-7s to Cubi Point in the Philippines. The carrier's captain ordered the move to reduce the number of fixed-wing aircraft to make room for Marine helicopters, which might be needed for the evacuation of Saigon.

On the 29th, with Saigon about to fall to the communists, Navy and Marine Corps helicopters from the Seventh Fleet started to evacuate American citizens from the capital. The military situation around Saigon and its Tan Son Nhut airport made evacuation by helicopter the only way out. President Gerald Ford ordered the evacuation when Viet Cong shelling forced the suspension of transport aircraft flights from Tan Son Nhut. With fighter cover provided by carriers, helicopters landed on Saigon's rooftops—including the U.S. Embassy and at Tan Son Nhut—to evacuate the Americans. The airport became the main helicopter landing zone because it was defended by Marines from the 9th Amphibious Brigade flown in for that purpose. All but a handful of the 900 Americans in Saigon were evacuated.

When Macdonald returned to the *Midway* on April 29 the evacuation was in full swing. Upon their arrival on board the *Midway,* Marine helicopter crews had been briefed about evacuation sites and told to stand by for immediate launch. By April 29th, most helicopter crews on board the carrier had spent the last twenty-four

A deck full of salvaged planes. The USS *Midway* returns to Guam in May of 1975 with aircraft flown on or placed on board by cranes following the fall of South Vietnam to the communists. (Courtesy of Tony Macdonald.)

hours strapped in their aircraft waiting for the order to take off and evacuate key personnel from Saigon. Once the order was given, the helicopters began a series of flights to and from these designated sites. Navy and Marine helicopter crews actually spent more hours in the air than they did sleeping.

Simultaneously, South Vietnam's air force helicopters began to arrive aboard the *Midway* unannounced with evacuees. Among the first of the arrivals was ex-Premier Nguyen Cao Ky. Carrier Air Traffic Control on the *Midway* picked up his pilot's distress call and vectored his helicopter to the carrier. After an official greeting by Seventh Fleet flag and staff officers, Ky broke down and wept. Undoubtedly his emotion was triggered by his earlier public announcement that "any Vietnamese who deserts his country is a traitor."

Macdonald watched with disbelief as helicopters continued to land, depositing hundreds of South Vietnamese. Some copters arrived with people clinging to the outside. The carrier's deck soon became so overcrowded with American helicopters and South Vietnamese aircraft that the captain ordered some Vietnamese aircraft pushed over the side so that more helicopters could land on the flight deck.

The *Midway*'s police form, the Master-At-Arms, was assigned to categorize, store, and safeguard valuables that the arriving evacuees had brought with them. It was an enormous task, and soon several fifty-gallon drums were filled with tagged items, such as precious jewels, gold bars, and stacks of large-denomination bills from various countries. The first arrivals, both Americans and Vietnamese, were the main contributors. They were well dressed, only slightly in disarray, and usually carried expensive luggage. Later arrivals, mostly Vietnamese, were visibly shaken, disheveled, and carried very little of value.

Many of the evacuees were neither dependents of American citizens nor themselves American citizens. But due to their close ties to the American government during the war, abandoning them now in Saigon was tantamount to a death sentence once they were captured by the communists. If there was room in a helicopter, they were evacuated along with American personnel and their dependents. Other Vietnamese simply forced their way into pickup sites and onto helicopters. Late at night on the 29th, panic prevailed at all sites and American crews had to use brute force to prevent their helicopters from becoming so overloaded that they would be unable to lift off and possibly crash. There had been estimates of the number of evacuees who would have to be accommodated, but the actual figure surpassed all expectations.

During a lull in the arrival of waves of helicopters to and from Saigon earlier that day a lone South Vietnamese observation plane, an O-1E, began circling the *Midway*. At first the carrier was unable to establish contact with the pilot. Then, several small objects were thrown out of the aircraft that either missed the flight deck or rolled off the deck into the sea. Finally, a piece of fruit landed on the dock with a note folded inside. In broken English the pilot's note said that he was a good pilot with seven people on board. He wanted to land, and asked that the captain make room for his tiny plane on the flight deck. The note was passed from enlisted deck crews to the air boss and finally to the ship's captain. He immediately ordered that some helicopters be pushed over the side to clear a landing space. Once the deck was available, the O-1E pilot quickly approached the *Midway*. After several passes the little tail-dragger landed, and due to a closing speed of less than thirty knots, it rolled slowly to a halt abeam of the island. Deck crews frantically threw chocks under the little plane's wheels as the more than twenty-knot wind caused the O-1E to roll backward. After the plane stopped rolling, seven people, adults and children, emerged. As a gesture of goodwill, the crew of the *Midway* "passed the hat," donating more than $7,000 to these escapees to aid them in their search for a better life in the United States. The O-1E was later placed in the Naval Aviation Museum at Pensacola, Florida.

Throughout the night of April 29–30 the helicopter flights continued, with hundreds of American and Vietnamese evacuees arriving each hour. The *Midway*'s kitchens began around-the-clock operations to feed them. The lines to the *Midway*'s mess decks wound through the hangar dock from fore to aft several times as these hungry and frightened people were fed. Macdonald noted that the food barely raised the spirits of the Vietnamese, whose weary and terrified faces reflected their despair after leaving families and relatives to an uncertain fate in Saigon.

Berthing facilities had been at a premium aboard the carrier even before the influx of so many people. The *Midway* had a smaller number of crew than were assigned by later supercarrier standards, totaling approximately 3,500. Now with several thousand evacuees, the ship's population almost doubled. The problem of where to put these people became a monumental headache. Ordinary crew bunkrooms were filled up quickly, forcing many Navy personnel to sleep in noisy repair shops, ready rooms, and passageways. Next, storage rooms were used to house the arrivals, and finally even the "heads" (bathrooms) were converted into temporary living quarters. Families were kept together whenever possible, and the cry of infants—so out of place aboard a Navy warship—added a further note of unreality to the scene.

One final chapter ended the U.S. Navy's participation in the long war in Southeast Asia. The *Coral Sea* participated with other Navy and Air Force units in the recovery of the American merchant ship U.S. *Mayaguez* on May 14. The ship and her thirty-nine crewmen had been seized illegally on May 12 in international waters by a Cambodian gunboat controlled by the Communist Khmer Rouge. Protective air strikes were flown from the carrier against Cambodian naval and air installations as Air Force helicopters with 288 Marines from Battalion Landing Teams 2 and 9 were launched from Utapao, Thailand, to rescue the crew and secure the merchantmen. Eighteen Marines, airmen, and Navy corpsmen were lost in an action that demonstrated ineptitude at various levels, including the Commander of the Pacific Fleet.

The *Oriskany,* last of the *Essex*-class carriers, was decommissioned this year on September 30 and placed in the mothball fleet. It had served in Korea and Vietnam.

The Navy's new F-18 strike fighter was named Hornet on March 1, 1977. This was an unusual action, because the name Hornet had previously only been applied to ships. For replacement of the Phantom II in the early 1980s, and the A-7 Corsair II, the F-18 was sorely needed to maintain the Fleet's operational capabilities.

In the Mediterranean, the *America* and other elements of Task Force 61, with the *Nimitz* standing by, supported the evacuation on July 27 of 160 Americans and 148 other nationals from Beirut, Lebanon. Since the first of the year, the contingency evacuation force involved with the Lebanon civil war crisis had drawn upon the services of six carriers. The ongoing Lebanese civil war continued to generate unrest throughout the region for the remainder of the 1970s.

★ ★ ★

The Navy took delivery of the last A-4 Skyhawk from the McDonnell Douglas Corporation on February 27, 1979, setting a record for the longest production run of any American military aircraft. Built as an attack bomber and as a two-place trainer, the A-4 had been in continuous production for twenty-six years. The final Skyhawk off the production line was an A-4M attack bomber built for the Marine Corps. It was the 2,960th Skyhawk manufactured by the company, and it was delivered to VMA-331.

One naval aviator and fourteen Marines were among the more than sixty Americans taken hostage on November 4, 1979, when the U.S. Embassy in Tehran was seized by a mob of Iranian hoodlums. A spokesman demanded that the United States return the deposed shah to Iran. He was in a New York hospital at the time suffering from incurable cancer.

In response to this new crisis the *Midway* and her escorts were transferred from the Indian Ocean on November 18 to the northern part of the Arabian Sea. The *Kitty Hawk* was directed three days later to join the *Midway* in the same area. Their attack and fighter aircraft then would be in a position to respond to a variety of situations if called upon during the Iranian hostage crisis. This was the first time since World War II that the United States had positioned two carrier task forces in the area.

The Defense Department announced that it was sending a three-ship, nuclear-powered carrier battle group from the Sixth Fleet for deployment to the Indian Ocean on December 31 to relieve the Seventh Fleet's battle group led by *Kitty Hawk*. The Sixth Fleet was composed of the nuclear-powered *Nimitz* and her nuclear-powered escort ships.

The situation in the area worsened on December 24 when 5,000 Russian airborne troops and equipment landed in Kabul, Afghanistan. Despite American protests, the Soviet Union claimed that the government of Afghanistan had sought their assistance. So the commanders of the two task forces in the northern Arabian Sea had an additional concern.

Throughout the 1970s the American people became increasingly aware of their country's critical dependence upon oil from foreign sources. During this time an acute consciousness of the United States' position as a two-ocean nation reemphasized its reliance upon the U.S. Navy to keep the sea-lanes open.

Despite this need for vigilance, the Navy was faced with a declining material inventory, and experienced difficulty in maintaining trained personnel. As the surplus of equipment from the war in Indochina was used up through constant use, money for replenishment became less available. The high inflation rate that beset the world's industrial nations reduced defense budgets and depressed the Navy's purchasing power. Fortunately, research and development funds remained available to keep the Navy moving ahead on an efficient basis.

During 1979 the Navy's carrier forces responded to five crisis situations around the world: The *Constellation* responded to the political situation involving North and South Yemen; the *Saipan* was called upon during the Nicaraguan civil war; the *Nassau* was involved in the response to Russian combat troops in Cuba; the *Kitty Hawk* was summoned to the alert in Korea; and the *Midway* and the *Kitty Hawk* conducted contingency operations during the Iranian hostage crisis.

Navy surface and aviation forces of the Seventh Fleet continued their patrols, and in the Far Pacific they frequently rescued Vietnamese "boat people" following the president's order to do so. During the last six months of 1979, Navy ships picked up more than 800 Vietnamese refugees. Merchant ships, with the aid of Navy P-3 patrol planes, picked up another 1,000.

The *Nimitz* and escorts joined the *Midway* and the *Kitty Hawk* on station in the Arabian Sea on January 22, 1980, as the Iranians continued holding the fifty-three American hostages. The *Kitty Hawk* departed the next day after 64 days on station. The *Eisenhower* relieved the *Nimitz* task force April 16. Carrier operations in connection with the Iranian hostage crisis involved ten tours by eight attack carriers. Two of them served two tours.

On April 24 eight RH-53D Sea Stallions operating from the *Nimitz* in the Arabian Sea took part in a joint task force to rescue the American hostages in Tehran. But the attempt was aborted later at a desert refueling site. One helicopter collided with a C-130 Hercules, resulting in the loss of eight men. All others were evacuated. It was an incredibly botched performance by all involved.

Part X

A Shift in
Soviet Grand Strategy

25

Budget Restraints

When Admiral Sergei G. Gorshkov assumed command of the Soviet Navy in 1956, it was largely a waterborne adjunct to the ground forces. He changed the thinking of the Soviet Union's military high command by insisting upon a truly massive effort to increase the size and potential of the Russian Navy. By the early 1970s it was still not on a par with the U.S. Navy in all categories of combat readiness, but the Soviets were rapidly closing the gap. Such a fundamental change in Soviet strategy posed a threat to the free world that far exceeded the confrontational dangers of the American involvement in Southeast Asia. By 1973 it was apparent that the Soviet Navy was a well-balanced modern force that was equally at home on the high seas and in its coastal waters.

The significant growth of the Soviet nuclear submarine force, and the introduction of a wide variety of antiship cruise missile systems that could be launched at long range from air, surface, and subsurface platforms was of particular concern to military and civilian officials of the U.S. government.

By the mid-1970s the Soviet submarine fleet was the largest in the world, with 340 undersea craft. This number was two and a half times the size of the United States submarine fleet. Most significantly, about 110 of these Soviet submarines were nuclear-powered, and the Soviets were producing three times as many nuclear submarines as the United States.

Vice Admiral H.E. Shear, director of ASW Programs for the U.S. Navy, said in the fall of 1973, "The Soviets are operating their submarines with a high degree of confidence in a total ocean environment. They are no longer tied to the homeland by a logistical umbilical cord and are no longer thought of as a defensive measure against allied sea power projections; rather they operate and make frequent extended patrols throughout the world's oceans.

"The potential challenge posed by Soviet submarines is both strategic and tactical," Admiral Shear said. "On the strategic side, the Soviets have a construction program for 'Yankee Class' submarines which are similar to the United States Polaris-type submarines. They have some 31 in operation already and they are maintaining this type of submarine operation off both coasts of the United States. And, the Soviets are in the process of constructing a follow-on class to the 'Yankee' that will carry extended range ballistic missiles. It is called the 'Delta' and may be described as the Soviets' 'Trident.'"

The admiral said the "Yankee Class" submarine was equipped with a missile of about 1,300-nautical-mile range, corresponding roughly to the first-generation Polaris A-1, which the United States put into service in 1969. "We followed that down the years with the A-2, then the A-3, the current *Poseidon* and then, in the future, the Trident," Shear said.

The admiral said he was certain the Soviets were proceeding along the same developmental lines because they were testing a missile with a 4,000-mile range. "They have, in effect," Shear said, "leap-frogged us by four generations of strategic missiles."

On the tactical side, Admiral Shear said, the Soviets possess an attack submarine force that surpasses the United States in many respects. "They have about 280 anti-shipping submarines," he said, "and 65 are nuclear-powered. Of the 280 attack submarines, 65 are equipped with advanced anti-shipping cruise missiles."

American Navy officials became increasingly disturbed about antiship cruise missiles because they posed a threat to both merchant and naval ships. The ability of Russian submarines to fire these missiles while submerged and attack surface forces from a great distance made all ships of the U.S. Navy vulnerable anywhere in the world.

In the Atlantic Ocean during World War I a small force of German submarines— no more than 200—nearly brought the British Empire to its knees. In World War II, the Germans had only fifty-seven U-boats in service at the start of the war. They almost destroyed the Atlantic Alliance.

Soviet naval strategists learned profound lessons from these experiences in both wars, and the strength and efficiency of their modern nuclear-powered submarines demonstrated how well they had profited from the misfortunes of others.

As Admiral Shear pointed out, "When one considers the relative economy of the submarine, it is evident how the investment of relatively minor resources has yielded large dividends. This is the reason the Soviets find the submarine such an attractive weapon."

"When we add the destructive power of Soviet anti-shipping missile capability to the large number of attack submarines she possesses, we can see the magnitude of the challenge facing us."

Although the problems posed by the Soviet Union's navy in the 1970s were complex and difficult to solve, U.S. Navy officials knew they had to overcome the obstacles. The security and welfare of the United States were at stake, as well as the security of its allies, who have long depended on the American Navy to meet any and all threats on the high seas.

When it became clear that the Soviet Navy was seeking mastery of the seas, it offered a global challenge such as the U.S. Navy has seldom faced. Despite a move toward détente with the United States in 1973, it must be remembered that Soviet leaders had not discarded their long-range goal of world domination.

In pursuit of this aim it became Soviet policy to maintain a nuclear weapons capability sufficient to provide a shield under which their large and growing non-nuclear forces could operate. To achieve that goal, the Soviet Union began to employ political, economic, and military means to advance its national interests in the international arena, using the inherent mobility of seapower as one of the best tools to pursue its objectives.

Fundamental to the employment of seapower is the traditional "Freedom of the Seas." Along with other nations, the Soviet Union can freely utilize the world's oceans.

It can deploy ships almost anywhere without committing an act of war or encroachment upon another nation's sovereignty. Thus their seapower was a useful means to promote or protect their national interests.

In the past, Russian efforts to develop a modern navy were deterred by a number of causes, including economics, purges, war, leadership changes, and the rapid development of technologies which, until the 1970s, they had had difficulty in keeping up with.

But in this decade the Soviet Navy emerged as a versatile and capable force. It was a threat to the United States and its allies because the Russians were capable of applying force far from their own borders.

In addition to their large and expensive research fleets, intelligence fleets, and fishing and merchant fleets, the Russians were building surface units to fire missiles either for defense against air attack or to strike surface naval forces well beyond the range of naval gunfire. They began to build missile patrol boats in significant numbers, and were providing some to their allies.

One of the new types was the *Moskva,* a combined-capability ship with some of the most modern weapons, including a large helicopter platform on its stern. Its major capabilities include antisubmarine warfare and amphibious warfare.

The Soviet Navy's Commander in Chief Gorshkov had long advocated nuclear submarines as the basis of Soviet naval might. There was no reason to doubt that the Soviet submarine had emerged in the 1970s as a weapons system posing the most serious challenge to the supremacy of the U.S. Navy.

In March of 1973 Secretary of the Navy John W. Warner told the Senate Defense Subcommittee of the Committee on Appropriations, "In an era when the perception of naval power will carry great weight at the negotiating tables of the world, the military portion of that power must remain strong. Our economic and political interests dictate we retain a stature such that no adversary will question our ability to protect the vital sea lanes of our commerce, or our resolve to support our allies in pursuit of their sovereign interests, or our intention to work for world peace from a position of benevolent strength."

"To achieve these goals we must have a modern naval force of adequate size and composition to offset the growing capabilities of our major political adversary."

Rear Admiral Herbert S. Ainsworth, Commander, Patrol Wings, Pacific, said on July 30, 1973, that the Commander of the Sixth Fleet in the Mediterranean and the Commander of the Seventh Fleet in the Western Pacific have both indicated that the most serious threat to their ability to exercise sea control was the nuclear attack submarine, particularly those that fire antiship missiles.

"In World War II," Ainsworth said, "submarines were essentially surface ships that could submerge to conduct an attack. They transmitted on the surface, where they were vulnerable to ASW forces. The modern nuclear attack submarine is a very different threat of a new order of magnitude. It is quiet, fast, mobile, and lethal."

He said Americans sometimes forget how important freedom of the seas is to the prosperity of the United States, to our economic well-being, and to our way of life. "There are 72 raw materials," the admiral said, "that are vital to the production of automobiles, newspapers, typewriters, washing machines, television sets, just about everything that goes to make up our American way of life. Sixty-nine of these raw materials must be imported from overseas by ship. By the end of this decade, by 1980, 50 percent of our oil will come from overseas.

"Now let's look at the other side of the coin. Our exports. Today we are exporting about $45 billion a year and with the free-floating dollar I would expect this figure

Close-up of the Russian ship *Moskva.* (Courtesy of U.S. Navy.)

to increase as American products become better bargains. Each billion dollars of exports equates to about 100,000 jobs for Americans. The $45 billion worth of exports represents about four million jobs and is vital to the welfare and prosperity of a great many of our working people. These imports and exports depend upon our ability to control the seas and to maintain freedom of the seas."

As the potential threat posed by the Soviet submarine force increased throughout the 1970s, the U.S. Navy stepped up its antisubmarine warfare capabilities. Both land- and carrier-based aircraft continued to be vital. The P-3 Orion landbased ASW patrol aircraft had been a vital member of the ASW team since 1959. Since that time the Lockheed-California Company had produced over 400, including a new version, the P-3C.

There was no equivalent to the P-3C, an offensive aircraft capable of searching for and detecting a hostile submarine before it reached a target.

It was now realized that the most effective ASW platform for large ocean areas was an aircraft. Its great advantage lies in its speed, its quick reaction to tactical intelligence, its capability for immediate attack, and its relative immunity to counterattack by a hostile submarine.

Aircraft are not the only means to counter enemy submarines. Other submarines and destroyers play an important role.

Admiral Ainsworth said of the P-3C during the acceptance of the 400th Orion, "If there is one motto for our P-3 community it might be 'anywhere in the world in 30 minutes.'" He explained, "If we were to draw the perimeter of P-3 bases, we would start at Siganolla and Souda Bay in the Mediterranean, go to Norway, to Iceland, the United States, Alaska, Japan, Okinawa, the Philippines, Thailand, Australia, New Zealand, Guam, and Hawaii. This perimeter includes all the broad reaches of the Atlantic and Pacific Oceans and borders on the Indian Ocean. The P-3 is used not only by the United States Navy but by Australia, New Zealand, Norway, and Spain. It is used in all sea-controlled missions, surveillance, tracking, search and rescue, ASW, and Medevac. The P-3 in other versions is also used for electronic surveillance, weather reconnaissance, and oceanography."

Lockheed had done an outstanding job in the perfection of ASW systems since 1942, starting with the diversion of twenty Hudson bombers scheduled for the Royal Air Force. Recognizing the success of the Hudson, the Navy selected this land-based aircraft to counter the threat of enemy submarines in World War II.

Since then over 3,500 ASW aircraft were produced by Lockheed, including the PV-1 Ventura, the PV-2 Harpoon, and the P2V Neptune, which made the most significant contribution to the future of airborne antisubmarine warfare by combining a group of independent systems into a single, integrated airborne system. Succeeding ASW systems were based on the electronic systems integration conceived for the P2V.

The P-3C uses an advanced avionics system developed by the Naval Air Development Center, Johnsville, Pennsylvania. It is a fully integrated system built around an airborne digital computer. The P-3C's weaponry includes torpedoes, depth bombs, mines, rockets, nuclear devices, and air-to-surface Bullpup missiles.

Grumman's venerable S-2 Tracker was later phased out, an action that was accelerated in 1974 when Lockheed's new jet-powered S-3A Viking joined the fleet.

The Viking is used on carriers, and it is equipped with surface and subsurface search equipment. A computer collects, processes, interprets, and stores ASW data, thereby reducing the time required to identify a hostile submarine so it can be expeditiously attacked with a variety of weapons.

U.S. Navy officials became increasingly concerned by the mid-1970s with the apparent shift in Soviet grand strategy—a shift with emphasis on possible offensive action anywhere in the world. There was little doubt that the Soviet Union had changed her strategic concept of seapower to take an expanded role on the high seas. The fact that she was building her first aircraft carrier was added proof. The *Kiev* is approximately 900 feet long, with an angled flight deck two-thirds of its total length. By American carrier standards the *Kiev* is small, used mainly for helicopters and vertical-landing or short-takeoff aircraft.

In the past, the Soviet Union's intent was to deny the United States and its allies freedom to use the seas during wartime. In the decade of the 1970s, her intent appeared to be to contest control of the seas. This was a fundamental change and a grave threat to the future of the United States and the free world.

The S-3A Viking. This underview shows bomb bays and sonobuoy chutes.
(Courtesy of Lockheed.)

After the war in Southeast Asia ended for the United States in 1975, defense budgets were slashed, and there was no provision for replacement of all aircraft lost there.

The F-4 Phantom made up the major part of the Navy's fighter aircraft inventory, but its technology was almost twenty years old. To maintain force levels at an adequate strength, funds were included in budgets to extend the service life of older F-4B aircraft.

Obviously a new fighter was needed. Despite early cost overruns that almost eliminated the Grumman F-14A Tomcat, the program was continued. For many years after its introduction into the Fleet, the Tomcat was vital to the maintenance of the Fleet's operational readiness. It was ideal for countering multiple aircraft or missile raids, and hopefully, for helping to achieve local air superiority against any potential aggressor.

The F-14A became the first major innovation in naval fighters since 1953, when initial design efforts of the F-4 began. An all-weather, carrier-based fighter, it was capable of performing air-to-air combat as well as surface attack missions.

For most of the rest of the century—while Soviet fighter technology, missile technology, and strategy advanced dramatically—the Tomcat had the capability of

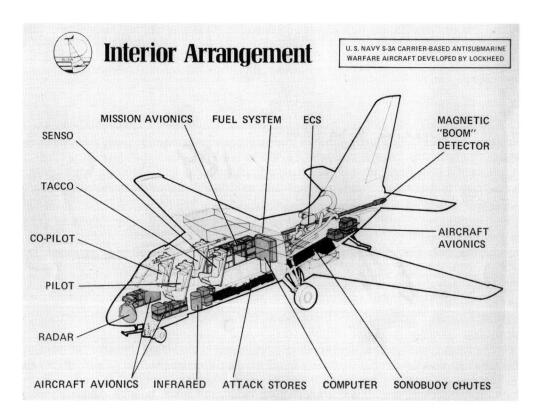

Interior Arrangement

U. S. NAVY S-3A CARRIER-BASED ANTISUBMARINE
WARFARE AIRCRAFT DEVELOPED BY LOCKHEED

SENSO

MISSION AVIONICS   FUEL SYSTEM   ECS

MAGNETIC
"BOOM"
DETECTOR

TACCO

CO-PILOT

AIRCRAFT
AVIONICS

PILOT

RADAR

AIRCRAFT AVIONICS   INFRARED   ATTACK STORES   COMPUTER   SONOBUOY CHUTES

Interior arrangement of the S-3A Viking. (Courtesy of Lockheed.)

countering the new Soviet threat of trying to saturate an opposing navy's defenses with coordinated aircraft and antiship missile attacks. Such a threat required an aircraft with a large radar search volume, and a longer-range multiple missile–firing capability than had ever been needed previously. The Tomcat, with its Phoenix missiles, was designed to meet that threat.

The F-4 has done yeoman service through the years, but was now an old airplane that should have been replaced. Budget constraints kept it in service long past its prime.

In modern naval warfare, air supremacy requires an acceptable level of effectiveness in all tactical fighter roles because the limited space available aboard carriers does not permit different types of aircraft for different missions. The ideal is an aircraft that is capable of performing all fighter missions.

The F-14A, with its variable-sweep wings and Mach 2 speed, proved to be a weapons system capable of detecting and tracking multiple targets simultaneously at distances over a hundred miles with the ability to engage such targets at altitudes from sea level to over 80,000 feet. It proved itself capable of doing these multiple jobs. The Phoenix missile has a fifty-mile range and the F-14A can fire six missiles at six different targets. For years it was the only weapons system that could counter existing Soviet capabilities against American naval operations.

The F-14A has been flying since December 1970. The first production aircraft was delivered in May 1972, with initial deployment aboard the nuclear-powered

Enterprise in the spring of 1974. It proved to be three to five times as effective as the F-4 in defending against bomber and cruise missile attacks, and ten times as effective against the Soviet Union's most modern fighter of that year, the MiG-21.

During the conflicts between Israel and the Arab countries in 1967, and again in the fall of 1973, the limited availability of land bases for American forces pointed up again the importance of seapower if the United States had become involved. Air bases that had been available around the Mediterranean in the past were either unavailable for political reasons or were of limited use because of the distances involved. Therefore the only forces that could be put into position immediately were seaborne forces.

At the end of World War II the United States had 1,100 air bases overseas. By 1973 the total was down to fifty, and it has been declining ever since.

With the nationalism of newly emerging nations being felt all over the world, it became increasingly clear that the availability of bases the United States could rely on for routine military missions and possible war contingencies would continue to diminish.

In the sustained type of operations that were necessary in Southeast Asia, it became apparent that nuclear power for ships was imperative. The *Nimitz,* first of a new class of nuclear carriers, was commissioned on May 3, 1975. It carries more ordnance than conventional carriers, and with its added jet fuel storage tanks, it is able to conduct sustained operations for ten days or longer without replenishment of supplies. Aircraft from this new class of carriers have a potential strike range almost five times as great as that of comparable land-based aircraft.

Since 1958 the U.S. Navy has almost continuously maintained five carrier task groups in forward areas with its fifteen-carrier force. It was demonstrated off Vietnam that six or seven carriers could be maintained on station for long periods of time.

The Soviet Union built over 400 combat ships between 1963 and 1975. The United States built only 100, but the superiority difference was made up by construction of new carriers, particularly those with nuclear power.

In terms of utilization, experience has shown that three nuclear ships can do what four conventionally powered ships can accomplish in a similar six- to seven-month deployment. This factor is significant in extended combat situations—such as the war in Southeast Asia—where rotation of ships on and off the line, and from one task group to another, was required to sustain the desired level of pressure. The future course of nuclear-powered ship construction should be judged against this background of experience.

In the last quarter of the twentieth century America's overseas bases outside of Europe shrank, and the same can be said within Europe. Therefore the aircraft carrier has become increasingly central to the whole idea of adequate U.S. military forces.

Navy officials believe that in many situations aircraft carriers are the only way to bring immediate air support to the nation's allies in time of need. That is why it so desperately fought critics throughout the 1970s when the Navy sought funds for a third nuclear-powered carrier of the *Nimitz* class, in addition to the *Dwight D. Eisenhower.*

"The Fustest with the Mostest." This painting by R.G. Smith portrays the old
USS *Langley* and her biplanes along with the new USS *Nimitz*
and her modern jets. (Courtesy of McDonnell Douglas.)

They pointed out that the aircraft carrier is a vital part of America's sea control forces because the Soviets have increased their ability to launch air attacks over wide ocean areas and bring surface combat strength to bear.

The Navy won that battle, and the *Carl Vinson* was commissioned in 1982. By then none of World War II's carriers was suitable for first-line operations, and the oldest carriers were thirty-three to thirty-five-years old.

With four nuclear carriers (the *Enterprise* was commissioned in 1961)—two based in the Atlantic and two in Pacific coastal ports—rapid reinforcement and response to any new crisis is possible.

The construction rate to maintain a modern carrier force of twelve to fifteen carriers, with thirty-year life cycles, calls for the construction of a new carrier every two and one half years. By the mid-1970s, construction forecasts called for only one carrier every four years, and even that schedule could not be met due to budget restraints.

The Navy launched its first fully guided Tomahawk cruise missile over the White Sands Range in New Mexico on June 5, 1976. It was airborne for sixty-one minutes after it was released from under the wing of an A-6 Intruder. This was the first in a series of flights intended to test the functional operation of the test vehicle's capability to perform while undergoing low-terrain following maneuvers. It was also the first test flight with a turbofan engine. Previously, missiles had been equipped with turbojet engines.

The next day an Intruder successfully fired a tactical version of the Tomahawk. It was designed to be launched from tactical and strategic aircraft, surface ships, submarines, and land platforms. It made its first submarine launch from the USS *Barb* on February 20, 1978.

The new SH-60B Seahawk helicopter demonstrated its potential in antisubmarine warfare in January of 1981. It provides surveillance and target information of surface targets, performs search and rescue operations, and is used for vertical replenishment and gun support.

Later that year the three-engine CH-53E Super Stallion became operational with the Marine Corps. The largest helicopter in the free world, it can carry sixteen tons of cargo or ferry fifty-five fully equipped Marines. It can also deliver aircraft on board carriers.

Fighter Squadron 125 became the first unit to receive the new F/A-18 Hornet in February of 1981. Thus began the transition from F-4s and A-7s to the new fighter. For use also by the Marines, the Hornet is designed for interdiction and close air support without compromising its fighter capability.

New aviation ships joined the fleet in the early part of 1980. The amphibious assault ship *Peleliu* and the nuclear-powered *Carl Vinson* were commissioned in 1980 and 1982.

The first of a new class of amphibious assault ships, the LHD *Wasp* class, was delivered in 1989. The combination of this ship with AV-8B Harriers and other aircraft, and an air cushion landing craft will assure success of future amphibious assault operations. The Harrier II met the Marine Corps' requirement for a light-attack aircraft to provide responsive firepower while operating from the forward sites of a battlefield in direct support of ground troops.

The F-14 Tomcat has been consistently upgraded throughout the 1980s for its role as an air superiority fighter. Its Phoenix air-to-air missile is the only one in the world that can hunt a target more than 120 miles away at speeds up to Mach 5 (more than a mile a second). It also has an M-61 gun for close-in attacks.

Older airplanes were also updated. The A-6E Intruder, the Navy's only all-weather attack aircraft, remained operational into the 1990s. Newer series incorporate a solid-state weapons release system, a simple integrated track and search radar with a moving-target indicating system, plus an inertial navigation system. The A-6 continues to provide pathfinder and strike leader functions for attack groups. The EA-6B Prowler, a derivative of the Intruder, supports strike aircraft and ground troops with a radar countermeasure capability for battle groups. It accomplishes this mission by suppressing and degrading an enemy's electronic systems. It is the only carrier-based aircraft that provides all these capabilities. The E-2C Hawkeye continued to provide service as an all-weather, carrier-based airborne early warning aircraft.

Since 1987 the E-CA has provided a communications platform to serve command links for the Fleet's ballistic missile submarine force. This Boeing four-engine aircraft has also been adapted to such roles as transport and aerial tanker. It is a much-improved version of the original 707.

Women made naval aviation headlines with several firsts in 1983. An all-female crew in a C-1A Trader from VRC-30 conducted an operational mission that terminated in a carrier-arrested landing. Lieutenant Leslie Provo became the first woman to be designated a landing signal officer, and Lieutenant Colleen Nevius was the first female graduate of the U.S. Navy's Test Pilot School.

★ ★ ★

During the invasion of Grenada on October 25, 1983, by United States forces and its Caribbean allies, aircraft and crews from the carrier *Independence* supported Marine combat amphibious assault operations. The first day was marked by poor intelligence, inadequate communications, and interference by higher headquarters. Although the conflict was concluded in a few days, men died unnecessarily due to poor planning.

Naval aviation engaged more actively in America's space programs during the decade of the 1980s. An all-Navy crew manned the space shuttle *Columbia* in April 1981, on her maiden voyage. In November of that year astronaut Captain Richard H. Truly rode aboard *Columbia* to become one of the first men to fly into space and return in a reusable spacecraft. The following year in November former Navy and Marine Corps aviators were on board the space shuttle during its first operational flight.

In recognition of the Navy's expanded role in space, the Naval Space Command was established in 1983, headed by Captain Truly, to consolidate the Navy's space-related activities. On February 7, 1984, astronaut Captain Bruce McCandless II made history when he took the first untethered walk in space.

Prior to the 1980s the Pacific Ocean was left largely to operations by the U.S. Navy, but with the global-ranging Soviet fleet, it became one of the world's most dangerous spots. Soviet and American ships keep constant watch on one another, ready to move to war status on a moment's notice if ordered to do so by the president of the United States or the head of the Soviet Union. The U.S. Navy had built up its forces and was prepared to fight in the air, on the surface, and deep within the ocean in an area covering half the world's surface. By the mid-1980s, the United States had forty-four fast attack submarines armed with sonars and computers that listened constantly for Russian ships through a system of undersea hydrophones. Overhead 120 P-3C Orions maintained their constant vigil, while destroyers with bow-mounted sonar domes sought the Russians all over the Pacific Ocean, from Asia to the American mainland and north to Alaska and the Aleutians.

In addition to their own submarines and warships, the Soviet Union used so-called fishing boats bristling with radar cones and antennas to flash word of any American presence to the huge base at Vladivostok while listening in on the thousands of radio and microwave transmissions between U.S. stations. Russian trawlers routinely cruised four miles off Pearl Harbor, just outside Hawaii's territorial limits. Normally, six trawlers were positioned between California and Midway Island to monitor American naval ships or to observe missiles launched from Vandenberg Air Force Base in California. Before 1980, units of the Soviet fleet were seldom found so far from their bases. But when they took over operation of the abandoned American base at Cam-ranh Bay in Vietnam, they began to expand throughout the Pacific now that they had reached the decision to develop a navy second to none.

In 1985, when the United States positioned Pershing-2 nuclear missiles in West Germany, drastically reducing the flying time of missiles to Russian targets, officials of the Soviet Navy moved their missile-carrying submarines closer to the West

Coast of the United States. Now their submarines could hit targets in California in six minutes or less.

The United States countered this threat by positioning eight Trident ballistic missile submarines in the Pacific, each capable of firing twenty-four ICBMs. These deadly submarines lurk in deep water, ready to launch whenever the president so directs.

America's attack submarines are designed to detect, pursue, and destroy Russian submarines. Crews of these submarines train endlessly, engaging in a secret war of maneuvers to prepare themselves for a real war that they devoutly hope will never happen. But they are prepared to respond quickly if necessary by maintaining a constant state of readiness.

World War II submarines were not very accurate, because torpedoes were not reliable. Forty years later torpedoes had become acoustically guided and wire-guided. A thin copper wire between the submarine's control room and the torpedo permits an operator to guide it to its target. Any change in direction is instantly registered and taken into account.

American submarines sank fifty-five percent of Japan's merchant fleet in World War II, and thirty-eight percent of its warships. But they were highly vulnerable to attack because they had to surface to make sightings. As a result, one out of every five American submarines was sunk. They are much less vulnerable in the latter part of the century because they rarely come to the surface.

The strains on the Russian economic and political system of maintaining such a huge military organization were not evident at first to officials of the United States. Even when it became obvious in 1987 that the Soviet Union was cutting back its shipbuilding program and reducing naval exercises throughout the world, the information was greeted with skepticism. This was such a departure from past practices that American officials were suspicious. But in the following years there was growing proof everywhere that Russian seapower had scaled back its grandiose plan to challenge the U.S. Navy everywhere in the world and to project itself into Third World conflicts. Instead, defense of the homeland was stressed. Several indicators of this drastic change in strategy began to display themselves. The largest aircraft carrier in the Soviet Union, the *Leonid I. Brezhnev*, later renamed the *Kuznetsov*, became the first true aircraft carrier after it was commissioned in the Black Sea. During sea trials in 1989 it was revealed that the new carrier had no catapults, and used a ski jump bow over which conventional aircraft performed rolling takeoffs. Along with cuts in shipbuilding programs begun by President Brezhnev, and a reduction in naval exercises around the world, there was a fifteen percent drop in deployments outside of the Soviet Union's own adjoining oceans. This was the first such decline in decades, due in part to Soviet leader Mikhail S. Gorbachev's insistence that the Russian Navy's huge fuel bills had to be sharply reduced. There was another reason, of course, that brought about this reduction. The U.S. Navy's new ability to take any future war right to the Soviet Union's home waters had become a frightening prospect for Russia's leaders.

When Admiral Gorshkov took command of the Soviet Navy in 1956, he said, "The flag of the Soviet Navy flies over the oceans of the world. Sooner or later the United States will have to understand it no longer has mastery of the seas."

In the early 1970s Gorshkov, in conjunction with other military leaders in the Soviet Union, had promoted an "interventionist" policy, supporting forces of other nations around the world. Huge amphibious landing ships were designed for its new role to permit the landing of Soviet naval infantry, while provision ships transferred fuel and supplies to warships under way. But by the early 1980s, these programs were largely scrapped, and only one ship of each kind was completed.

In 1967 and 1973 the Soviet Union sent huge fleets into the Mediterranean during the wars between Israel and the Arab nations. When war threatened on September 14, 1982, as the Phalangists killed hundreds of Palestinians in refugee camps, the Soviets did not send in their fleets. Leonid I. Brezhnev's death later that fall, followed by the succession of two incompetent leaders, further weakened the Soviet Union's once great military machine.

American military chiefs viewed the changes going on in the Soviet Union almost with disbelief because new Russian submarines were being reduced in number. The Typhoon class, as large as a World War II aircraft carrier and capable of carrying ballistic missiles, was reduced to a rate of one per year by 1987 instead of the original requirement for two. The same proved true of the Oscar class of cruise missile submarines. They had been projected for three per year but production was now down to one. There was also a strong indication that the allocation of nuclear reactors for ships had been cut in half.

The reduction in the presence of the Soviet Navy was most marked in the Indian Ocean. In 1983 the Soviet Union had twenty-eight ships there, but four years later the number was down to seventeen ships, despite the fact that Cam-ranh Bay in Vietnam was available to the Soviet Navy.

The Soviets also reduced their presence in the Atlantic Ocean, from an average of forty-five ships in 1984 to thirty-three in 1987. Ballistic missile submarines also were reduced in the Atlantic.

These were mind-boggling changes that were greeted with disbelief until a growing body of evidence indicated that drastic changes were under way in the Soviet Union.

Unfortunately the Reagan administration could not seem to grasp the fact that these incredible changes were going on, and authorized a 600-ship Navy at enormous cost at a time when the worldwide threat to the United States was receding at an accelerating tempo.

As part of the buildup, the $3.5 billion, 95,000-ton aircraft carrier *Abraham Lincoln* was commissioned on November 11, 1989. This huge carrier was designed to carry the war to an enemy's homeland and to help to control small Third World conflict. Since then two other huge aircraft carriers have been commissioned. The *John Stennis* was commissioned in 1995, and the *Harry S Truman* on July 25, 1998. The *Ronald Reagan* will be the eighth *Nimitz*-class carrier when it is commissioned in 2000. These ships were part of President Reagan's $350 billion Navy buildup to 600 ships, which was later scaled down during President Bush's administration. (This number was never achieved. It fell thirty-three ships short of its goal.) Budget restraints fortunately compelled the Navy to reduce its size to a level more in line with its responsibilities. The forty-three-year-old *Coral Sea* was decommissioned in April 1990 as part of this reduction.

Defense Secretary Dick Cheney explained the situation publicly on October 29, 1989: "We'd all like to have 15 carriers, but I made a decision not to build up to 15 for budget reasons."

The Navy had developed its "maritime strategy" of waging conventional war against the Soviet Union during a period when it posed a substantial threat to the United States. This war strategy called for the U.S. Navy to seize the offensive against the Soviet fleet, attack and destroy it, or bottle it up in its home ports. Once that was accomplished, American convoys could rush reinforcements and supplies to Europe with little fear of reprisals.

Admiral Frank B. Kelso II, Supreme Allied Commander, Atlantic Command, said the Navy planned to protect the convoys by using submarines and carriers to trap and destroy Soviet surface ships and submarines in Soviet ports and coastal waters before they could slip out into the open ocean. From stations near Soviet harbors, carriers could then threaten strikes on key Soviet ports such as Murmansk in the Barents Sea and Petropavlovsk in the northern Pacific.

But times had changed by 1989, and the Joint Chiefs in the Pentagon now believed that the likelihood of a superpower conflict was "perhaps as low as it has been at any time in the postwar era."

Deployment of accurate naval weapons, such as the Tomahawk missile, made carriers less important as the Navy's chief offensive weapon. With introduction of long-range cruise missiles, all Navy ships can conduct strikes on land that once required a carrier. With the proliferation of missiles, carriers were now more vulnerable. But carriers still have a role to play. Their flexibility and mobility for dispatch to Third World flash points permit them to operate despite irresolute allies and a lack of land bases.

A carrier battle group—with a complement of eighty to ninety multimission aircraft—with its accompanying destroyers, cruisers, and submarines provides a four-dimensional war capability because it can control an area more than a thousand miles around it.

Meanwhile, the Soviet Union's 65,000-ton *Tbilisi,* the first of three new carriers, conducted sea trials in late 1989. Its angled and "ski-jump" decks indicate that it is for use primarily against Allied submarines. Although it is capable of carrying about sixty combat jets and helicopters, its principal aircraft are short-takeoff-and-landing craft and helicopters. Her sister ship, the *Riga,* was being fitted out in a shipyard, while a third ship of this class was in an early stage of construction.

Despite Lockheed's long record of producing excellent antisubmarine warfare aircraft, the Navy cancelled its P-7A on July 20, 1990. A replacement for the P-3C, it was scheduled to go into full-scale development, but the Navy said Lockheed had "failed to make adequate progress toward completion of all contract phases."

Another Navy program faced cancellation in 1990 when serious flaws were discovered in the A-12 program. When the Navy's program managers failed to report to Defense Secretary Dick Cheney that there would be a $1 billion overrun and a one-year delay in the development phase of the program, the Navy was ordered to remove three senior officers in charge of the secret program. Vice Admiral Richard C. Gentz, Commander of the Naval Air Systems Command, retired in early 1991, and the A-12's executive officer, Rear Admiral John F. Calvert, and its program manager, Captain Lawrence G. Elberfeld, were reassigned to other duties and given letters of censure from Navy Secretary H. Lawrence Garrett III.

This was the most serious punishment meted out to top Navy officials in recent years. This action was taken because they had testified to Congress in late April that the program was in good shape. On their evidence, Cheney had supported the program, although two companies developing the A-12—McDonnell Douglas and General Dynamics—had warned the Pentagon the previous June of huge overruns.

Cheney ordered a full investigation, and a new Navy report said the two firms were "overly optimistic" in their cost projections to the Navy before their disclosures.

It further stated that the companies' program managers "perceived significant pressures from upper management" to keep up the cash flow with optimistic reports.

The two firms were developing the A-12 under a fixed-price contract awarded in June of 1986 with a ceiling price of $4.78 billion. Any costs above this ceiling would have to be borne by the companies.

The A-12 Avenger's stealth technology was supposed to help it evade enemy radar. The A-6 Intruder, which it was designed to replace, had almost outlived its useful operational life, so a new plane was needed for the Navy's attack mission.

After a thorough review, Secretary of Defense Cheney ordered the $57 billion production program canceled on January 7, 1991, claiming that McDonnell Douglas and General Dynamics had badly mismanaged the contract—a charge they denied. He said that in his opinion these companies could never meet the government's contract terms.

The government had already spent $5 billion on the A-12, involving 10,000 workers in forty-two states. The companies sought a restructuring of the contract, but Cheney refused to consider their request. "This program cannot be sustained unless I ask Congress for more money and bail the contractors out. But I have made the decision that I will not do that. No one can tell me exactly how much more it will cost to keep this program going. And I do not believe a bailout is in the national interest. If we cannot spend the taxpayers' money wisely, we will not spend it." Cheney said the contract was being terminated for "default" based on the "inability of the contractors to design, develop, fabricate, assemble, and test A-12 aircraft."

The defense secretary agreed with the Navy that it needed an all-weather, carrier-based attack plane to replace the 25-year-old A-6 Intruder. The Navy had planned to build 620 A-12s, down from the 858 planes originally sought, because the Navy's carrier fleet was being reduced from fifteen to twelve ships. Now it had to decide on a new airplane called the AX, and Cheney authorized the Navy to begin the contract selection process all over again. Like the A-12, the AX would be designed for both fighter and bomber missions, and would replace the aging A-6 with a longer-range, more versatile aircraft using stealth technology.

Mikhail Gorbachev was named president of the Soviet Union on May 25, 1989, signifying that the vast changes in his country made superpower military confrontation between east and west more and more unlikely. With the Warsaw Pact disintegrating as Russia's satellite countries each broke away from the alliance, it was imperative that the massive nuclear armaments possessed by the United States and the Soviet Union be sharply reduced.

This seemed more likely to happen after President Gorbachev and President Bush met in Washington on June 1, 1990. This was their second conference, and they reached a broad understanding, which included a reduction in nuclear arms and chemical weapons. Bush also signed a trade treaty that Gorbachev had long sought, although "most favored nation" status had to await Senate ratification.

The two presidents signed several pacts and joint statements in which both countries agreed to cut strategic nuclear arsenals by a third. Under terms of these pacts, each nation would be limited to a total of 6,000 warheads in their missile, submarine, and bomber organizations. They both agreed that their nations would cease production of chemical weapons and reduce their current stockpiles to 5,000 tons—roughly twenty percent of the U.S. stockpile. They also reaffirmed their nations' commitment to a treaty that would reduce conventional forces in Europe.

While the Soviet Union and the United States agreed to reduce their nuclear arsenals, the U.S. Department of Defense pushed ahead with nonnuclear weapons, including adaptations of the standard cruise missile. These vehicles can fly long distances, and can circle a target repeatedly, if necessary, before homing in on an enemy's radar electronic signal and destroying whatever target it represents. "Tacit Rainbow" has enormous implications for seeking out and destroying hidden targets. Another missile, the Long-Range Conventional Standoff Weapon, can fly 2,000 miles to destroy railroad yards, bridges, and major industrial targets.

The U.S. Navy began to develop the Tomahawk cruise missile in 1984 with nuclear and nonnuclear warheads. During the early 1990s the Navy planned to order several thousand, most of which would have nonnuclear warheads. The Tomahawk has a range of up to 1,500 miles and sells for $1 million to $1.4 million each, depending on its warhead.

Dean Wilkening, a military analyst for the RAND Corporation in Santa Monica, which makes long-range studies for the Department of Defense, gave the Tomahawk a realistic evaluation. "As weapons of terror, these new weapons will never replace nuclear weapons. But as weapons of warfare, these high-tech conventional weapons can pick up some of the nuclear weapons' jobs."

Cruise missiles with or without nuclear warheads can be carried by ships and aircraft, giving them the ability to respond quickly to an emergency. Their use against massed troops, communications centers, and dug-in armies gives an advantage to the United States. But they are one-shot weapons, and their cost is extremely high compared to a reusable aircraft. They are not a substitute for manned aircraft, but rather a valuable weapon for use in conjunction with modern warplanes. For example, a manned bomber on just one mission can carry between fifty and sixty times as much destructive power as one cruise missile. They are not a panacea for reducing defense costs by eliminating manned aircraft.

Part XI

Desert Storm

26

A Classic Air/Land Battle

President Saddam Hussein of Iraq became an increasingly disturbing influence in the Middle East during July of 1990 as he pressured the Organization of Petroleum Export Countries (OPEC) to raise oil prices. Despite his own vast reserves of oil, the high cost of Hussein's eight-year war with Iran had almost destroyed his nation's economy.

Despite Iraq's protestations of innocence, Hussein moved 30,000 troops to the Kuwaiti border and accused the heads of that nation and of the United Arab Emirates of undermining oil prices by producing more oil than OPEC had approved. At first Western leaders believed these accusations were just more bellicose talk from a man known for such threats.

Although Middle Eastern diplomats agreed to defuse the situation by boosting oil prices $3.00 a barrel, Hussein's true intentions became evident on August 2 when Iraqi troops, spearheaded by tanks, invaded Kuwait and occupied it in a matter of hours. The emir was forced to flee his unarmed country as hundreds died, and Iraq announced that Kuwait no longer existed because it had been annexed to Iraq.

Suddenly there was a full-blown crisis in the Middle East again, and Saudi Arabia's leaders feared they would be next to be invaded. Her border with Iraq and Kuwait was fortified, but Saudi leaders knew they were no match for Iraq's armed forces.

Two days after Iraq's armed forces invaded Kuwait, planners with the Joint Chiefs of Staff met in the basement of the Pentagon in Washington to draw up a war plan to free Kuwait and eliminate Iraq as a military threat to the Middle East. The plan they developed—literally written on the backs of envelopes—called first for an air war in four phases. Phase 1 would involve deep strikes into Iraq, particularly in and around Baghdad, to destroy strategic targets. Phase 2 called for clearing the skies over Iraq and Kuwait of Iraqi planes to establish air supremacy over the battlefield. Phase 3 called for the destruction of bridges and lines of communication to reduce the supplies moving up to the war fronts. This phase called for the destruction of bridges by laser-guided bombs as well as old-style unguided iron bombs. Its goal was to achieve at least a fifty percent destruction of the Iraqi Army, and the

Kuwaiti oil fields, set on fire by the Iraqis during the Gulf War, are examined by an F-14A from the *Abraham Lincoln's* VF-114. (Courtesy of U.S. Navy.)

total destruction of its elite Republican Guard divisions. Once these phases were completed, Phase 4 would begin, with a strong, highly mobile ground offensive.

Prior to this crisis, there had been no specific contingency plans for Iraq, because Iran was considered the primary threat to the free world.

At the time of Iraq's invasion of Kuwait, the American presence in the region was limited to two Navy carrier battle groups, with the USS *Independence* in the North Arabian Sea and the USS *Dwight D. Eisenhower* in the eastern Mediterranean. These two carriers, even with the aid of Saudi Arabia's military establishment, would have been insufficient to prevent Saddam Hussein from moving beyond Kuwait. (There is every indication that he planned to invade Saudi Arabia, thereby controlling forty percent of the world's known oil reserves.) But statements made by President Bush and other world leaders, plus the presence of the two carrier battle groups, warned Hussein that the cost to his nation might be prohibitive. But he refused to heed these warnings. Then, too, Hussein's army had outrun its logistics support, and for several days his armies had to remain in position in Kuwait until they were resupplied.

These days were precious to the coalition forces that began to move into the theater, giving them time to prepare for the defense of Saudi Arabia. When American forces were rushed to the region, Hussein's indecision cost him an even greater victory.

Now Iraq's armed forces were faced with a defensive war. After the Navy and Marine amphibious forces were positioned offshore, Hussein ordered ten of his best divisions to meet this threat along the southern Kuwaiti coast. It was learned after the war that Hussein was so certain of an amphibious landing (it never became necessary) that some of his most experienced units were placed in defensive positions, where they were particularly vulnerable to naval gunfire, and when the main thrust of the ground war was made farther west, these troops were not in a position to oppose the coalition's advancing armies.

While most major nations placed an embargo on Iraqi oil, the UN Security Council on August 6 approved a sweeping trade and financial boycott of Iraq and Kuwait. The following day President Bush ordered troops, planes, and tanks to Saudi Arabia to protect it as part of a coalition. Two days later the Arab League voted to send troops to defend Saudi Arabia. Some American reservists were called up, starting on August 20. The Iraqis—stunned by the UN action and the prompt response by members of the Arab League—countered by moving Americans held in Iraq and Kuwait to strategic sites as human shields. Even more serious for Iraq, the United Nations on August 25 gave the United States and other nations in the coalition the right to enforce the embargo. Iraq's President Hussein, after most of the world condemned his actions, permitted foreign women and children to leave.

President Bush's forthright action was approved by the vast majority of the American people, and at this stage he received overwhelming support from Congress. It was clear that the question of war or peace lay in Hussein's hands. If he withdrew from Kuwait, as demanded by the coalition's leaders, there would be no war. If he did not, and the embargo failed to convince Hussein of the futility of resistance, then war was inevitable. President Bush repeatedly reiterated this stance. Meanwhile, coalition forces began to build up their armed strength to contain Hussein's threat of further conquests, and possibly defeat his armies on the battlefield if that should become necessary.

As the buildup began in the Persian Gulf, Navy carriers were soon on the scene. This was the Navy's fifty-third crisis situation since 1980. The *Independence* arrived in the Gulf of Oman on August 7, while the *Dwight D. Eisenhower* moved through the Suez Canal (Egyptian approval was necessary for this transit, and it was given without argument) and headed for the Persian Gulf. Between them they had 164 warplanes that were ready to fight. Their presence undoubtedly proved a deterrent to Iraq's President Hussein, whose next goal apparently was to annex Saudi Arabia.

The U.S. Army and Air Force, however, needed at least six months to get on the scene and be prepared to fight, while the Navy's two battle groups were prepared to fight at once. If Hussein had moved quickly—a possibility that was a nightmare for American commanders—only carrier aviation would have been available to resist his massive ground and air forces (except for Saudi Arabia's small air force).

While the buildup continued, the U.S. Navy and its coalition partners intercepted Iraq-bound shipping under the United Nations embargo so that Iraq's armed forces soon began to feel the pinch as they lost most of their supplies and spare parts. Navy sealift ships carried ninety-five percent of the American military cargo to Saudi Arabia, while Marine equipment was prepositioned aboard Navy warehouse ships in the first month of the crisis.

General H. Norman Schwarzkopf, in command of American forces in the Persian Gulf, worked with other leaders of the coalition to draw up detailed plans to defend Saudi Arabia and, if ordered to do so, to defeat the Iraqis. It was a complex plan that relied heavily on airpower and deception. Simply, the plan was to deceive the Iraqi command about the coalition's true intentions by destroying the electronic eyes and ears of its communications system, to bomb its key installations into rubble,

MK-83 laser-guided bombs are loaded on a VA-52 A-6E from the *Kitty Hawk* during "Operation Southern Watch" in January, 1993. (Courtesy of U.S. Navy.)

and then to encircle its armies and destroy them. One of its major goals was to destroy Iraq's airfields and air defenses to deny its air force an opportunity to challenge the Allied air assault.

A major deception was the use of diversionary actions, such as amphibious exercises and saturation bombing, to confuse the Iraqis about the primary focus of a land assault. It was hoped that these moves would keep numerous Iraqi divisions along the Kuwaiti coast to guard against an amphibious assault, and to deploy additional infantry divisions along the Kuwaiti–Saudi border to block an anticipated allied land attack. This ruse proved highly successful.

The goal of the plan was to freeze the Iraqi army in Kuwait while the coalition army made a huge, high-speed maneuver with two full Army corps 200 miles west of the Kuwaiti–Iraqi border. These army units would then sweep to the Euphrates River and trap the Iraqis in Kuwait, and further instill doubt in the minds of the Iraqi commanders as to whether the coalition army was headed for Baghdad.

The First Marine Expeditionary Force was airlifted to Saudi Arabia while prepositioned ships—five from Diego Garcia and four from Guam—brought their supplies and heavy equipment. The first ship arrived on August 15. These were combat-ready troops with thirty days of supplies, and they were ready to fight.

While sealift ships brought ninety-five percent of the troops and their equipment to Saudi Arabia, C-5s, C-141s, and C-130s of the Military Air Transport Command flew around-the-clock to airlift the highest-priority men and equipment.

An F-14A Tomcat strike/fighter of AC Tophatters is loaded with MK-83 bombs.
Flown by Lieutenant Commander Thomas Prochilo, with
Commander Dana Potts as *Rio* over the Adriatic Sea in
March, 1993. (Courtesy of U.S. Navy.)

The Schwarzkopf plan was a variation of the air/land battle developed by Army planners to distribute firepower over the whole depth of a battlefield for a hundred miles or more. Tanks, armed helicopters, antitank aircraft, multiple rocket–launchers, self-propelled artillery, and "deep-strike" bombers were all assigned to meet an enemy strong in tanks.

Analysis of the Iran-Iraq war had noted that the Iraqis fought a static, trench-type war such as had been fought in World War I. Against the ill-equipped Iranians this strategy of fighting a war of attrition was effective. Schwarzkopf and his planners knew the Iraqis were dug into fixed positions to fight the coalition army, and he was confident they were inviting disaster. With his powerful mobile forces, Schwarzkopf gave orders to pound these fixed positions with bombs and ship bombardment until they were, in effect, impotent to withstand his rapid-moving armies.

The theory behind a classic air/land battle is that successful military campaigns are founded on a few clearly defined concepts. The most important is that mobile armies almost always defeat static defenses (the Nazis used their blitzkrieg tactics to gain incredible early victories in World War II), and that control of the air and control of communications are essential. Through the synchronization of air, land, and sea forces, overwhelming force can be concentrated against an enemy's weakest point.

An AIM-7 Sparrow missile is fired by an F-14D of VF31 in 1993 over
Southern California. (Courtesy of U.S. Navy.)

From the outset, General Schwarzkopf's plan was "to put the Republican Guards
out of business." Supposedly, Hussein's elite force would be the adversary to beat
before victory in the Gulf could be assured.

President Bush then decided in November not to wait for the embargo to bring
Iraq to its senses, and to liberate Kuwait by force. Schwarzkopf and General Colin
Powell, chairman of the Joint Chiefs of Staff, insisted that they needed 514,000
American troops to do the job effectively, and the president supported their request.
Both generals insisted that the policy of "incrementalism" that governed the war in
Indochina must not be repeated, because they knew from personal experience that
it was a strategy for defeat. The president readily agreed.

For five and a half months the Maritime Interception Force—involving ships from
several coalition nations—had played a key role in enforcing the United Nations'
embargo, as they challenged more than 7,000 merchant ships and boarded more
than 800 others to search them for war materiel. This force gave Saddam Hussein a
clear signal that the United States and its coalition partners were serious about
driving him out of Kuwait. It also cemented the coalition. In the early days, the
organization was fragmented and ill-directed prior to the start of hostilities. Military people from each nation were doing their own thing, and were reluctant to
participate in the work of the coalition as a whole. The Maritime Interception Force
made them work as a team. This gave President Bush a foundation for laying out

the strategy against Iraq among all twenty-eight countries involved and the 800,000 military men and women they contributed to the war, thereby assuring him the support that he needed from the American people and the world community of nations. The war could have been fought successfully with the armed forces from Saudi Arabia, Great Britain, and the United States. But the war could not have achieved a political as well as a military victory without the other members of the coalition. They did not add that much militarily, but they gave the United States a moral platform on which to stand that was of incalculable value.

On January 12, 1991, the U.S. Congress voted President Bush the authority to use "all necessary means to drive Iraq out of Kuwait."

UN Secretary General Javier Perez de Cuellar left Baghdad the next day after a last attempt to convince Hussein that his armed forces should evacuate Kuwait peacefully and avoid a war. There had been four days of meetings in Geneva, Switzerland, prior to de Cuellar's meeting with Hussein between Secretary of State James A. Baker, III and Iraqi Foreign Minister Tariq Aziz, but all had ended in failure.

Coalition forces were outnumbered on the ground, so much was expected of a preinvasion air war to destroy most of Iraq's ground armies prior to an invasion. To run the air war Lieutenant General Charles Horner of the U.S. Air Force was assigned as the Joint Forces Air Component Commander under General Schwarzkopf. Under him Captain Lyle G. Bien, Commander of Carrier Air Wing 15 on board the USS *Carl Vinson,* was appointed as the Navy's senior strike planner.

After they were sent to Riyadh in Saudi Arabia, they refined the plan agreed to in Washington on August 4, 1990. The most important strategic targets, to be hit first, were Iraq's nuclear, biological, chemical, air defense, and communications capabilities. They were ready when the president reached his historic decision.

Although there was a high level of confidence that the war would be fought swiftly and decisively, measures had been taken to prepare for the worst. Medical facilities at bases throughout the world contributed doctors, nurses, and technicians in case casualties proved unexpectedly high. Many military hospitals in the United States were stripped of personnel to send to the Persian Gulf. Now facilities were available to treat as many as 20,000 casualties. U.S. forces had 18,000 hospital beds in place at sixty-five facilities staffed by 41,000 medical personnel. This obviously was a worst-case scenario, but Schwarzkopf was being prudent, and rightfully so.

Marlin Fitzwater, White House press secretary, read a statement at 7:00 P.M. Eastern Standard Time, on January 16, 1991. It was from President Bush and it announced the commencement of "Operation Desert Storm." He said, "The liberation of Kuwait has begun. The offensive action against Iraq is being carried out under provisions of 12 United Nations Security Council resolutions and both houses of the United States Congress."

The attack began seventeen hours after expiration of the UN Security Council's deadline for Iraq to withdraw from Kuwait. Missiles and bombs hit targets in Baghdad starting at 6:30 P.M. (EST) or 2:30 A.M. the following day in Iraq.

It was early morning in the Middle East when the president addressed the nation. "This is a historic moment. We have in this past year made great progress in ending the long era of conflict and cold war. We have before us the opportunity to forge for ourselves and for future generations a new world order, a world where the rule of law, not the law of the jungle, governs the conduct of nations. When we are successful, and we will be, we have a real chance at this new world order, an order in which a credible United Nations can use its peacekeeping role to fulfill the promise of vision of the UN's founders."

Rio in full gear, taken by Lieutenant Tony Curran in Fallon, Nevada in an F-14B of VF-101. (Courtesy of U.S. Navy.)

The president assured the nation that the action against Iraq would not be another Vietnam type of war. "Our troops will have the best possible support...and will not be asked to fight with one hand tied behind their back."

U.S. Navy ships in the Arabian Sea and the Red Sea launched more than one hundred Tomahawk cruise missiles to start operations, and approximately 1,000 aircraft sorties were flown by Americans, British, French, Saudi Arabian, and Kuwaiti aircraft. The cruise missiles attacked Iraq's air defenses because they posed a threat to manned aircraft, but only fifty percent of these sorties were effective. Those that failed to hit their targets did so because of mechanical and weather problems. American planes faced limited attacks by Iraqi air force planes, but Iraqi Scud missiles landed in Israel that first day, causing seven civilian injuries, and another landed in Saudi Arabia. Militarily speaking, the Scud was a nonweapon; it was used primarily as a terror weapon. The American Patriot missile destroyed one Scud the first day over Saudi Arabia, and this was the start of a successful series of interceptions.

Pilots over Baghdad reported heavy flak at first, saying it turned night into day. Several pilots said it was so heavy that they could have stepped out of their airplanes and walked across the city on it.

Air Force F-15 pilot Captain Steve Tate quickly dispatched an Iraqi F-1 Mirage in the first kill of the war. The Iraqi aircraft was locked on the tail of another F-15 when Tate fired a Sparrow air-to-air missile. The Mirage blew up, lighting up the

An EA-6B of VAQ-135 is shown over the USS *Abraham Lincoln* off Southern California. (Courtesy of PHAN Sean Linehan, U.S. Navy.)

whole sky, and it continued to burn as it exploded on the ground. As Tate turned for home base, he could see the city of Baghdad in the distance..."like a huge blanket of Christmas lights.... The entire city was just sparkling at me."

Then American, Saudi Arabian, British, and French pilots dropped their bombs on the city. Lights flashed as bombs and missiles exploded while tracers lit up the night sky as guns on the ground went into action.

Throughout the night, and during daylight hours, U.S. Air Force F-15s, F-16s, F-4 Phantoms, A-10 tank killers, F-111s, F-117 stealth fighters, Navy AV-8s, F/A-18s, A-7s, A-6s, and EA-6B electronic jamming aircraft joined French Jaguars and British Tornado jets to sweep Iraq from one end to the other. Cruise missiles and F-117 fighters sought out Iraq's high command headquarters to cut communications links, destroy key government ministries, and eliminate Iraq's control network. That first night, an F-117 placed a bomb down the airshaft of the Iraqi Defense Ministry, using a hardened laser-guided missile. They came over Baghdad in two waves— with ten planes in the first—to knock out key communications centers. The second wave arrived an hour later with twelve aircraft. Pilots found ground firing haphazard after Iraq's control network was largely destroyed. Iraq's airfields were struck repeatedly, but the Iraqi air force remained largely on the ground, unable or fearful to contend the air space above their own country. The coalition air forces from the United States, Great Britain, France, Saudi Arabia, and Kuwait established air superiority within forty-eight hours, and they eventually achieved air supremacy when Iraqi pilots refused to fight and dozens of military and civilian aircraft were

An F/A-18C of VFA-131 flies over "Star Wars Canyon" in Southern Oman
February 11, 1991. (Photo courtesy of Lieutenant Steve Krieger, U.S. Navy.)

flown to Iran requesting asylum. In the most intensive aerial offensive in history,
Iraqi's air force was eliminated from combat, and antiaircraft defenses were largely
nullified, while ballistic missile launchers were attacked.

Iraq's nuclear research facility, used for chemical and biological weapons production, was attacked by seventy Air Force planes in daylight, including fourteen
F-16s, with the rest as escorts—such as jammers and tankers—to support them.
Heavy ground fire and smoke generators obscuring the target prevented these aircraft from scoring hits. F-117s went in at 3:00 A.M. the following morning and knocked
out three of the four nuclear reactors and heavily damaged the fourth.

Among other special targets attacked were those storing or producing nuclear,
biological, and chemical warfare agents. After the war, it was learned that key seg-
ments of these nuclear and chemical complexes emerged unscathed.

The F-117 stealth airplane performed well that first night, but without its two
2,000-pound guided bombs it would have been just another airplane. Subsonic, with
no radar, radio, or lights, it was incapable of flying long distances without refueling.
It is limited to night attack missions because it has no guns or air-to-air missiles.
The engines are muffled to eliminate noise, and its range is limited to 1,200 miles. It
was designed to attack extremely high-value targets to neutralize enemy defenses.
Fifty-nine F-117s were built at a cost of $43 million each.

The F-117s flew 1,271 missions, or one percent of the total of coalition air sorties, but it accounted for forty percent of the damage to targets—although it had
only a sixty percent direct hit score. Its "smart bombs" deserved the credit, and the
F-117's abilities have been grossly exaggerated, although none were lost.

Strike Team-VAW-121. An E2C Hawkeye and two VFA-136 F/A-18A Hornets
head out February 11, 1992 against targets off the coast of Oman.
(Photo courtesy of Commander J. Leenhouts, U.S. Navy.)

Other planes—such as the Navy's old A-7, and the Air Force's older F-111—flew
the same missions with a few minor exceptions. It is true that no F-117s were lost,
but neither were there any A-7s or F-111s lost. There were only two places in downtown Baghdad that were prohibited to nonstealth aircraft. In each instance enemy
defenses would have had to be eliminated first, and this involved too much of an
effort because of the need to protect civilian areas from extensive damage.

Lieutenant Commander Mark Fox launched from the *Saratoga* on January 17
in his F/A-18 as a "spare" but another plane aborted and he was on his first combat
mission. Operations officer for VFA-81, Fox was flying one of four fighter-bombers
assigned to attack an Iraqi airfield.

An E-2C Hawkeye from *Saratoga*'s Carrier Airborne Early Warning Squadron
(VAW) 125 provided strike control and battle management. Lieutenant John Joyce,
the Hawkeye's air control officer, radioed an alert to the inbound Hornets. Then the
squadron's skipper Commander Howard McDaniel called, "Hornets, bandits on your
nose at fifteen."

Fox spotted the Soviet-made MiG-21 Fishbeds in front. He switched from bomb
release to air-to-air and locked on the MiGs on the right. He quickly noted that
Lieutenant Nick Mangillo was aligned on the MiG on the left. Meanwhile the other
two Hornets acquired radar locks.

The MiGs came at them, nose on, at a 1,200-knot closing speed without maneuvering. Fox released a Sidewinder. Believing at first that he had missed, he then

Desert Storm. An F-14A of VF-1 over Kuwait and burning oil fields.
(Courtesy of U.S. Navy.)

fired a Sparrow. Then Fox saw a flash as the Sidewinder struck, followed by a brilliant orange-yellow flame and a puff of black smoke. His Sparrow hit the flaming aircraft seconds later, engulfing the Fishbed's rear half in flames. Fox noted that the canopy of the Iraqi plane was still in place when it went past him. Fox never learned, but he expected the Iraqi pilot was killed. The incident lasted only forty seconds.

Mangillo's Sparrow missile hit the second MiG, but it evidently escaped.

The Hornets proceeded to the airfield and dropped their bombs, but what was notable was that for the first time an airplane made an "air-to-air kill" while carrying four 2,000-pound bombs!

As the air war pounded away at targets in Iraq and Kuwait, the level of destruction rose to incredible heights. Air Force B-52 bombers concentrated on Republican Guard divisions, carpet-bombing areas where they were deployed. Later it was learned that some defensive lines were left virtually unmanned. The psychological impact on Hussein's armies was explosive, causing two out of every five Iraqi soldiers to desert. At first it was believed that the bombings had killed a great many Iraqis, but this was later disproved. Thirty-three thousand civilians may have died and 75,000 to 120,000 troops. But the impact on Hussein's forty-two divisions and their 540,000 men in Kuwait and southern Iraq was as severe as if they had been killed in battle, because at least 200,000 soldiers deserted their units and went home, and counting casualties, only 183,000 soldiers faced coalition forces when the ground action began.

On January 18, attacks were initiated from Turkey by USAF bombers to start a second front against Iraq.

Seven days later Iraq released millions of barrels of Kuwaiti oil that fouled the Persian Gulf.

In the first days of the war Iraqi antiaircraft fire was radar-guided, but once their ground bases began to swallow HARM (antiradiation) missiles that homed in on their radar signals, surface-to-air missiles were fired unguided, which dramatically reduced their effectiveness. But even on that first night of 1,000 sorties the United States lost only two aircraft (an F/A-18 and an A-6) with one crewman listed as missing, while other coalition air forces lost only one aircraft.

From the first bomb or missile dropped at 3:00 A.M. January 17 (Baghdad time), it was evident that those that provided the greatest effectiveness were precision weapons, or "smart" bombs guided by "seeker" heads.

The most critical Iraqi targets were nuclear facilities, biological and chemical factories and storage facilities, air defense systems, and electrical and communications systems. Many were struck massive blows in the war's early hours by coalition aircraft and the U.S. Navy's Tomahawk cruise missiles. Iraq's air defenses were quickly made ineffective by high-speed antiradiation missiles, and the electronic jamming by the EA-6B and the Air Force's EF-111.

The decision to amass the coalition's airpower against Iraq proved sound, because the government was thrown so seriously off balance that it was never able to organize its defenses. When the pressure increased in the following days, the war was all but won in the air. The massive bombing in the first seventy-two hours so stunned the Iraqis that they were never again able to put up a worthwhile fight.

With sensitive consideration for world opinion, as well as the moral issues involved, the air war planners tried to avoid damage to civilian areas and to limit civilian casualties.

For many years the United States had been criticized for its extensive use of high-tech hardware. Now that costly decision paid off. The need to destroy selected targets—especially "hardened" targets—was acute in Iraq because the most formidable of such targets were in Baghdad deliberately surrounded by civilian buildings. The United States' arsenal of weaponry proved ideal for hitting targets with an accuracy measured in feet by its Tomahawk, Walleye, and Maverick missiles and its laser-guided bombs.

Even more exotic weapons—those with a "standoff" capability (out of range of antiaircraft fire)—were not used much because Iraq's air defenses were so disrupted in the early days of the war that they were not needed. With coalition bombers roaming at will throughout the breadth of Iraq, bombers firing missiles from great distances were not needed.

Tomahawk cruise missiles and F-117 stealth bombers repeatedly struck Iraq's surface-to-air defense systems in Baghdad and quickly destroyed them. Otherwise medium-range weapons like the standoff land-attack missile (SLAM) and Walleye would have been essential, not just nice to have as a backup.

The Iraqi air force's decision not to contest the coalition air forces was due, in part, to the loss of its ground control facilities on the first day. Its commanders also had no illusions about contesting the skies over their country. Their pilots were not equal to the challenge, although they had some of Russia's most modern airplanes.

Prior to the start of hostilities it was fashionable among some so-called military experts to credit Saddam Hussein's military establishment with superhuman capabilities. After the Iran–Iraq war ended in such a debacle for Iran, it seemed ridiculous to label his armed forces second-rate. Neither was true. Iraq had a 1.3-million-man army equipped with the latest and most modern Russian aircraft and tanks. Many in the army, particularly its officers, had experience in the eight-year war with Iran. The truth is that Iraq's military and civilian leaders never recovered from those stunning early air blows. Like a boxer who takes a right hook on the chin and continues to fight although he never recovers from that hard blow, Iraq was knocked out of the war within a few days. Once the nation's command and control system was destroyed in those early hours, the war was won. The United States was aided by data provided by some of the Western firms who built the hardened sites, and this information proved of great value to determine the type of weapons to use.

"Smart" bombs proved more effective than was believed possible. Inside hardened shelters there was total and massive destruction, although quite often the outer framework of the building remained intact. Often these bombs penetrated fourteen feet of reinforced concrete. If these sites were used as underground aircraft hangars, not a single part of an aircraft was recognizable. Although only nine percent of the weapons dropped during the war were precision-guided (smart bombs), the destruction they caused was way out of proportion to their numbers. Ninety-one percent of the bombs were old-style iron bombs. Most were carried by B-52s that carpet-bombed areas with large troop concentrations.

In Baghdad the Ministry of Air Defense building was located in the center of a block housing civilian buildings. The Ministry of Agriculture building was across the street from it. When smart bombs knocked out the Defense Ministry, the Agriculture building emerged not only unscathed, but without even one of its windows broken. Nonprecision weapons could have destroyed the Ministry of Air Defense, but they probably would have also destroyed everything else around it. In the modern world of instant communications, such wanton destruction would have been intolerable to most people—although it was common in World War II.

The F-117 is not an invisible aircraft. It can be picked up by high-powered low-frequency radar sets of World War II vintage that are still readily available. But it does reduce the time of acquisition, which is a mark in its favor. If an enemy radar can track a bomber for a hundred miles and plot its track all the way to the target, that bomber is at increased risk when it enters the defense envelope. Any reduction in that acquisition time is desirable, but the same situation can be achieved by a terrain-following aircraft flying close to the ground with a supersonic dash capability.

Some stealth technology is incorporated in all modern combat aircraft, but consideration must be given to its serious limitations. Such airplanes are three times as costly as a comparable conventional aircraft, and many of the materials used in full-stealth aircraft are very susceptible to damage. Some of the radar-absorbent material is actually soft to the touch. This is hardly the type of material to stand up to the bump and grind world of sailors, toolboxes, tie-down chains, and fender benders so common in the close quarters of a carrier. Such bumps and scrapes on metal aircraft pose little penalty aside from the man-hours needed to repair them. On stealth aircraft, many of these small bumps will result in fairly complex repairs due to their exotic materials. If not repaired properly, much of the "stealthiness" will be lost. Even more important is the need for a controlled environment. In part, this is due to the bonding material used on the skin of these aircraft. Like fine furniture, it is subject to accelerated deterioration when exposed to the elements. Even more importantly, the manufacture of stealth aircraft invariably involves the bonding of materials with vastly different qualities of expansion and contraction characteristics due to heat. Over a period of time, these stresses weaken the bond qualities of the airframe. Keeping a stealth aircraft in an environmentally controlled hangar is intended to minimize these stresses. This condition can be easily maintained on land, but borders on the impossible on ships at sea.

Should the Navy abandon stealth technology? Of course not. But its officials should proceed slowly in embracing the concept for its carriers at a time when it cannot even afford to fill its decks with first-rate conventional aircraft. Industry may someday develop the capacity to build a suitable, sturdy, and affordable Navy stealth aircraft. Until then it would be wise for the Navy to procure conventional aircraft and not spend naval aviation into oblivion with a handful of aircraft of questionable capability. One of the lessons of the Gulf War should be kept in mind. Although no F-117's were lost, neither were any A-7s or F-111s that flew similar missions, and they were two of the least-defensible aircraft used in the war.

During the first three weeks of the war, Navy and Marine planes contributed a third of the more than 42,000 sorties. Six carriers were positioned in the Red Sea and the Persian Gulf. The *America* and the *Theodore Roosevelt* had departed Norfolk, Virginia, on December 28, 1990, arriving at their battle stations just in time to participate in the air war. The *Midway,* the *Saratoga,* the *Ranger,* and the *John F. Kennedy* had preceded them. This was the first time that carriers had conducted combat operations from either the Red Sea or the Persian Gulf. These are narrow, restricted waters normally used by large numbers of commercial ships. As a result, carriers were highly vulnerable, and twenty-five percent of their planes had to be used to protect their task groups and other shipping in the area from possible air or missile attack.

At ground bases, there were large numbers of Marine tactical aircraft on the Saudi Arabian Peninsula or aboard amphibious warfare ships offshore. Marine F/A-18 Hornets, AV-8B Harriers and OV-10 Broncos and AH-1 helicopter gunships attacked Iraqi positions in Kuwait. These planes used a variety of air-to-ground ordnance, such as the Standoff Land Attack Missile or SLAM, the high-speed antiradiation missile HARM, the Shrike antiradar missile, Walleye bombs, regular iron bombs, Rockeye cluster bombs, and Harpoon antiship missiles.

The venerable A-6E Intruder and the F/A-18 Hornet bore the brunt of strikes by carrier planes. The last two A-7E squadrons from VA-46 and VA-72 were on board the *Kennedy,* and Corsair IIs were on their 26th year prior to their retirement.

An S-3 Viking antisubmarine plane on a routine surveillance flight from VS-32 on February 20 became the first such aircraft to destroy an enemy ship by bombing. As the only plane aloft capable of tracking a high-speed hostile vessel, it dropped three bombs on the ship in the first ever use of the aircraft's high-altitude bombing equipment. Flown by Lieutenant Commander Bruce Bole and his crew in the northern part of the Persian Gulf, it had to remain high to avoid antiaircraft fire.

The Central Command in Riyadh announced on February 14 that the coalition forces had destroyed 1,300 of an estimated 4,280 Iraqi tanks (a figure raised to 2,100 tanks a week later) 1,450 out of 2,870 armored personnel carriers, and 1,100 of Iraq's 3,110 artillery pieces.

Five days later General Schwarzkopf told the press that the Iraqi forces were "on the verge of collapse."

27

"Like Shooting Fish in a Barrel"

With atrocities mounting in Kuwait, and most of the nation's oil wells set on fire by the Iraqis, the stage was set for the ground invasion. Schwarzkopf ordered it to begin at dawn on February 24. He believed that the interdiction of supply lines by coalition aircraft had weakened frontline Iraqi troops to the point of starvation, while the bombing of bridges had cut off routes for reinforcement or escape.

With amphibious forces threatening to land on the shores of Kuwait's Persian Gulf, Schwarzkopf ordered the main thrust of his forces to advance into southern Kuwait. While a Saudi–Kuwaiti mechanized task force moved into southern Kuwait along the gulf, the American 1st Marine Division, in combination with the 1st Brigade of the U.S. Army's 2nd Armored Division and the 2nd Marine Division, moved forward on their left flank.

Schwarzkopf ordered another massive movement of troops on the extreme western end of the battlefield because the vast majority of Iraq's forces were in Kuwait. With coalition forces in control of the air, he knew the Iraqis would be incapable of countering this move, even if they were aware of it. He was convinced that Iraq's "eyes and ears" had been destroyed.

Prior to the ground offensive, Schwarzkopf had ordered his heavy armored forces to Iraq's border, where the front was lightly defended. (Iraqi commanders had massed their forces in Kuwait in anticipation of an amphibious assault that was never needed.)

After Marine and Arab forces moved rapidly into Kuwait in the east, the American and British armored corps swept rapidly northward into Iraq before turning eastward in a sweeping enveloping maneuver that caught the Iraqi Republican Guards in a trap inside Kuwait. Meanwhile, even farther to the west 2,000 troops of the 101st Air Assault Division, transported by helicopters, moved deep into the Iraqi desert and established a supply base. Women flew some of these helicopters. These troops moved up to the Euphrates River and blocked any escape to the west by Iraqi forces.

The coalition's main force, the U.S. VII Corps with its strong armored units, supplemented by British and French tank divisions and two American airborne divisions, covered 200 miles in two days against little opposition. Then they attacked the positions of the vaunted Republican Guards.

The French 6th Armored Division set up a screen farther to the west to prevent escape or reinforcement, and to confuse the Iraqis as to whether they might be headed for Baghdad.

In one of the great armored sweeps of all time, troops from the 101st Airborne Division secured airfields and blocked escape routes across the Euphrates River. The 24th Mechanized Division then closed a noose around the Republican Guard. There was a brief pause while coalition troops stopped to rest, refuel, and rearm.

When the coalition forces attacked, the forces of the Republican Guards were too demoralized to put up much of a fight, and their resistance was broken within forty-eight hours.

The preinvasion bombing campaign had effectively quarantined the battlefield, and now coalition air forces shifted to the south, to the immediate theater of ground operations. Bombs were dropped in front of advancing troops to destroy barriers and to burn off oil-filled trenches the Iraqis had dug to slow the advance of coalition troops. Bombs, rockets, and missiles destroyed thousands of Iraqi tanks, armored vehicles, and artillery guns.

Tens of thousands of Iraqis began to desert as casualties mounted. Some surrendered to coalition units, while others fled northward, abandoning their positions.

In eastern Kuwait, two Marine divisions broke through the lines despite minefields, barbed wire, and the constant threat of chemical weapons fired by massed artillery. Schwarzkopf said of their advance, "If I use words like brilliant, it would really be an under-description of the absolutely superb job that they did in breaching the so-called 'impenetrable barrier.'

"It was a classic—absolutely classic—military breaching of a very, very tough minefield with a barbed-wire, fire trenches–type barrier. They went through the first barrier like it was water, then went across into the second barrier line."

American, French, British, Saudi, and Egyptian forces followed through the breaches, fortunately meeting unexpectedly light resistance and encountering masses of Iraqis who begged to surrender.

As the Iraqis clogged the roads into Iraq's interior, coalition airplanes hounded them every mile of the way. The roads north of Kuwait City became clogged with retreating Iraqi trucks and armored vehicles. One Navy pilot said attacks against fleeing Iraqis on the highway north of Kuwait City was like "shooting fish in a barrel."

Captain Ernest Christiansen Jr., skipper of the *Ranger,* told the air crews, "It looks like the Iraqis are moving out, and we're hitting them hard. It's not going to take too many more days until there's nothing left of them." One pilot, Lieutenant Brian Kasperbauer, said, "It was the road to Daytona Beach at spring break. Just bumper-to-bumper. Spring break's over."

The *Ranger, Midway, Roosevelt,* and *America* focused their A-6 attacks on two roads leading north from Kuwait City to Basra, the stronghold of the Republican Guards. While they bombed the roads, B-52 Air Force bombers devastated the area with 1,000-pound bombs.

Navy jets went below low clouds and dropped antitank and antipersonnel Rockeye cluster bombs. Once clusters explode they send a deadly shower of armor-piercing bomblets that shred cars and trucks within their range.

Allied troops were now advancing so fast that pilots had to be given new targets as they boarded their planes.

On the extreme left flank of the battle line a Franco-American component moved to the Euphrates River and captured a key Iraqi air base only one hundred miles from Baghdad by driving off an Iraqi army division.

Now there were no Iraqi forces between the coalition forces and Baghdad. Allied forces also developed a defensive line to protect coalition forces in the east.

Schwarzkopf provided details of the ground offensive three days later. He compared the brilliant flanking maneuver to a "Hail Mary" pass play in American football, saying, "I can't recall any time in the annals of military history when this number of forces have moved over this distance to put themselves in a position to attack."

Schwarzkopf said the quick collapse of the Iraqi Army was due to poor leadership. When asked if he considered Saddam Hussein a great military strategist, he said, "He is neither a strategist, nor is he schooled in the operational arts, nor is he a tactician, nor is he a general, nor is he a soldier. Other than that, he is a great military man."

Schwarzkopf denied that the high-tech war was fought like a video game. "It is a tough battlefield where people are risking their lives at all times, and great heroes are out there, and we ought to be very, very proud of them."

On February 25, with the war coming to an end, a Scud missile killed twenty-eight American soldiers and wounded at least eighty-nine others in a Dhahran barracks.

President Bush announced the suspension of military operations in a television address at 9:00 P.M. Eastern Standard Time on February 27 after Iraq agreed to withdraw from Kuwait, and committed itself to comply with all twelve UN resolutions. The president said, "Kuwait is liberated. Iraq's army is defeated. Our military objectives are met."

The president recalled his words seven months earlier, "America and the world drew a line in the sand. We declared that the aggression against Kuwait would not stand, and tonight America and the world have kept their promise.

"This is not a time of euphoria, certainly not a time to gloat, but it is a time of pride, pride in our troops, pride in the friends who stood with us in the crisis."

With operations ordered to end at midnight, the ground offensive had lasted only 100 hours.

Iraq agreed to release all prisoners of war, Kuwaiti internees, third-party nationals, and "the remains of all who have fallen." Further, Iraq pledged to inform Kuwaiti authorities of the location of all land mines, and to meet with coalition military leaders to arrange the military aspects of the surrender.

The allies flew 106,000 air sorties (the Navy and Marines flew 30,000) in forty-three days of fighting and lost thirty-six planes in combat. One hundred and forty-seven Americans were killed in action (thirty-five due to friendly fire) and six of them were Navy airmen, while the number of wounded totaled 357 (seventy-two due to friendly fire). Forty-five coalition prisoners of war were released on March 5. Another hundred American service people died prior to the war in noncombat accidents.

Navy crews shot down only two fixed-wing Iraqi airplanes, and an F-14A off the *Ranger* destroyed an MI-8 helicopter. It was hit by a Sidewinder missile fired by Lieutenant Stuart Broce and Commander Ron McElroft, the VF-1 Wolfpack's commander. The Navy had only one chance to make "kills," and that was during the war's first thirty hours.

The Air Force destroyed thirty-two fixed-wing aircraft and the Saudi air force another two.

Although there were some glitches, such as incompatible radio codes among the various air forces, the air war was run with remarkable expertise. But losses due to friendly fire were intolerable. Their unusually high number was due to confusion on the battlefield, poor visibility, a lack of positive identification of ground vehicles, the intensity of the ground combat, and the rapid advance of coalition forces.

Coalition forces had a priceless five and a half months to prepare for the war with Iraq, a circumstance not likely to be repeated in future emergencies. Although coalition forces were inadequate to counter an early Iraqi invasion, Saudi Arabia had the vital infrastructure of airfields, ports, fuel, and transportation resources to permit a rapid buildup. It was soon apparent that the realistic training given American and ground personnel in the preceding years paid off to an unprecedented degree. Aside from their exposure to live fire, crews said their stateside training exceeded combat conditions in complexity.

The Air Force supplied about 200 aerial tankers, but they were inadequate to refuel the large number of sorties flown by all air forces. The need for more carrier-based tankers, such as were used in Vietnam, became acute.

The need for more expeditious processing of intelligence film to strike leaders was quickly apparent after war broke out. Unfortunately, the need for real-time video data was not recognized prior to the war.

Despite the achievements of the Marine Corps, it's AV-8 Harrier proved highly vulnerable to attacks by heat-seeking missiles because its large engine is located in the center of the aircraft. It emits a high level of heat, which attracts such missiles to a critical structural area of the aircraft. An unusually large percentage of Harriers were lost—compared to other aircraft—because of this problem. F/A-18s were also hit by such missiles, but no planes were lost. They survived because their engines are located in a less vulnerable area.

Marine ground troops lacked adequate night-vision equipment for all its men, which could have hampered their activities if they had had to do more night fighting. This was a serious oversight, and could have had tragic consequences.

The first troops in the area in the summer of 1990 would have been at the mercy of Iraq's chemical and biological weapons, because the Pentagon had no policy regarding vaccinations against such agents. Enormous casualties could have resulted, which could have been prevented by protective masks that filter out biological agents. But the services at first had not developed an adequate warning system.

The success of the Tomahawk cruise missile, originally developed by the U.S. Navy, has been overestimated. The psychological impact of its accuracy often exceeded the devastation caused by its 1,000-pound warhead. Two hundred and eighty-eight Tomahawks were fired from surface ships and submarines in the Persian Gulf. Only half of them hit their intended targets. Unlike FLIR (infrared) and laser systems, which cannot be used successfully on cloud-covered targets, the Tomahawk is not dependent upon good weather.

The six U.S. Navy carriers that participated in the war performed magnificently. Typically, workdays were twelve hours long. Pilots proceeding to and from their targets were often forced to fly in formations of dozens of aircraft clustered around Air Force tankers at night and often in bad weather awaiting a chance to refuel. After eight hours in the air, and exposure to some of the heaviest flak ever encountered in war, crews concluded their flights with night-carrier landings. That they completed their missions with only six fliers killed in action, and two listed as missing, with only thirteen Navy and Marine aircraft lost speaks volumes for their professionalism. The fact that they were involved in an extremely small number of operational accidents (two Navy fliers and seven Marines died in ten accidents) is well nigh incredible.

Of equal importance, the six carriers on duty had fewer outstanding material requisitions (planes grounded for lack of parts) than was common on a single carrier just a few years previous, and often in a single squadron a few years before that.

Pilots feared running into one another more than they feared encountering Iraqi fighters, antiaircraft fire, or missiles, particularly on days when 2,000 sorties were sent out. The fact that these daily sorties were flown over a twenty-four-hour period helped to reduce the chances of midair collisions. Strict traffic control measures were instituted at the start, and good air discipline was maintained. As a result, there was not one midair collision, although an F-111 hit a tanker while it was being refueled.

Iraq had an effective antiaircraft system, but it was destroyed in the war's early days. The Iraqis made extensive use of hardened shelters to protect their lines of communication, but deep penetrating bombs were used against these hardened sites. Once they were knocked out, Iraq's armed forces became blind and deaf, permitting coalition air forces to fight with a minimum of losses. Once Iraq's ability to defend itself was dissipated during the early days the end was inevitable.

Even on the battlefield the extreme accuracy of allied bombs made the difference between success and failure. F-16s that dropped 1,000-pound bombs near Iraq's fifty-two-ton tanks merely moved them laterally as they continued to move forward. It was soon learned that a ten-foot miss—exceptionally good for unguided bombs—was not good enough. Tanks had to be hit to render them useless. Two-thousand-pound guided bombs, with their spun-steel casings and fourteen-inch titanium noses, could penetrate steel or concrete. When these bombs were dropped with a laser seeker to give them a deadly accuracy, nothing could survive their impact.

During and after the war the Navy was criticized by newsmen for its apparent failure to be as effective with its air strikes as the U.S. Air Force. Some of this misconception was due to the Navy's inability to prove what they had accomplished. Inevitably, however, the Navy's air arm had only a third of the planes assigned to the Air Force theater, with 1,800 available aircraft. On any one day the Navy flew an average of 400 strikes, while the Marine Corps and other coalition air forces flew another 400, and the Air Force averaged about 1,200 sorties. In actual numbers the Navy had a three-to-one imbalance. A total of 88,000 tons of bombs were dropped during the war by all air forces. Only a small fraction of them were guided bombs.

But the war inevitably became a media event, with strong demands by worldwide television networks for spectacular footage of strikes. Unfortunately, the Navy could seldom produce them. U.S. Air Force planes—such as the $100 million F-117, the F-111, and F-15E—were all equipped with very high-quality laser recorders. The Navy had only one such plane—the A-6. The Air Force extensively used the 2,000-pound penetrating bomb, and they had the video proof of how deadly it was against hardened sites. The Navy's old A-6 did an extraordinarily good job with its laser guidance system, but it had far fewer of such strike aircraft than the Air Force. And, the A-6's recording equipment was an archaic two and a half-inch magnetic tape, whose quality was almost unusable when converted to a VHS television tape. In effect, the Navy had nothing to prove to television audiences what they had accomplished, although the success of their strikes was on a par with similar strikes by the Air Force.

Actually, the best strike flown in the war was made by an A-6 off the *Theodore Roosevelt*. The crew put four of its Mark 84 20,000-pound bombs across Iraq's largest power plant. That one Navy airplane destroyed twelve percent of Iraq's electrical power, but the quality of the plane's tape was so poor it could not be used on television.

The problem was created much earlier when a high-level decision was made within the Department of the Navy to put all intelligence for weapons delivery in the F/A-18 and not in the bomb. These officials also decided to put greater reliance on old-fashioned iron or "dumb" bombs with "smart" delivery systems. The system was upgraded only for the A-6, an old airplane dating back to 1958.

The Air Force made a different decision, putting some intelligence into their weapons. Air Force officials theorized that if you're going to hit a small target with devastating accuracy, you can't put enough intelligence into the airplane delivering the weapon because it cannot accurately predict the bomb's environment and characteristics. In other words, with present-day technology it is impossible to integrate all factors in a precise manner without putting some intelligence into the weapon. The primary systems used in the Gulf War were FLIR (forward-looking infrared) and laser systems with some television systems.

The Air Force's ability to record the results of a mission on a high-resolution tape gave it an advantage in justifying its equipment. And the cost of doing so was relatively small. The equipment is available on an off-the-shelf basis at any radio shop. Originally the Air Force bought this equipment to train their pilots. It proved a marvelous training aid for pilots during debriefing sessions.

The Navy's modern F/A-18 does not have FLIR or laser systems, which reduces its effectiveness because it has to rely primarily on "dumb" bombs. Although the A-6 can drop the more sophisticated bombs with great accuracy, its two-bit recording equipment makes it impossible to develop a usable video. And, with only one hundred A-6s remaining in the fleet, the entire number of Navy strike airplanes is seriously compromised. Whoever made the decision demonstrated incredibly poor judgment.

The Navy was also criticized for using twenty-five percent of its planes to protect its carriers and supporting ships, while the Air Force chose to support its bases with fewer fighters. The Navy also provided greater protection for its bombers, in particular the subsonic A-6 Intruder, because of its vulnerability to modern fighters in daylight. It was thirty-three years old in 1991 and long since should have been retired, but there was nothing to replace it. The Air Force's F-15E could largely take care of itself on the battlefield, and the Navy needs a comparable airplane.

Much of the criticism about the use of large numbers of planes for protection was not justified. The 200 coalition ships at sea—half of them American—demanded protection. Throughout the war there was always concern about what action Iran might take, so an F-14A Tomcat aerial barrier was placed between the ships and Iran. Ship batteries can offer some protection, but the problem of identification of 2,000 aircraft returning over the gulf from missions proved insurmountable for shipboard gunners. Aircraft flew such a route to avoid flying over trigger-happy coalition ground forces. The extensive use of ground-to-air missiles by these troops made such flights particularly dangerous. With Navy aircraft overflying the entire area, identification of friend and foe proved a simple matter.

In the early days Iraq had a substantial Navy, including missile ships and patrol boats, which posed a threat until it was liquidated.

Throughout the war the Navy put up a minimum of four strike aircraft on a twenty-four-hour-a-day basis to be responsive to attacks by small patrol boats. If they found no boats, they were instructed to bomb targets in Kuwait. These missions were listed as defense sorties, which was really a misnomer. For example, every Air Force AWACS (Airborne Warning and Control System) plane had four F-15s flying protective cover, but these planes were called strike aircraft. Actually there was no difference between Navy strike aircraft flying fleet protection and Air Force fighters protecting their early warning planes. What counted in the Navy's disfavor was how their aircraft were designated.

Actually, Navy combat air patrols covered all coalition bases in Saudi Arabia and provided a massive air defense for the northern and eastern flanks of the coalition's ground, sea, and air forces.

Air Force General Horner never criticized the Navy's efforts, saying, "I save myself thousands of combat air patrol sorties because the Navy provides a big barrier up there between Iran and my most northern force."

The Navy–Marine amphibious feint off Kuwait's coast was the largest such operation since the Korean War. Prior to the air and ground war, amphibious forces and other naval forces in the Persian Gulf exercised extensively to convince Saddam Hussein that such an assault was imminent. As a consequence it tied up Iraq's entire southern flank.

After the war ended General Schwarzkopf explained their presence. "We continued heavy operations out in the sea because we wanted the Iraqis to believe that we were going to conduct a massive amphibious operation. The Iraqis believed that we were going to take them head-on in their most heavily defended area. We launched amphibious feints and naval gunfire so that they continued to think that we were going to be attacking along the coast, and therefore fixed their forces there. Our hope was that by fixing the forces in this position and with a ground attack, we would basically keep the forces here [in the south], and they wouldn't know what was going on in this area," he said, pointing to a map of the area west of Kuwait. "We succeeded in that very well."

Iraq's Scud missile proved to be a nonweapon. Mobile Scuds were hidden beneath bridges, culverts, and hangars until they were moved into firing position. Therefore they were almost impossible to locate from the air. Although the Scud was not effective as a military weapon, politically it became a monster. At the height of the Scud campaign, coalition air forces were putting up 300 strikes a day against its sites. Half of this anti-Scud campaign was to bolster President Bush's plea to the Israelis to remain on the sidelines despite Scud attacks on their country. There is little doubt that fear that Israel might join the war against Iraq was overdramatized. It is doubtful they would have done so despite their belligerent talk. Coalition forces were fighting their war, and there was nothing more that Israeli forces could have done that was not already being done.

The mobile Scud proved to be a true needle in a haystack because it was largely invisible in the daytime. At night, Iraqi crews drove their trucks out of their hiding places, fired their missiles, and quickly backed the trucks into underground culverts or beneath bridges. Crews particularly took advantage of cloudy nights when FLIR and laser systems could not be operated. Even when a combat air patrol was maintained above them at all times the Scud's truck could fire and be back in hiding before a plane could attack it.

Reports of successful attacks against Scuds proved inaccurate. Actually, very few Scuds were destroyed on the ground. At one time it was thought the use of night-vision goggles would be helpful to crews, but they achieved few successes. The Scud became a terror weapon with its 2,200-pound warhead because although its accuracy was erratic, psychologically and politically it was a potent weapon easily capable of stampeding civilian populations.

Patriot missiles accomplished little in the anti-Scud campaign, even though they intercepted forty-three out of forty-five missiles. The only time a Patriot missile destroyed a Scud's warhead was when it hit head on and exploded it in the air. In most cases the Scud was destroyed, but its warhead fell to the ground and exploded. Those aimed at Saudi Arabia for the most part landed in vacant desert land, but this was not true in Israel. The same effect would have been gained if the Patriot had never been launched. Patriots need to be redesigned with a warhead of sufficient size to destroy a missile's warhead if they are to justify their high cost of $1 million each.

Unlike the war in Indochina, where North Vietnam was divided into route packages and targets were assigned to the Air Force and the Navy, this could not be done in the war against Iraq. The combat theater was too small, and the forces were too interdependent. For example, the Navy relied upon the Air Force AWACS and tankers, while the Air Force depended to a large extent on the Navy's EA-6B for electronic information. Although it was not practical to divide Iraq and Kuwait into assigned zones, in some instances this was done on an informal basis. Targets around Basra in Iraq's northeast corner were assigned to the Navy because they were geographically closer.

Air Force Lieutenant General Horner, the Joint Forces Aerial Commander responsible for all strike aircraft, kept reminding those who represented the various air forces under him that "there are plenty of targets out here." He inspired them to think not merely of their own service, but as a team.

At first the Navy was assigned responsibility to destroy Iraq's key bridges. Due to bad weather and insufficient forces equipped to release laser-guided bombs (only the A-6 had this capability) the Navy fell behind during Phase 3 in their assignment to destroy approximately forty bridges. Horner typically faced the problem, telling Captain Lyle G. Bien, the Navy's senior strike planner, "We need to get on with getting this war over in a hurry, and one of the things we have to do before preparing the battlefields is to get these bridges cut. It's going to take too long for the Navy by themselves to take out all those bridges. We'll do it as a combined effort."

To some members of the press covering the war this appeared to be a Navy failure—that the Air Force had to be called in to complete the job. Such was not the case. The F-18's accuracy with unguided bombs was excellent. Navy planes, despite their limited equipment, had done a fine job, and they continued the same missions, but as part of a joint effort. If there had been more time, Navy planes could have dropped every bridge in the theater.

Coalition leaders were surprised by how quickly the Iraqis completed bypasses around downed bridges, and by their effective use of pontoon bridges. They should not have been. The identical pattern was seen during the Korean War and in Vietnam.

Bridges have always been devilish structures to destroy. Some Iraqi bridges could not be knocked down by 2,000-pound unguided bombs. In such cases the Air Force used their penetrating weapons, which were not available to the Navy.

Throughout history coalition armies have a record of fighting among themselves more than against their enemies. There was almost none of that in the Persian Gulf war. General Horner made it clear from the start that he did not care which service gave him the weapon that could do the best job on a target. He told his staff, "I don't care which service or nation is involved. I don't care if the Navy doesn't fly a single sortie on a particular day if it doesn't fit the schemes and I don't care if the Air Force doesn't fly."

Horner never worried about keeping everybody equal. Instead he focused on whatever was needed to prosecute the war in the fastest and most efficient manner.

Air Force officials after the war estimated that it took 9,000 bombs in World War II and 300 unguided bombs in Vietnam to destroy a point target such as an aircraft shelter. The Air Force claims that only one precision bomb was needed in the Gulf War. Actual figures later doubled this number. It is still a vast improvement over previous wars.

Officers like Horner and Schwarzkopf, and all other top men in the 800,000 coalition force, instilled in each man and woman that they must remain almost unconscious of their specific services or even nationality. At the front, positioning of people relative to one service versus another was almost lacking. Schwarzkopf set the pattern that all services were as one, with identical insignia on their uniforms (only wings, name, and rank in Horner's command), and without squadron or unit patches. These conditions contributed to a sense of oneness that is vital to success in a coalition involving diverse people from twenty-eight nations.

Prior to the war, Arab-speaking nations had little contact with or knowledge of the problems to be faced while working with American and British or French in a military undertaking. But they developed a singleness of purpose that focused primarily on winning the war as quickly as possible with as few casualties as humanly possible. No one worried that someone's feelings would get hurt. For Horner, who managed the air war, this former Iowa farmer achieved the impossible—he got the various air forces to work as one.

Unfortunately, after the war the old service rivalries were renewed, as the American Army, Air Force, and Navy jockeyed for position as appropriations began to be allocated for each service. Such sentiments are detrimental to all members of the Department of Defense, and to America's military position in the world.

Although American Navy aviators fought with superb competence, they were poorly equipped to fight a modern air war. Surface admirals in the Navy had long fought against airpower, insisting during the first forty years of the twentieth century that the battleship was the supreme expression of military power.

The debacle at Pearl Harbor on December 7, 1941, appeared to have destroyed this argument forever, particularly after World War II, when carrier admirals took charge of the Navy and made carrier battle groups preeminent.

But then a series of admirals in the top positions of the U.S. Navy became obsessed with nuclear warfare, and submariners assumed command of the Navy. Although the insurance value of the nation's nuclear retaliatory capability helped to prevent World War III, a number of wars have been fought without the use of nuclear or thermonuclear weapons. For moral and political reasons, no nation wanted to be the first to use such awesome weaponry. But wars continue to be fought, and between World War II and the Persian Gulf War, large numbers of American men and women died needlessly because the nation's armed forces were not prepared to fight a nonnuclear or conventional war.

In the 1980s, the Navy became a house divided, particularly after John F. Lehman became Secretary of the Navy under President Reagan.

Naval airmen, who believed that the battleship was finished in the Navy, were appalled when Lehman ordered that four battleships be returned to service and refitted at enormous cost. He justified his decision by saying they were needed for cruise missiles, although *Spruance*-class destroyers can carry an equal number mounted vertically and at far less cost. Lehman fought for a 600-ship Navy at a time when there was no justification for so many ships. As a result, appropriations for aircraft were reduced. Lehman actually wanted to build battle groups around the four battleships to relieve the extraordinary operational commitments assigned to the Navy's carriers. This action would have committed the Navy to tactics that had long been discredited and cost the nation billions of scarce dollars over the years. Fortunately, Lehman did not succeed, but his policy of downgrading airpower in favor of surface and undersea ships came close to destroying naval aviation. The results of his actions were all too apparent in the Persian Gulf War. Fortunately, the dedication of the men and women of the U.S. Navy made up for his incredible shortsightedness.

One tragedy of the Gulf War is only now emerging, despite efforts of the Department of Defense to deny its implications since the war ended. On October 23, 1996 the department conceded that Iraqi chemical weapons may have caused medical problems for at least 20,000 service personnel who were exposed to vapors from the demolition of the huge Khamisiyah Iraqi weapons depot in 1991. A spokesman said that battlefield logs suggest that soldiers may have been involved in another demolition in the area as far as thirty miles away. He admitted that the figure could be much higher, and efforts are now under way to screen all military personnel who could have been exposed to these chemicals. Thousands have charged they are suffering from the ill effects caused by fumes discharged from the depots, but until now their urgent pleas for recognition that at least some of their problems are service-connected have been largely ignored.

Initially only 5,000 troops were believed to have been within 12.5 miles of the weapons complex when it was destroyed March 4, 1991. Now it is admitted that a nearby complex was destroyed close to Khamisiyah, and that even a third plant making Iraqi artillery shells may have caused problems. In this event 840 artillery shells were destroyed on March 12.

When one considers that these thousands of American men and women, who fought so well for the free world, should be so shabbily treated, it is evident that a thorough investigation is called for. Let's start at the top with former President Bush and his administration.

Captain Lyle G. Bien, the Navy's senior strike planner on General Horner's air staff, put the successful ending of the war effort against Iraq in true perspective when he said, "No discussion is complete without mention of the fighting hearts of the troops of Desert Storm. Under brutally sparse living and working conditions, they maintained their equipment and their spirit on a plane that made the term troop morale obsolete—we called it troop euphoria. They were bolstered throughout by compassionate leadership, a rousingly supportive nation and a clear-cut understanding of their sworn duty. But in the end they were simply a half million proud, hardworking, determined kids who went to war believing in their nation's cause, like generations of proud Americans before them. The American people thrust the banner of our nation's freedom in their hands, and added the charge to guard it with their honor. They could not have chosen more wisely."

In sharp contrast to the bravery and honor exhibited during Desert Storm was the behavior on September 7, 1991, when the Tail Hook Association gathered at the Hilton Hotel in Las Vegas, Nevada, for its annual meeting. Several thousand Navy fliers and their families attended various meetings, but the majority did not stay at the hotel. A few officers drank too much and made perfect asses of themselves. There were several instances of sexual harassment against female pilots, who were groped and fondled in the hotel's hallways despite their protests. The officers who took part disgraced themselves for conduct unbecoming an officer.

When female pilots protested to their superiors, and word of these deplorable actions leaked to the press, the situation developed into a full-blown scandal that has haunted the U.S. Navy ever since. It has become a witch hunt that has lowered morale and helped to create a serious shortage of pilots in the late 1990s as hundreds of pilots have resigned from the Navy in protest.

The chief of naval operations was there, along with many four-star officers who could have put a stop to this nonsense before it got out of hand. Instead, they avoided their responsibilities and sought scapegoats among lower-ranking officers who were totally innocent of any wrongdoing. Hundreds of officers were found guilty by association and their records were "flagged," which often denied them promotions. Squadron commanders were especially singled out when offending officers were in their units.

High-ranking officers failed to act quickly to punish those responsible, and to take steps to assure that such conduct is never repeated. Instead they left the problem in the hands of lower-ranking officers without the authority to act decisively. As a result, the media kept harping on the subject, spurred by members of the feminist movement, who tried to use the situation to force the Navy into giving women more flying jobs. Unfortunately, their comments have only served to inflame emotions, to the detriment of women serving as pilots and those who hope to become Navy pilots.

It is imperative that steps be finally taken to rectify the failure of the Navy to resolve this situation, and for all parties to cease inflammatory rhetoric that has caused so much damage to morale and naval aviation's effectiveness.

After coalition forces soundly defeated Iraq's armed forces in 1991, President Bush exercised incredibly bad judgment in not insisting on the removal of Saddam Hussein as president of Iraq. He continued to survive, in effect "thumbing his nose" at the United States and its allies. When Hussein refused to permit further inspection of his military capability, the United States and Great Britain participated in four days of aerial strikes under "Desert Fox" on December 16–19 of 1998. A thousand sorties were flown and 400 cruise missiles were launched, including some missiles with 2,000-pound warheads, which are more effective than smaller warheads. Twelve Navy ships were involved, along with Air Force bombers such as the B-52 and the B-1B. The F-117 was not used because its iron bombs would have destroyed its effectiveness as a stealth airplane. This aircraft was designed primarily to drop nuclear weapons. Every effort was made to avoid hitting civilian targets, although the Iraqis took newsmen to a hospital where civilians had been killed or wounded. This apparently was an isolated case. Bridges, water, and electric sources were not bombed, and there was no interruption of service.

As one large bomb was attached to an F/A-18 Hornet on the carrier *Enterprise,* its graffiti included "Lewinsky's Lip Trainer," a reference to Monica Lewinsky's affair with President Clinton which caused him so much anguish.

The Pentagon claimed after the attacks that eighty-five percent of the one hundred targets listed for strikes were hit, with a seventy-four percent success rate. However, these were not vital targets that would seriously impair Iraq's war-making capability.

Secretary of Defense William Cohen said the strikes have delayed ballistic missile production by a year, and sharply reduced Iraq's threat to its neighbors. These targets included twenty command posts and nine sites linked to the Republican Guard. The Pentagon claimed 600 to 2,000 Republican Guards were killed. Iraq claimed only sixty-two, but evidently this number included many of the Guard's leaders.

Factory damage was described as heavy by a Pentagon spokesman. He said weapons factories were special targets and there was heavy loss of irreplaceable machinery, setting back war production by two years. He claimed that the attacks were more successful than in any previous aerial strikes. His claim appears to be substantiated as far as accuracy, but such an effort should have been expended on more worthwhile targets. Fortunately—and this is unusual—no lives were lost.

When President Clinton cancelled the strikes he said he was confident "we have achieved our mission."

Attacks against biological and chemical plants were restricted to research and development facilities and not against production facilities to avoid releasing huge clouds of lethal gases that could kill people and animals in areas outside Iraq. Years earlier the Soviet Union had experienced an accidental release of anthrax that spread over much of Siberia and even into Manchuria. The decision to avoid production facilities was undoubtedly a difficult but correct one in view of the possible disastrous consequences.

Bombing barracks is an exercise in futility. The allied forces should have destroyed bridges and water and electrical facilities, which would have caused havoc throughout the country. Such a condition might have led to Hussein's overthrow. Again, like Vietnam and Desert Storm and other such operations, the attacks caused no appreciable long-term damage to Iraq and Hussein retained his hold over the nation.

The cruise missiles used on Iraq depleted the Navy's stock to almost zero and will have to be replaced at a cost of more than half a billion dollars.

President Clinton's 1999–2000 military budget calls for a sharp increase in spending to rectify serious shortcomings in the armed forces. It will be the largest increase in military spending since the Cold War ended.

Epilogue

After years of devotion to the concept of mobile airpower, can Navy airmen now relax, feeling secure that the nation appreciates the need for continuance of carrier aviation?

Events in recent years have proved otherwise, as the years ahead should also prove. The United States defense establishment rightly believes that no concept should be perpetuated unless there is a vital need for it. Times change, and old ideas must give way to new.

Has this time come? The Navy firmly believes that the attack carrier striking force is an essential weapons system in support of the nation's military objectives. Some call it the sea-based striking force. How large a part should attack carrier aviation play in the years ahead?

Arguments are strong that carrier aviation is obsolete, that it represents an expensive duplication of effort. Critics reason that atomic and thermonuclear missiles now take the place of most aircraft missions, but they forget that all wars since World War II have been nonnuclear, as future wars probably will be for political and moral reasons. A conventional explosive in a single missile does relatively minor damage. It takes thousands of such weapons to cause significant damage, due to relatively small warheads, which rarely exceed 1,000 pounds.

Air Force proponents believe a force of strategic bombers can do a more effective job than carrier aircraft because of their electronic suppression equipment. Carrier planes also use such equipment, and both carrier planes and Air Force bombers have proved their ability to penetrate heavily defended targets, such as were found in North Vietnam and in Iraq.

If Air Force bombers are properly used against large strategic targets that need a number of bombs to destroy them, and carrier aircraft are used against specific pinpoint targets, they will complement one another.

Both the U.S. Navy and the U.S. Air Force would serve the nation more sensibly if they solved their differences in private and finally agreed to roles that each can perform best. Airpower would be strengthened and billions of dollars would be saved each year in the defense budget.

For the foreseeable future, naval aviation is an indispensable part of America's defense establishment. The responsibilities of the United States are worldwide. They will remain so in the years to come.

Although the free world must be prepared to wage a nuclear war, the possibilities of such a conflict have receded because most countries appear to realize the futility of such a war. But limited wars will always pose a threat to mankind's security.

The U.S. Navy's role in a nuclear war has been overemphasized in the past. In such a war, with ballistic missiles delivering the massive assault, naval aviation's role will be vital but limited to specific target systems.

In limited wars, manned tactical aircraft take on a new significance. Their quick reaction time and ability to support amphibious and ground operations on distant battlefields are unlimited.

After World War II, emphasis was placed on the Soviet–European land mass. With the dramatic changes within the Soviet Union, and the collapse of the Warsaw Pact as Russia's satellites gained their freedom, Europe is no longer the focal point for "hot spots" that it was in the past.

In the future, the underdeveloped areas of Africa, the Middle East, and Southeast Asia will continue to be areas for exploitation by enemies of the free world. In these areas, where limited wars could occur, sea-based tactical aviation can play a decisive role.

Nationalism has reared its proud head in many new countries. In most it is a source of satisfaction to democratic institutions. In others, it bodes ill for the free world, because nationalistic ardor is often tinged with fascism or communism.

No one can foresee how many hot spots will occur throughout the world, where they will be, or when they will erupt. We only know from tragic experience that there will be conflicts somewhere.

In a world constantly in ferment, it is imperative that the United States provide suitable protection against all types of known and unknown enemy weapons.

The attack carrier deserves its rightful place. Its maneuverability is an asset because ballistic missiles would find it an impossible target, since they rely upon pretargeted information.

Thirty-knot carrier speeds also make them extremely difficult submarine targets and—with modern missile defenses—extremely hard targets to attack from the air. As missile systems improve in quality, reliability, accuracy, and survivability, American attack carriers will undoubtedly be released from general war alert commitments. They will retain, however, a dual capability for conventional and nuclear weapons and remain in the general war strategic reserve as long as aircraft are capable of delivering weapons.

In a general war, precision strikes after the initial nuclear bombardment will be vital for the support of ground troops and in the destruction of targets missed by missiles.

The missile threat in 1991 was reduced, by agreement with leaders of the United States and the Soviet Union to a strategic reduction treaty. It was signed by the United States and Russia December 5, 1994. This treaty calls for a forty-eight percent cut in Soviet ballistic warheads, and the United States agreed to a thirty-nine percent reduction. Most importantly, the Soviet Union agreed to sharply reduce all land-based missiles (eighty percent of her missiles were in that category that year) from 6,500 to 2,000. Such a treaty has a strong stabilizing influence, because the extreme vulnerability of these ballistic missiles makes them tempting targets for would-be aggressors. A major stumbling block for a time withheld ratification of the treaty. The Republic of Ukraine, which had many missile sites in its territory, refused at first to turn them over to the Russian republic, although they were among the missiles that Russian leaders had promised to destroy. The leaders of the Ukraine finally agreed to do so in early 1994.

Since 1988 the Soviet Union has sought to include navies in strategic arms reduction talks, but the United States has refused, claiming America's global economy requires a strong Navy to protect its shipping lanes. But each nation, on its own, had started to reduce the size of its fleet. In mid-1991, Admiral Vladimir N. Chernavin, Commander of the Soviet Fleet, announced an intriguing reduction in numbers of men and ships amounting to approximately twenty-five percent, thus making official what had already been taking place. He said *Typhoon*-class guided missile submarines, third generation attack submarines, *Kirov*-class cruisers, and the new *Sovremenny*-class destroyers would now form the basis of his nation's fleet.

Despite this reduction, the Soviet Navy remains a formidable force for fighting a defensive war. Its combat capabilities will be maintained because this decision involves the decommissioning of older ships that have outlived their usefulness. (The United States is following the same course.) Both navies now emphasize quality instead of quantity. Prior to Chernavin's announcement it was believed that the Soviet Union had 323 submarines, 227 surface ships, 395 patrol and coastal vessels, 331 mine warfare ships, and 77 amphibious craft.

American Navy strategists now maintain that a force of twelve attack carriers is necessary to provide quick reaction by tactical aircraft to meet any war emergency. They point out that their presence alone can prevent a war from starting. Despite tight defense budgets, the U.S. Navy has been authorized twelve carriers, although one will be used only for training. This is highly significant, because the 1993 authorization of a 452-ship Navy has been reduced to 346 ships of all types. A battle line of eleven carriers appears to be the minimum for the Navy to fulfill its worldwide commitments into the twenty-first century.

Along with the reduction of carriers from fifteen to twelve, the Navy plans to cut back its active air wings from thirteen to ten by 1999. Three new *Nimitz*-class carriers are scheduled to replace older carriers by the end of the century.

The F/A-18E and F will replace the A-6, although fewer aircraft will be procured than originally proposed. The F-14 will undergo modest improvements to make it a more efficient air-to-air fighter and give it the capability to drop smart bombs as a strike aircraft. With the Navy's cancellation of the P-7 antisubmarine warfare aircraft, the P-3C is expected to undergo extensive upgrading.

After the heavy losses of aircraft at the Battle of Midway during World War II, it was apparent to top Navy officials that future plans must allow for such losses with assignment of two air wings to each carrier. In the late 1940s, the Navy reduced that requirement to one and a half air wings.

When the Korean War broke out, AD Skyraiders and F4Us had to be taken out of storage to provide five squadrons from each air wing. Since *Essex*-class carriers could operate only four squadrons, the spare squadrons were made into air task groups that were group equivalents. The new plan calls for a reduction from the thirteen-fighter air wing strength in 1990 to ten, with only one reserve fighter wing. Ideally there should be a reserve wing for each carrier.

The Navy has fought three wars—Korea, Vietnam, and the Persian Gulf conflict—with little or no air opposition. This was fortunate, because in each case there were insufficient spare aircraft in the system to provide for heavy losses, such as were incurred during World War II.

Although the Navy's two reserve air wings were not called up in the Persian Gulf War, some support groups were given active-duty assignments. This was a departure from the policy in the Vietnam War, when President Johnson refused to call up the reserves. But aviation units were limited to C-9A Nightingale medical evacuation aircraft and four units of HH-60 helicopters that were sent to the war front as search and rescue aircraft. The latter were needed because the Navy had no such units on active duty. The two tactical air reserve wings, CVWR-20 and CVWR-30, fly frontline aircraft such as F-14s, F-18s, and EA6Bs—the same as squadrons on active duty—and they are manned by experienced and competent aircrews. Unlike their U.S. Air Force counterparts, they must have a carrier on which to deploy, and all of the Navy's carriers have their own assigned wings. These organizations could have been deployed from shore bases, but they would not have had the support infrastructure that is provided by carriers. They were also available as attrition replacements for the active wings, but that is a costly way to provide such replacements.

The need for only two reserve air carrier wings needs to be reassessed. Unless definitive plans are made for their ultimate utilization, they should be abolished as a cost-saving measure. But eliminating such an experienced pool of talented professionals would have tragic consequences. The answer may be to abolish the wings and assign their squadrons to specific carriers. Such an action would provide these carriers with the nucleus for rapid expansion in time of war. The present status is costly and unrealistic, and nonuse of these reserve air wings in Desert Storm is added proof that a change is needed.

Retired Rear Admiral James D. Ramage, who served with distinction in World War II and Vietnam, has long been concerned by the Navy's failure to assign an adequate reserve of airplanes to the fleet.

"In order to provide for losses," he says, "and give naval aviation the capability it must have to meet any war threat, it is apparent that more aircraft must be assigned to each carrier. The ideal would be two wings per carrier." He reminded critics that "a carrier without its air wing becomes an expensive transport."

Further changes are in store for the U.S. Navy in its organization and mission in the years ahead. Some of these changes were revealed when President Bush disclosed a bold new plan on September 26, 1991, to reduce the nation's reliance on nuclear weapons. He announced that he was immediately withdrawing short-range nuclear weapons from Europe, abandoning the mobile intercontinental missile, and cancelling the short-range attack missile SRAM-2. He also called for the elimination of all multiple-warhead missiles.

President Bush said, "We can now take steps to make the world a less dangerous place than ever before in the nuclear age." He called on the Soviet Union to join the United States in building only one kind of a less-destructive, ground-based missile. Some of these steps, the president said, will be taken unilaterally. But he called upon the Soviet Union to "go down the road with us" by making the same reductions. He warned that if such a step is not taken "a historic opportunity will have been lost."

Prior to his historic speech from the Oval Office the president talked to Soviet President Mikhail S. Gorbachev and Russian Federation President Boris Yeltsin, both of whom praised the president's initiative. They said his proposals were too complex for them to reach an early decision, but were "very positive."

The unsuccessful coup by hard-line communist conservatives to oust Gorbachev from power caused the Bush administration to reevaluate its position in regard to short-range nuclear weapons because thousands of them were in artillery shells and battlefield weapons in some of the republics trying to break away from the USSR. The president's greatest fear was that some of them might fall into the hands of terrorists.

Therefore, at the president's request, Defense Secretary Cheney was charged with development of a new program to reduce America's nuclear arsenal to enhance the world's stability and reduce the chance of another war. The president requested a proposal that would cover all tactical nuclear weapons, which, he said, were no longer needed to defend Europe as a counter to the Soviet Union's former advantage in conventional weapons and manpower. Bush claimed that the collapse of the Warsaw Pact, and the Soviet Union's internal problems, practically eliminated the Soviet Union as a threat to peace.

Bush also called for a sharp reduction in other nuclear weapons, but he continued to urge production of the B-2 bomber and the Strategic Defense Initiative. In an otherwise wise and provocative plan, Bush's support for these weapons systems demonstrated his mindset about two unneeded and outrageously expensive programs that long since should have been abandoned.

Nuclear cruise missiles were ordered to be removed from the Navy's surface ships and attack submarines. Many will be destroyed, while others will be placed in storage. Nonnuclear cruise missiles, such as the Tomahawk, which was used successfully in the Persian Gulf War, will continue in use.

Strategic bombers, which have been on around-the-clock alert for years, were ordered removed from their alert status. Intercontinental ballistic missiles, scheduled for deactivation under START, were also taken off alert. These actions were taken two days after the president's speech, when Defense Secretary Cheney ordered the crews of 280 strategic bombers and 450 Minuteman II missiles to cease their twenty-four-hour alert status.

In challenging the Soviet leaders to "match their words with their deeds," Cheney announced that he had signed an executive order implementing the first part of the president's decision. At a Pentagon briefing the defense secretary said, "It is, in my opinion, the single biggest change in the deployment of U.S. nuclear weapons since they were first integrated into our forces in 1954." He noted that the Soviet Union had never placed their bombers and missiles on continuous alert.

He warned that such a relaxation in the operation of strategic weapons must be followed by similar moves by the Soviet Union. This decision could be reversed in twenty-four hours, he said.

General Colin Powell, chairman of the Joint Chiefs of Staff, provided additional details. He said that only forty of the Strategic Air Command's 280 B-52 and B-1 bombers had been on continuous alert. He said they could carry "several hundred" nuclear bombs and missiles, although he refused to divulge the exact number for security reasons.

Powell said the president's new plan was a "historic turning point" in the United States' military presence around the world. "If the Soviets respond fully and in kind,

I think we'll go a very, very long way to allowing both nations to finally begin to step down the thermonuclear ladder after some 40 years."

The Soviet Union had approximately 27,000 warheads, including 10,000 ground-launched, short-range weapons such as have been recommended for elimination by President Bush. An estimated 7,000 other nuclear weapons were available for Soviet aircraft.

President Bush's proposal would eliminate 3,000 American nuclear weapons, leaving the United States with approximately 18,000 such weapons. But two-thirds of those recommended for destruction have already been declared obsolete and due for retirement. Two thousand, one hundred tactical nuclear weapons will be destroyed. They include about 1,000 eight-inch and 155-millimeter artillery shells, and 850 Lance single-warhead missiles. Seventeen hundred of these warheads have been assigned to bases in Europe and South Korea.

Approximately 500 tactical nuclear weapons on board Navy ships will be returned to the United States for storage. They include nuclear bombs and depth charges and nuclear Tomahawk cruise missiles that have been carried on surface ships and attack submarines. Half of them will be maintained for future emergencies, while the rest—including all nuclear depth charges—will be destroyed. General Powell said that there were 350 nuclear Tomahawk cruise missiles in the Navy's inventory, with an average of one hundred at sea at any one time.

Both Bush and the Defense Department heads said they expected no short-range savings, but that eventually a budget reduction of $20 billion could be expected. Cancellation of the mobile ten-warhead MX missile, scheduled to be based on railroad cars, was expected to cost $6.8 billion over the life of the program. Possibly $11 billion may be saved by cancellation of the small mobile intercontinental ballistic missile, while the short-range attack missile known as SRAM-2 will result in saving an additional $2.2 billion.

The United States' nuclear deterrent will remain massive, because there will be no change in the status of the Trident and *Poseidon* ballistic missile submarines. Twelve of them will remain on alert at all times. They are almost invulnerable.

In Europe, the United States plans to maintain medium-range bombers equipped to launch nuclear cruise missiles and to drop conventional bombs.

"The world has changed," Cheney said, "but insurance is still a good idea. Under this plan, we believe we will have enough."

"I am absolutely confident, based upon the work that we've done, that we can say with confidence that our security and that of our allies is protected. Even with these initiations, we will retain sufficient nuclear forces, and we are committed to keep them up to date and effective."

President Bush continued his efforts to reduce the nuclear inventories of the United States and the Soviet Union through negotiations designed to eliminate all ground-based, multiple-warhead missiles under the guidelines of the Strategic Arms Reduction Treaty (START).

Two basic differences earlier divided the superpowers. The Soviet Union had long sought a reduction in sea-based systems before they would agree to cut their ground-based multi-warhead missile systems. The Reagan and Bush administrations repeatedly rejected any limits to American submarine-based multiple-warhead missiles. They argued that submarines are considered impossible to target, whereas land-based missiles of this nature are considered destabilizing because they present such tempting targets for a preemptive strike. The other disruptive factor was their insistence on deployment of SDI against ballistic missiles. The

1972 Antiballistic Missile Treaty, signed by the United States and the Soviet Union, would have to be amended if such a defense system were to be permitted. President Clinton resolved this impasse by canceling SDI in 1993 after $35 billion had been spent on it. This was a wise decision, because SDI could never hope to destroy all incoming warheads, thus making it ineffectual as a deterrent to nuclear holocaust.

President Bush and Russian President Boris N. Yeltsin reached another accord on June 16, 1992, that neither nation would ever make war on the other. (The treaty is subject to ratification by each nation's legislative bodies—which as of mid-1999 still had not occurred.) Most importantly, they agreed to destroy two-thirds of their nuclear arsenals within an eleven-year period. This agreement would significantly reduce the number of missiles agreed to in the previous year's Strategic Arms Reduction Treaty. Under terms of the new agreement, neither nation will have sufficient nuclear weapons to make a disabling strike against the other.

The United States agreed to cut its submarine-launched ballistic missiles—the core of its retaliatory power—from 3,840 to 1,750, and its attack submarines from eighty-eight to forty-five, while retaining twenty-two ballistic missile submarines. Russia agreed to give up all land-based heavy missiles with multiple warheads and scrap them. This has long been their most feared and potentially devastating arsenal.

In early 1994 Presidents Clinton and Yeltsin agreed to stop targeting one another with missiles. The United States approved this treaty January 26, 1996, but opposition in Russia is strong in its parliament and it is still unratified. Meanwhile, START 1 remains in force.

The Cold War has lasted almost half a century, but with this agreement it came to a quiet end, necessitating dramatic changes in the nation's armed forces. Before he left office in early 1993, President Bush ordered the creation of a new Strategic Command. Under it, all American strategic nuclear forces—including the Navy's submarines—will be placed under control of a joint services staff answerable only to the Secretary of Defense and the president. This change was long overdue.

In late 1992, the U.S. Navy announced it was shifting the emphasis from open-ocean conflicts to development of expeditionary forces to fight "brush wars" anywhere in the world.

Emphasis is now placed on using Marines for beach landings, and establishing task forces of small warships to operate in restricted waters such as the Persian Gulf. A new war plan, devised by the Clinton administration, recognizes the Marine Corps' increasing importance by recommending only a one-percent reduction of active duty personnel, for a total of 174,000. The Marine Corps now has three active and one reserve expeditionary forces.

In contrast, the Army will be cut to ten divisions from its recent eighteen divisions on active duty, and the National Guard will be reduced from ten divisions to five in 1999. The Air Force had twenty-four active fighter wings in 1990, and they will be cut to thirteen by 1999, while reserve fighter wings will be reduced to seven from a peak of twelve.

Admiral Frank B. Kelso, Chief of Naval Operations in 1993, described the Navy's role in the post–Cold War era this way: "We've been trying for some time to come to grips with how the world has changed, and we've refocused from the idea of a global confrontation to how we can operate with one foot on the sea and one on land."

Kelso's comments mark a strategic shift in emphasis of vast importance, probably heralding the end of the Navy's primary focus on aircraft carriers. Although the Navy's carrier strength will be reduced by approximately twenty-seven percent, the remaining eleven carriers will provide a potent backbone for the nation's naval forces, with a fleet of more than 800 carrier airplanes.

These fundamental changes in the U.S. Navy's role in the 1990s and into the twenty-first century stress that battle groups designed around amphibious vessels, specially equipped cruisers, or even attack submarines will be tailored to developing events. "The answer to every situation," Kelso said, "may not be a carrier battle group."

He stressed that global trouble spots are developing increasingly along the world's coasts. In the next twenty years, he said, fully eighty percent of the populations of developing nations in Latin America, Africa, and the Pacific Rim will migrate to or be concentrated in urban coastal regions.

The overall size of the U.S. armed forces will be sharply reduced from its total of approximately two million men and women in all the services in the 1980s to 1.4 million by 1999. The Navy will have 194,000, for a twenty-three-percent reduction since 1989. The Navy's reserves will have about 893,000 selected reservists, for a reduction of twenty-seven percent. The practice of having one civil service employee for each two personnel on active duty will be retained, but the total number will be reduced to 718,000. With an all-volunteer armed forces it is essential that they become almost self-sufficient in handling their responsibilities. In World War II active duty personnel were quickly trained to support themselves in the field with only a small number of civilian specialists, primarily those representing companies whose hardware was in use by the unit. Ideally, no more than five percent of the total personnel should be civilians. Any other percentage is an admission that the uniformed personnel have not been adequately trained to handle their responsibilities. This is an area where vast savings can be achieved with no detriment to the nation's defenses.

When Secretary of Defense Les Aspin revealed the nation's new defense plan on September 1, 1993, he claimed it would enable the United States to maintain its "forward presence" and to fight two major regional conflicts—such as the Persian Gulf War and Korea—almost simultaneously.

Aspin said the plan fulfills President Bill Clinton's campaign pledge to restructure the U.S. armed forces for the post–Cold War era. "We have a highly mobile, high tech...force here that will deal with the dangers in the new world." Under the plan, the U.S. Army would have sufficient combat troops to maintain 100,000 men and women each in Europe and Asia.

In 1993, the Clinton administration agreed to build a new F-22 Stealth fighter for the Air Force and the Navy. The Navy was also authorized to build twenty new versions of the F/A-18 Super Hornet, but it was forced to drop plans for the more advanced AF/X. The existing F-14 will be upgraded because the aging A-6 attack plane has outlived its usefulness. The Department of Defense will also provide $1 billion for development of the Joint Strike Fighter, and $1.1 billion to procure five Marine Corps V-22 aircraft and to continue research and development work.

Many critics have condemned further production of stealth aircraft, claiming there is no conceivable threat from foreign fighters in the next forty years to warrant such an aircraft, whose cost is three times that of conventional fighters. No aircraft today is completely "stealthy." Although low-frequency radar, used in World War II, is no longer in service, stealth aircraft can be readily picked up by it. Today's high-frequency radar cannot detect stealth aircraft readily, but a potential enemy could easily develop low-frequency radar to counter such a threat.

The Air Force F-22 Raptor (bird of prey) by weight is thirty-six percent titanium and sixteen percent aluminum, with composite materials covering most surfaces. The latter adds another twenty-four percent of unnecessary weight and raises the cost of manufacturing to an excessive degree. Composite materials are also easily damaged in service, necessitating costly repairs. Such an airplane would be expensive to maintain on carriers because rough treatment of aircraft is at times unavoidable. The F-22's price tag is already $192 million each.

The use of suppression devices to knock out enemy radar gun or missile sites is an added safety factor for all aircraft. This technique was used successfully during attacks on Baghdad in the Gulf War. Terrain-following aircraft can also avoid most defensive weaponry, particularly those with a supersonic dash capability.

Few people realize it, but the Russians explored stealth technology after World War II and decided it was too costly for the limited advantage it gives attacking aircraft. It is also little known that aircraft for years have used "stealth" technology. The famed DeHavilland Mosquito bomber in World War II was made predominantly of wood to make it less visible on radar, although its engines, equipment, and crew could not be eliminated in a radar signature. The Mosquito, of course, was also built to conserve metal, which was in limited supply in England at the time.

Radar-absorbing materials have been incorporated in the design of otherwise conventional airplanes to reduce their cross section and to isolate multiple antennas. The Lockheed U-2 reconnaissance aircraft is a typical example.

In the 1960s, research activities indicated that great improvement could be achieved by using different shapes and components. For example, engineers found that a flat plate at right angles to a radar beam had a very large signal, and that a cavity, similarly heated, also gave a large return. Inlet and exhaust systems of jet aircraft are large contributors to the radar cross section in a nose-on viewing direction, while a vertical tail dominates the side-on signature.

These facts were used by engineers to design aircraft with more curved surfaces and composite materials to reduce their radar cross sections. The result was an aircraft like the SR-71 Blackbird, which could fly Mach 3 in sustained flight. It reduced its radar signature, but not to an appreciable extent, because it had to use afterburners to achieve supersonic flight.

A further expansion of stealth technology was used to create the Lockheed F-117A and the Northrop B-2 stealth bomber through the use of extensive composites.

There are two basic ways to reduce an aircraft's radar cross section. The first is by shaping the configuration and by coating exterior surfaces with energy-absorbing composites to cancel out most radar beams. The F-22 incorporates both approaches to achieve low observable levels when exposed to high-frequency radar.

At a normal right angle, a flat plate acts like a mirror, and its radar reflection is excessively high. If surfaces avoid right angles the advantage in stealthiness is enormous. It is claimed that the F-22 has a radar return equal to birds and bees.

Engineers at Lockheed learned that it was relatively easy to decrease the F-22's radar cross section by avoiding obviously high-return shapes and angles, but that the cost is high.

The F-22 air superiority fighter was developed by a Lockheed-Martin–Boeing team and unveiled August 30, 1990, at Palmdale, California. Northrop had designed a competing fighter, but Lockheed won the flyoff. It is a more sensible fighter than its competitor because it relies on maneuvering and dogfighting capability rather than on stealth technology. It has a shorter fuselage than the Northrop design, a smaller wing, and two horizontal and vertical stabilizers to increase its maneuverability. The two F119, 35,000-pound-thrust Pratt and Whitney fan engines power it to Mach 1.8 without use of afterburners. This is a major breakthrough, because an afterburner "wake" increases the radar exposure. The F-22 is the first operational fighter engine with vectoring exhaust nozzles, which improve aircraft maneuverability during flight.

Originally, the winners of the competition hoped to sign a contract worth $80 billion for 750 Air Force aircraft, and an additional $40 billion for 450 Navy aircraft. The program for the Air Force alone is now projected to cost almost $66 billion, with a per-airplane cost of $138.5 million. No decision has been reached yet on a Navy derivative.

The Raptor's integrated avionics system is one of the key elements to giving the fighter a tactical advantage against enemy threats. It is designed with zones of operational interest. These zones, based on the enemy's and the aircraft's own capabilities, determine the information requirements for each object encountered on a mission.

Today's fighters have some of the same sensing capabilities and subsystems, but each avionics function in the F-22 has its own processor, and essentially works independently. The pilot is the integrator of data and the manager of all supporting subsystems.

The AN/APG-77 radar is the key to the F-22's integrated avionics and sensor capabilities. It will provide pilots with detailed information about multiple threats before an adversary's radar even detects the F-22.

The Raptor's internal weapons include an M-61 twenty-mm cannon, six AIM-120 air-to-air missiles or two AIM-120 missiles, and two 1,000-pound GBU-32 Joint Direct Attack Munitions. Externally, the fighter can carry two fuel tanks and four missiles or, in a ferry configuration, four fuel tanks and eight missiles.

In 1997, General Ronald R. Fogleman, chief of staff of the Air Force, said that without the F-22 there was no viable overall tactical aircraft plan. He stressed that the Navy's F/A-18 E and F and the multiservice Joint Strike Fighter needed the F-22 to complement them. "Without air superiority," he reminded critics, "you are not going to function. By dominating the airspace over a hostile area, the F-22 will provide the environment in which you will be able to operate 18s and Joint Strike Fighters. If the F-22 is cancelled, the Joint Strike Fighters will need costly redesign to make them survivable."

His arguments are valid, but neither he nor anyone else in government will admit that one of the primary reasons for these costly new aircraft is to counter the hundreds of American-built fighters and bombers sold throughout the world in the past two decades to countries that either have become or could become potential enemies. The worst example is the sale of F-14 aircraft to Iran while the Shah was in power. These aircraft probably are no longer flyable, but their technology can be sold to anyone for the hard currency so desperately needed by such nations. These sales of America's most modern combat aircraft were made for political reasons by both major parties and without military approval. If the situation were not so tragic, it would be laughable.

Secretary of Defense William S. Cohen revealed Clinton's defense budget on February 6, 1997, of $250.7 billion, to be effective October 1. With a real decline in spending of $5.8 billion in one year, it marks the thirteenth year that declines have occurred. The budget will remain about the same through 1999 and then increase slightly. Procurement of F-15s, F-16s, and A-10s in the mid 1970s took roughly six percent of the Air Force budget. The F-22 and the Joint Strike Fighter will take only four percent of the USAF budget. These are still controversial programs, because they are part of a six-year spending program projected to cost $1.57 trillion. The Pentagon has budgeted $2.2 billion for the F-22 program in fiscal 1998, and $2.4 billion in 1999. These funds will continue the plane's development and pay for the first two production aircraft. By the year 2004, the year the F-22 enters into service, the F-15—the world's top air-to-air fighter—will have been in use for thirty years.

The Air Force plans to commit $458 million of a Pentagon total of about $1 billion to continue development of the Joint Strike Fighter. This program is expected to produce 3,000 new fighters for the Air Force, the Navy, the Marine Corps, and the British Royal Navy at a cost of $220 billion over a twenty-year period.

Northrop-Grumman Corporation has joined Lockheed-Martin to compete against the Boeing Company to develop a prototype. The winner will be announced in 2001, and the first strike fighter is expected to be delivered seven years later.

The new Clinton budget contains $342 million to equip B-1 bombers with precision-guided munitions. Another $484 million is set for fiscal 1998 for procurement of four types of precision weapons—the Joint Air-to-Surface Standoff Missile, the Joint Standoff Weapon, the Joint Direct Attack Munition, and the Sensor-Fused Weapon.

The 1998 defense budget in real terms is forty percent below the inflation-adjusted sum of $418 billion voted in fiscal 1985. This was the peak year of post-Vietnam defense spending. Most Pentagon spending is for everyday activities. Procurement accounts for only seventeen percent of Pentagon spending. This is a huge decline from the Reagan era's peak.

The Navy and Marines now will get $79.1 billion, or 36.9 percent of the budget's authorizations.

The Navy's fleet is scheduled for a reduction to 335 warships, including eleven carriers. Among the ships being deactivated are nineteen nuclear-powered attack submarines.

Navy officials now concede that its large ocean-going ships encountered almost insurmountable difficulties while operating in the Persian Gulf. During escort operations for oil tankers, before and after the war, Navy ships faced mines, shore-based missiles, hostile aircraft, and small attack craft. Prior to that war, the Navy hierarchy refused to operate carriers in the Gulf's narrow waters, fearing they would have insufficient maneuvering room to ward off attacks. During the conflict with Iraq, four carriers did operate in the gulf, but it was an anxious period.

In the future, Navy officials must work with greater closeness with other services—something they have resisted in the past by demanding and often getting autonomous authority. Cooperation between the services is now essential, and the

Navy's primary need is to increase its transport capabilities for movement of Army soldiers and their equipment anywhere in the world on short notice. The Gulf War proved that this capability was sadly lacking.

As long as the Navy's attack carrier force provides a necessary alternative sea-based system for a portion of the nation's retaliatory capability, it will remain a vital part of the nation's defenses.

The day is not near when the carrier plane of the 1990s must take its place in history with the almost forgotten warplanes of yesteryear. In the final analysis, it will survive as long as it deserves to survive.

Bibliography

Agawa, Hiroyuki. *The Reluctant Admiral: Yamamoto and the Imperial Navy*. Translated by John Bester. Tokyo, New York, and San Francisco: Kodansha International, Ltd., 1979.

Buell, Thomas B. *Master of Sea Power: A Biography of Fleet Admiral Ernest J. King*. Boston: Little, Brown & Company, 1980.

Cagle, Malcolm W. and Mansons, Frank A. *Sea War in Korea*. Washington, DC: U.S. Naval Institute, 1957.

Commander, Naval Air Systems Command, United States Naval Aviation, 1910–1980. Washington, DC: Superintendent of Documents, 1981.

Conner, Howard M. "World War II History of the 5th Marine Division." Washington, DC: Infantry Journal Press, 1950.

Hough, Richard. *Death of the Battleship: The Tragic Close of the Era of Sea Power*. New York: Macmillan, Inc., 1963.

Jablonski, Edward. *Airwar*. Garden City, NY: Doubleday & Company, Inc., 1971.

Karig, Walter. *Battle Reports* (6 volumes). New York: Rinehart, 1946-1952.

Lavalle, Major A.J.C., Ed. *Airpower and the 1972 Spring Invasion*. Washington, DC: Office of Air Force History, U.S. Government Printing Office, 1976.

Lavalle, Major A.J.C., Ed. *The Tale of Two Bridges and the Battle for the Skies over North Vietnam*. Washington, DC: Government Printing Office, 1984.

Merrill, James M. *A Sailor's Admiral: A Biography of William F. Halsey*. New York: Thomas Y. Crowell Co. Publishers, 1976.

Millot, Bernard. *The Battle of the Coral Sea*. Annapolis, MD: Naval Institute Press, 1974.

Morison, Samuel Eliot. *The Two-Ocean War: A Short History of the United States Navy in the Second World War*. Boston: Atlantic Monthly Press, 1963.

Morrison, Wilbur H. *Wings Over the Seven Seas: United States Naval Aviation's Fight for Survival*. Cranbury, NJ: A.S. Barnes & Co., Inc., 1974.

Morrison, Wilbur H. *Point of No Return: The Story of the Twentieth Air Force*. New York: Times Books, 1979.

Morrison, Wilbur H. *The Elephant and the Tiger: The Full Story of the Vietnam War.* New York: Hippocrene, 1990.

Morrison, Wilbur H. *Donald W. Douglas: A Heart with Wings.* Ames, IA: State University Press, 1991.

Potter, E.B. *Nimitz.* Annapolis, MD: Naval Institute Press, 1976.

Reynolds, Clark G. *The Fast Carriers: The Forging of an Air Navy.* New York: McGraw-Hill, Inc., 1968.

Simpson, Albert F. *Aces and Aerial Victories.* Washington, DC: Historical Research Center, Air University and Office of Air Force History, 1976.

Van Wyen, Adrian O. *Naval Aviation in World War I.* Washington, DC: Chief of Naval Operations, 1969.

Winton, John. *The Forgotten Fleet: The British Navy in the Pacific.* New York: Coward, McCann, and Geoghegan, Inc., 1970.

Index

Note 1: Page numbers in boldface indicate appearance in a figure or map.

Note 2: Aircraft are listed by model (e.g., A4) and cross-referenced by name (e.g., Skyhawk) where applicable. Aircraft naming terminology is explained in the Introduction (page ix).

WELCOME TO
Hellgate Press

Hellgate Press is named after the historic and rugged Hellgate Canyon on southern Oregon's scenic Rogue River. The raging river that flows below the canyon's towering jagged cliffs has always attracted a special sort of individual — someone who seeks adventure. From the pioneers who bravely pursued the lush valleys beyond, to the anglers and rafters who take on its roaring challenges today — Hellgate Press publishes books that personify this adventurous spirit. Our books are about military history, adventure travel, and outdoor recreation. On the following pages, we would like to introduce you to some of our latest titles and encourage you to join in the celebration of this unique spirit.

Our books are in your favorite bookstore or you can order them direct at **1-800-228-2275** or visit our Website at **http://www.psi-research.com/hellgate.htm**

ARMY MUSEUMS
West of the Mississippi
by Fred L. Bell, SFC Retired

ISBN: 1-55571-395-5
Paperback: 17.95

A guide book for travelers to the army museums of the west, as well as a source of information about the history of the site where the museum is located. Contains detailed information about the contents of the museum and interesting information about famous soldiers stationed at the location or specific events associated with the facility. These twenty-three museums are in forts and military reservations which represent the colorful heritage in the settling of the American West.

BYRON'S WAR
I Never Will Be Young Again…
by Byron Lane

ISBN: 1-55571-402-1
Hardcover: 21.95

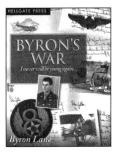

Based on letters that were mailed home and a personal journal written more than fifty years ago during World War II, *Byron's War* brings the war life through the eyes of a very young air crew officer. It depicts how the life of this young American changed through cadet training, the experiences as a crew member flying across the North Atlantic under wartime hazards to the awesome responsibility assigned to a nineteen year-old when leading hundreds of men and aircraft where success or failure could seriously impact the outcome of the war.

GULF WAR DEBRIEFING BOOK

An After Action Report
by Andrew Leyden

ISBN: 1-55571-396-3
Paperback: 18.95

Whereas most books on the Persian Gulf War tell an "inside story" based on someone else's opinion, this book lets you draw your own conclusions by providing you with a meticulous review of events and documentation all at your fingertips. Includes lists of all military units deployed, a detailed account of the primary weapons used during the war, and a look at the people and politics behind the military maneuvering.

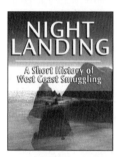

FROM HIROSHIMA WITH LOVE

by Raymond A. Higgins

ISBN: 1-55571-404-8
Paperback: 18.95

This remarkable story is written from actual detailed notes and diary entries kept by Lieutenant Commander Wallace Higgins. Because of his industrial experience back in the United States and with the reserve commission in the Navy, he was an excellent choice for military governor of Hiroshima. Higgins was responsible for helping rebuild a ravaged nation of war. He developed an unforeseen respect for the Japanese, the culture, and one special woman.

NIGHT LANDING

A Short History of West Coast Smuggling
by David W. Heron

ISBN: 1-55571-449-8
Paperback: 13.95

Night Landing reveals the true stories of smuggling off the shores of California from the early 1800s to the present. It is a provocative account of the many attempts to illegally trade items such as freon, drugs, sea otters, and diamonds. This unusual chronicle also profiles each of these ingenious, but over-optimistic criminals and their eventual apprehension.

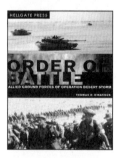

ORDER OF BATTLE

Allied Ground Forces of Operation Desert Storm
by Thomas D. Dinackus

ISBN: 1-55571-493-5
Paperback: 17.95

Based on extensive research, and containing information not previously available to the public, *Order of Battle* is a detailed study of the Allied ground combat units that served in Operation Desert Storm. In addition to showing unit assignments, it includes the insignia and equipment used by the various units in one of the largest military operations since the end of WWII.

PILOTS, MAN YOUR PLANES!

A History of Naval Aviation
by Wilbur H. Morrison

ISBN: 1-55571- 466-8
Hardbound: 33.95

An account of naval aviation from Kitty Hawk to the Gulf War, *Pilots, Man Your Planes! — A History of Naval Aviation* tells the story of naval air growth from a time when planes were launched from battleships to the major strategic element of naval warfare it is today. Full of detailed maps and photographs. Great for anyone with an interest in aviation.

REBIRTH OF FREEDOM

From Nazis and Communists to a New Life in America   ISBN: 1-55571-492-7
by Michael Sumichrast   Paperback: 16.95

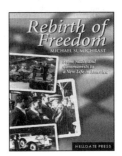

"...a fascinating account of how the skill, ingenuity and work ethics of an individual, when freed from the yoke of tyranny and oppression, can make a lasting contribution to Western society. Michael Sumichrast's autobiography tells of his first loss of freedom to the Nazis, only to have his native country subjected to the tyranny of the Communists. He shares his experiences of life in a manner that makes us Americans, and others, thankful to live in a country where individual freedom is protected."

— *General Alexander M. Haig, Former Secretary of State*

THE WAR THAT WOULD NOT END

U.S. Marines in Vietnam, 1971-1973   ISBN: 1-55571-420-X
by Major Charles D. Melson, USMC (Ret)   Paperback: 19.95

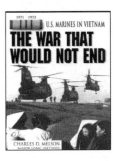

When South Vietnamese troops proved unable to "take over" the war from their American counterparts, the Marines had to resume responsibility. Covering the period 1971-1973, Major Charles D. Melson, who served in Vietnam, describes all the strategies, battles, and units that broke a huge 1972 enemy offensive. The book contains a detailed look at this often ignored period of America's longest war.

WORDS OF WAR

From Antiquity to Modern Times   ISBN: 1-55571-491-9
by Gerald Weland   Paperback: 13.95

Words of War is a delightful romp through military history. Lively writing leads the reader to an under- standing of a number of soldierly quotes. The result of years of haunting dusty dungeons in libraries, obscure journals and microfilm files, this unique approach promises to inspire many casual readers to delve further into the circumstances surrounding the birth of many quoted words.

WORLD TRAVEL GUIDE

by Barry Mowell

ISBN: 1-55571- 494-3
Paperback: 19.95

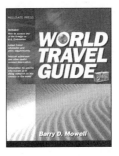

The resource for the modern traveler, *World Travel Guide* is both informative and enlightening. It contains maps, social and economic information, concise information concerning entry requirements, availability of healthcare, transportation and crime. Numerous Website and embassy listings are provided for additional free information. A one-page summary contains general references to the history, culture and other characteristics of interest to the traveler or those needing a reference atlas.

TO ORDER OR FOR MORE INFORMATION
CALL 1·800·228·2275

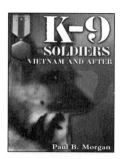

K-9 SOLDIERS
Vietnam and After   ISBN: 1-55571-495-1
by Paul B. Morgan   Paperback: 13.95

A retired US Army officer, former Green Beret, Customs K-9 and Security Specialist, Paul B. Morgan has written *K-9 Soldiers*. In his book, Morgan relates twenty-four brave stories from his lifetime of working with man's best friend in combat and on the streets. They are the stories of dogs and their handlers who work behind the scenes when a disaster strikes, a child is lost or some bad guy tries to outrun the cops.

AFTER THE STORM
A Vietnam Veteran's Reflection   ISBN: 1-55571-500-1
by Paul Drew   Paperback: 14.95

Even after twenty-five years, the scars of the Vietnam War are still felt by those who were involved. *After the Storm: A Vietnam Veteran's Reflection* is more than a war story. Although it contains episodes of combat, it does not dwell on them. It concerns itself more on the mood of the nation during the war years, and covers the author's intellectual and psychological evolution as he questions the political and military decisions that resulted in nearly 60,000 American deaths.

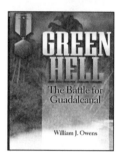

GREEN HELL
The Battle for Guadalcanal   ISBN: 1-55571-498-6
by William J. Owens   Paperback: 18.95

This is the story of thousands of Melanesian, Australian, New Zealand, Japanese, and American men who fought for a poor insignificant island is a faraway corner of the South Pacific Ocean. For the men who participated, the real battle was of man against jungle. This is the account of land, sea and air units covering the entire six-month battle. Stories of ordinary privates and seamen, admirals and generals who survive to claim the victory that was the turning point of the Pacific War.

OH, WHAT A LOVELY WAR
A Soldier's Memoir   ISBN: 1-55571-502-8
by Stanley Swift   Paperback: 14.95

This book tells you what history books do not. It is war with a human face. It is the unforgettable memoir of British soldier Gunner Stanley Swift through five years of war. Intensely personal and moving, it documents the innermost thoughts and feelings of a young man as he moves from civilian to battle-hardened warrior under the duress of fire.

THROUGH MY EYES
91st Infantry Division, Italian Campaign 1942-1945   ISBN: 1-55571-497-8
by Leon Weckstein   Paperback: 14.95

Through My Eyes is the true account of an Average Joe's infantry days before, during and shortly after the furiously fought battle for Italy. The author's front row seat allows him to report the shocking account of casualties and the rest-time shenanigans during the six weeks of the occupation of the city of Trieste. He also recounts in detail his personal roll in saving the historic Leaning Tower of Pisa.

Order Directly From Hellgate Press

You can purchase any of these Hellgate Press titles
directly by sending us this completed order form.

Hellgate Press

P.O. Box 3727
Central Point, OR 97502

To order call, 1-800-228-2275
For inquiries and international orders,
call 1-541-479-9464 or Fax 1-541-476-1479

TITLE	PRICE	QUANTITY	COST
Army Museums: West of the Mississippi	$13.95		
Byron's War	$21.95		
From Hiroshima With Love	$18.95		
Gulf War Debriefing Book	$18.95		
Night Landing	$13.95		
Order of Battle	$17.95		
Pilots, Man Your Planes!	$33.95		
Rebirth of Freedom	$16.95		
The War That Would Not End	$19.95		
Words of War	$13.95		
World Travel Guide	$19.95		
Memories Series			
After the Storm	$14.95		
K-9 Soldiers	$13.95		
Green Hell	$18.95		
Oh, What A Lovely War!	$14.95		
Through My Eyes	$14.95		

If your purchase is:	**your shipping is:**		
up to $25	$5.00	**Subtotal**	$
$25.01–$50.00	$6.00	**Shipping**	$
$50.01–$100	$7.00	**Grand Total**	$
$100.01–$175	$9.00		
over $175	call	*Thank You For Your Order!*	

Shipping Information

Name:

Address:

City, State, Zip:

Daytime Phone: Email:

Ship To: (If Different Than Above)

Name:

Address:

City, State, Zip

Daytime Phone:

Payment Method:

For rush orders, Canadian and overseas orders please call for details at (541) 479-9464

☐ **Check**   ☐ **American Express** ☐ **MasterCard** ☐ **Visa**

Card Number: Expiration Date:

Signature: Exact Name on Card:

Learn More About Hellgate Press

For more adventure and military history information visit our Website:

HELLGATE PRESS ONLINE
http://www.psi-research.com/hellgate.htm

With information about our latest titles, as well as links to related subject matter.

Hellgate Press Reader Survey

Send this survey to:
Hellgate Press
P.O. Box 3727
Central Point, OR 97502

or fax it: (541) 476-1479
or email your responses to
info@psi-research.com

Our readers and their opinions are important to us. By answering and mailing this survey, we can improve our book selection.

Did you enjoy this Hellgate Press title?
☐ Yes   ☐ No, I would improve it by:

_____

_____

_____

Would you be in interested in reading other Hellgate Press titles?
☐ Yes   ☐ No

Is there a period of history or event you would like to read about?

_____

_____

_____

How do you feel about the price of this book?
☐ Lower than expected   ☐ Fair   ☐ Too High

Where did you purchase the book?
☐ Bookstore   ☐ It was a gift
☐ Online (Internet)   ☐ Other: _____
☐ Catalog   _____
☐ Association/Club   _____

Do you use a personal computer?
☐ Yes   ☐ No

Have you ever purchased anything over the Internet?
☐ Yes   ☐ No

Would you like to receive a free Hellgate Catalog?
☐ Yes, please include information below: ☐ No

Name: _____

Address: _____

City, State, Zip: _____

Country: _____

Email Address: (optional) _____